TRANSDISCIPLINARY MARINE RESEARCH

Drawing on the expertise of marine researchers from both the natural and social sciences, this book examines how we, as both scientists and societies, can return to a sustainable co-existence with the ocean and use the tools of transdisciplinarity to bring together the diverse forms of knowledge needed to achieve this important task.

The marine sciences play a vital role in producing and providing the knowledge needed for a transition towards ocean sustainability. With a multitude of actors involved in using, exploiting, and safeguarding the seas, however, this task cannot be solved by science alone. Transdisciplinary research is needed, bringing together scientists and all other actors of society to jointly co-produce the knowledge and innovations that we so urgently need. In this context, this book examines and answers key questions at the forefront of transdisciplinary marine research: How can we provide approaches that integrate marine biodiversity and social systems in an appropriate relationship? What methodologies are most suitable to engage stakeholders in participatory processes providing new knowledge and tools for co-designing solutions with balanced socio-ecological embeddedness? How do we best integrate scientific with lay and local knowledge, and how are diverse knowledges valued in engagement activities? How can we reconcile socio-economic activities and the often divergent values attached to them to provide ethical principles for fair and equitable policy decisions? The book addresses these questions by combining an array of chapters about new theoretical approaches to transdisciplinary marine research, methodological considerations, as well as case studies from the nexus of the research and practices of engagement with a variety of stakeholder groups across the globe.

This book will be of great interest to students and scholars studying marine science and ocean research across a wide range of disciplines, including marine biology, environmental governance and policy, ocean resource management, oceanography, environmental anthropology, human geography, and sustainability.

It will also be of interest to those looking to build a greater understanding of transdisciplinary research and knowledge co-production, and practitioners working alongside academics.

Sílvia Gómez is Associate Professor and Researcher at the Universitat Autònoma de Barcelona, Spain. She is a social anthropologist whose research focuses on marine science from a transdisciplinary approach.

Vera Köpsel is Postdoctoral Researcher in the Institute for Marine Ecosystem and Fisheries Science at the University of Hamburg, Germany. Whilst her background is in human geography, she now focuses her work on marine social science and exchange with non-academic actors.

Earthscan Oceans

Marine Extremes
Ocean Safety, Marine Health and the Blue Economy
Edited by Erika Techera and Gundula Winter

Marine Policy
An Introduction to Governance and International Law of the Oceans
Mark Zacharias and Jeff Ardon

Conflicts over Marine and Coastal Common Resources
Causes, Governance and Prevention
Karen A. Alexander

Marine and Fisheries Policies in Latin America
A Comparison of Selected Countries
Edited by Manuel Ruiz Muller, Rodrigo Oyandel and Bruno Monteferri

Oceans and Society
An Introduction to Marine Studies
Edited by Ana K. Spalding and Daniel O. Suman

Transdisciplinary Marine Research
Bridging Science and Society
Edited by Sílvia Gómez and Vera Köpsel

For more information about this series, please visit: *www.routledge.com/Earthscan-Oceans/book-series/ECOCE*

TRANSDISCIPLINARY MARINE RESEARCH

Bridging Science and Society

Edited by Sílvia Gómez and Vera Köpsel

Routledge
Taylor & Francis Group
LONDON AND NEW YORK

earthscan
from Routledge

Designed cover image: Getty

First published 2023
by Routledge
4 Park Square, Milton Park, Abingdon, Oxon OX14 4RN

and by Routledge
605 Third Avenue, New York, NY 10158

Routledge is an imprint of the Taylor & Francis Group, an informa business

British Library Cataloguing-in-Publication Data
A catalogue record for this book is available from the British Library

Library of Congress Cataloging-in-Publication Data
Names: Gómez Mestres, Sílvia, editor. | Köpsel, Vera, editor.
Title: Transdisciplinary marine research: bridging science and society/ Sílvia Gómez Mestres and Vera Köpsel, editors
Description: New York, NY: Routledge, 2023. | Includes bibliographical references and index.
Identifiers: LCCN 2022040251 (print) | LCCN 2022040252 (ebook) | ISBN 9781032317601 (hardback) | ISBN 9781032317588 (paperback) | ISBN 9781003311171 (ebook)
Subjects: LCSH: Marine sciences. | Marine resources conservation. | Human ecology. | Aquatic ecology.
Classification: LCC QH541.5.W3 T73 2023 (print) | LCC QH541.5.W3 (ebook) | DDC 577.6—dc23/eng/20220831
LC record available at https://lccn.loc.gov/2022040251
LC ebook record available at https://lccn.loc.gov/2022040252

ISBN: 978-1-032-31760-1 (hbk)
ISBN: 978-1-032-31758-8 (pbk)
ISBN: 978-1-003-31117-1 (ebk)

DOI: 10.4324/9781003311171

Typeset in Bembo
by Apex CoVantage, LLC

CONTENTS

PART II
Methods and perspectives 83

PART III
Insights from the case studies 173

ABOUT THE CONTRIBUTORS

Mar Abbot-Jiménez is a PhD candidate in anthropology at University of Seville. Her project aims to evaluate participatory governance of marine reserves of fishing interest in Spanish waters. She aims to identify what political, economic, social, ideological, and symbolic factors are intervening in the management of these areas and what logics and powers are involved, and who are the main stakeholders in this context.

Katrina Anthony is the Inuit Research Coordinator (IRC) in Postville, Labrador for the Sustainable Nunatsiavut Futures project. She is originally from Makkovik. Katrina is interested in research that is guided by Inuit values and that will, in the end, benefit Inuit in Nunatsiavut, such as by ensuring all knowledge holders are respected and heard. She also enjoys that her work includes fieldwork on the land, water, and ice. Katrina struggles in large discussions at times but voices her concerns to ensure that the project ethically and respectfully aligns with Inuit values.

Megan Bailey is an associate professor in the Marine Affairs Program at Dalhousie University and a Canada Research Chair in Integrated Ocean and Coastal Governance. Her research spans the global to local, with study fisheries involving everything from transboundary tuna fisheries in the Indian Ocean, to lobster fisheries in Nova Scotia, to mixed commercial fisheries in Nunatsiavut (Labrador, Canada).

Stefan Baumgärtner holds the Chair of Environmental Economics and Resource Management at the University of Freiburg, Germany. His expertise is in ecological, environmental, and resource economics, with a focus on sustainability economics; biodiversity and ecosystem services; and risk, resilience, and insurance in ecological-economic systems.

Leandro Bergamino is a PhD researcher focusing his work on local ecological knowledge, coastal lagoon ecology, and coastal management. He works at the Eastern Regional University Center of the University of the Republic of Uruguay (Udelar).

Breanna Bishop is an interdisciplinary PhD student at Dalhousie University. She is interested in how Indigenous and western knowledge systems can contribute to understanding complex problems at the intersection of climate change and Indigenous rights. Her research is based in Nunatsiavut, exploring how Inuit understand indicators of oceanographic and climatological phenomena, including seasonality and change.

Hekia Bodwitch is a human geographer and a postdoctoral fellow with Marine Affairs at Dalhousie University. She studies how environmental governance can advance justice, with a particular focus on Indigenous fishery development. Currently, she is working with Inuit youth and leaders in the Canadian Arctic to explore how Indigenous-scientist knowledge co-production initiatives can support Inuit-led marine spatial planning.

Pamela M. Buchan is a marine social scientist and consultant who completed an inter-disciplinary PhD in marine citizenship at Exeter University in 2021. Alongside her PhD, she served in public office in Plymouth, UK. Her research goal is to transform the human-ocean relationship for sustainability.

Laura Brum Bulanti is assistant professor at the Interdisciplinary Department of Coastal and Marine Systems at the Eastern Regional University Center (CURE). Her work focuses on cultural landscapes, environmental history, socio-ecological conflicts, and participation in coastal and marine areas of eastern Uruguay.

Rachael Cadman is an interdisciplinary PhD candidate at Dalhousie University in Kjipuktuk/Halifax, Canada. Her research takes place in Nunatsiavut, where she works with Inuit government and community partners to advance Inuit visions for the future of natural resource management. Currently, she is focused on modelling inclusive governance strategies for the Nunatsiavut fishing industry and protected areas.

Jayne Carrick is currently a research fellow for the Knowledge Network on Climate Assemblies and a visiting researcher at Newcastle University. Her research interests lie in public engagement and participation in political decision-making, particularly regarding environmental issues such as climate change and renewable energy.

Kathryn Collins is a visiting fellow at the School of Architecture, Planning and Land-scape at Newcastle University and Principal Consultant at Howell Marine Consulting. Her marine spatial planning research focuses on the 'spatial' and brings urban design concepts into her work to consider how stakeholder values influence marine spatial understanding.

Susan de Koning is a PhD candidate at the Institute for Management Research at Radboud University Nijmegen. After graduating from her studies, she worked as a Marine Governance Researcher at Wageningen Marine Research, where she researched shellfish and seaweed farming, pulse trawling, and marine spatial planning. Currently, she examines the potential of landscape-oriented partnerships among farmers, governments, citizens, and NGOs to contribute to sustainability transformations via transformative governance.

Clare Fitzsimmons is a marine scientist, professor, and postgraduate degree programme director at the School of Natural and Environmental Sciences at Newcastle University. She has a background in environmental research, with commercial experience gained in defence and consultancy sectors. Her research interests lie in marine management and governance, primarily investigating human interactions with ecological systems.

Wesley Flannery is a reader in environmental planning. He leads the Marine Social Science Research Group at Queen's University Belfast. His research interests include power and participation in marine spatial governance, coastal risk management, coastal cultural heritage, the future of ports, social acceptance of marine renewables, and coastal transitions.

David Florido-del-Corral is a professor in the Department of Social Anthropology at the University of Seville. His main research subject is the analysis of fishing, mainly in the areas of management analysis and cultural heritage. In his research applies an interdisciplinary approach, collaborating with geographers, economists, biologists, archaeologists, and historians.

Sílvia Gómez is a marine social anthropologist working as a professor and researcher at the Autonomous University of Barcelona. Her research has focused on the socio-economic dimensions of fisheries, sustainable seafood markets, small-scale fishing rights, marine governance systems, and cultural heritage. She is currently working on the valuation of ecosystem services and the assessment of the impacts of maritime activities on human populations and the environment to provide ethical principles for policy decisions.

Tim Gray is Emeritus Professor of Political Thought and Senior Research Investigator at Newcastle University with a special interest in marine environmental governance and the principles of democracy, equality and justice. His recent publications focus on stakeholder participation in small-scale fisheries, normative analyses of regional fisheries politics, and the relation between co-management and fisheries science partnerships.

Felix Gross is a marine geoscientist at the Center for Ocean and Society. He is principal investigator of the team "marine and coastal geohazards". In his research, he focuses on the assessment, evaluation, visualization, and communication of marine and coastal geohazards.

Caroline Grünhagen is a PhD student at the Center for Ocean and Society (Kiel University). Her research focuses on the development of bio-economic models for the Baltic and North Seas, considering possible impacts of climate change and geo-engineering on fisheries.

Hugo Inda is a PhD researcher focusing on the long-term evolution of socio-ecological systems and climate variability, subsistence strategies, and adaptation to changing environments of prehistoric inhabitants and cultural heritage management. He currently works at the Eastern Regional University Center of the University of the Republic of Uruguay (Udelar).

Nathan Jacque is the Inuit Research Coordinator for the Sustainable Nunatsiavut Futures project from Rigolet, Labrador, where he was born and raised. Nathan greatly appreciates his culture and the land of Nunatsiavut which led him to pursue this career. The majority of his work in the project includes planning and taking part in on-ice fieldwork plus much more. Like everyone in Nunatsiavut, Nathan loves going off on the land for boil ups, hunting, fishing, and trips to his cabin. Nathan is very eager to learn and has really enjoyed his time in the project so far.

Vera Köpsel is a postgraduate researcher at the University of Hamburg, Germany. She is a human geographer currently researching how marine scientists perceive their engagement with stakeholders. Her practical work revolves around coordinating the stakeholder engagement activities of two large EU research projects on sustainable fisheries and marine nature-based solutions.

Marloes Kraan is a marine anthropologist working as a researcher at the applied research institute Wageningen Economic Research and at the Environmental Policy Group at Wageningen University. She is co-director of the Centre for Maritime Research (MARE) in Amsterdam, member of the Strategic Initiative on the Human Dimension in ICES and co-chair of the Working Group on Social Indicators. Marloes is interested in contributing to increasing inter- or trans-disciplinarity in maritime research and interactive fisheries governance, and improving the applicability of social science in fisheries science and policy.

Ximena Lagos is an associate professor at the Eastern Regional University Center of the University of the Republic of Uruguay (Udelar). Since 2014, she has been an advisor to the Association of Artisanal Fishermen of Coastal Lagoons (APALCO) in Rocha, Uruguay, and to Cocina de La Barra (Kitchen of La Barra), a cooperative enterprise of fisherwomen.

Cristina Larrea-Killinger is professor at the Department of Social Anthropology, the University of Barcelona, Spain. Her research is focused in environmental health in urban and rural settings, particularly on ethnographic work on urban sanitation in marginal urban neighbourhoods, and in health conditions of shellfish fishers from Afro-descendant communities in Brazil.

Louise Oliveira Ramos Machado is a PhD student at the Institute for Studies in Collective Health (IESC/UFRJ). Her work focuses on environmental contamination, climate and health, environment, and sustainability in artisanal fisheries. She is a member of the Health, Environment, Work and Sustainability in Communities Research Group and works with artisanal fishing communities in the Baía de Todos os Santos on implementation of strategies with populations exposed to environmental contamination by lead, cadmium, copper, and zinc.

Laura Marrero Beramendi is a graduated sociologist from the School of Social Sciences and holds a master's degree in Integrated Coastal Management of the University of the Republic of Uruguay (Udelar). She works as an advisor on gender and climate change for UNDP, providing technical assistance to the National Directorate of Climate Change of the Ministry of Environment of Uruguay.

Benedict McAteer is a research fellow in the School of Natural and Built Environment at Queen's University Belfast. Based in the planning department, his key research interests include participatory research, knowledge co-production, participation in marine governance, and sustainable development. His recent publications have explored the transformative potential and political dynamics of marine citizen science.

Paul McCarney is a transdisciplinary conservation social scientist with a PhD in environmental studies. He conducted research throughout northern Canada working on complex projects that bring together different academic disciplines, knowledge systems, and communities of people. His own perspectives on conservation are informed by hunting, fishing, hiking, and paddling, working with Indigenous communities in Canada and South-East Asia.

Megan McLaren is the project manager for the Sustainable Nunatsiavut Futures project. She draws on a background of applied transdisciplinary research in Southern Africa, ICT Systems Analysis and Design and Environmental Behavioural Economics, to address a range of social and ecological concerns – from agricultural and residential adaptation to climate change, to facilitating equal access to education and sanitation services. She loves solving problems, being in nature, and feels uncomfortable writing about herself in the third person.

Luize da Silva Rezende da Mota graduated in Nursing from the Federal University of Bahia (2018). She is a member of the Health, Environment, Work and Sustainability in Communities Research Group and works on projects with artisanal fishing communities in the area of quality of life with populations exposed to oil/petroleum spills. Currently she is a master's student in the Postgraduate Program in Health, Environment, and Work.

Josilan da Silva Nascimento is an artisanal fisherman. He is a member of the Associação Mãe dos Extrativistas da Resex de Canavieiras (AMEX) and has worked as a field research fellow in different projects at Universities and Institutes.

Caroline Nochasak is from Nain, Newfoundland and Labrador. She is currently an Inuit Research Coordinator (IRC) working on the Sustainable Nunatsiavut Futures Project with Dalhousie University. She conducts climate change research and acts as a connection between researchers and the communities. Caroline is also a co-founder of Inuusuktut Qaujisarnilirijut-Youth Working to Gain Knowledge, a research group run by and for Inuit youth. She has also worked on ice mapping and fieldwork, food security, and community interviewing.

Amanda Laura Northcross is an associate teaching professor focusing on assessing environmental exposures to understand their impacts on the health of communities. Combing community-engaged research approaches with her background in atmospheric chemistry and exposure assessment, she develops accessible tools, approaches, and methods to assess exposures to air and waterborne contaminants.

Eric C. J. Oliver is an assistant professor of physical oceanography at the Department of Oceanography, Dalhousie University, Halifax (CA). His research interests involve ocean and climate variability across a range of time and space scales including extreme events, and the predictability of climate variations. Eric is of Inuit descent with roots in Nunatsiavut and interested in understanding both Indigenous and scientific knowledge of the climate system.

Priscilla Andrea Orsi is a member of the Health, Environment, Work and Sustainability in Communities Research Group and works on a project with artisanal fishing communities in the Bahia coast affected by the oil spill of 2019. Moreover, she coordinates inter-municipal international cooperation projects focused on adaptation and resilience to climate change.

Michael A. Petriello is an interdisciplinary conservation social scientist. He is currently a postdoctoral fellow at the School for Resource and Environmental Studies (SRES) at Dalhousie University in Kjipuktuk/Halifax (CA), with the Sustainable Nunatsiavut Futures Project. He researches how knowledge co-production with and for Indigenous Peoples and local communities informs conservation objectives and redresses power imbalances towards transformative solutions to complex environmental problems.

Maira Ramos is a graduate student working on human impacts, local ecological knowledge, and coastal lagoon ecology. She works at the Eastern Regional University Center of the University of the Republic of Uruguay (Udelar).

Rodolfo Reboulaz is a graduate student at the Eastern Regional University Center of the University of the Republic of Uruguay (Udelar), working on participatory mapping, coastal environment and GIS analysis.

Rita de Cássia Franco Rêgo is an associate professor in environmental epidemiology. She currently coordinates participatory projects that assess the impacts of environmental exposure on the health of vulnerable communities. Her research results seek to support affected communities in their demands for reducing inequalities and accessing better health and environmental conditions.

Marie-Catherine Riekhof is professor of political economy of resource management and director of the Center for Ocean and Society (Kiel University). She works conceptually, empirically, and together with stakeholders to examine the effects of various institutional regulations in the field of marine and coastal resources.

Joana Sá Couto is a PhD candidate in climate change and sustainable development policies at the Institute of Social Sciences in Universidade de Lisboa. She holds a master's degree in anthropology and focuses her work on humans and nature, local knowledge, and transdisciplinary science – currently in the context of ocean sustainability.

Viola Schaber is a fisheries ecologist with a background in trophic interactions and stock dynamics. Her work focuses on environmental and anthropogenic effects on key fish species in an important Baltic Sea ecosystem and works at the Center for Ocean and Society (Kiel University).

Jörn O. Schmidt is working on social-ecological systems, marine and fisheries ecology, inter- and trans-disciplinary concepts, knowledge co-production, and games for communication (www.ecoocean.de, www.go-jelly.de). He has worked in the Baltic and North Sea as well as in Senegal, Cabo Verde, Haida Gwaii (Canada), Sitka (USA) and Peru. Jörn is now working with the ICES science network of over 6,000 scientists to implement ICES strategic and science plan. He is chair of the Science Committee of the International Council for the Exploration of the Sea (ICES).

Heike Schwermer is a researcher at the Center for Ocean and Society (Kiel University) focusing on transdisciplinary research with a strong interest in social-ecological systems including the conduction and analysis of interviews.

Michael Stecher is a PhD student at the University of Freiburg with a background in economics. His current research focus lies on quantifying the degree to which individual factors such as overfishing and climate change have been responsible for the current unsustainable state of the Western Baltic Sea, and what a transition to a more sustainable state might look like.

Nathalie A. Steins is a sociologist who has been working on fisheries in several roles. She currently works as a senior scientist and project manager at Wageningen Marine Research (Wageningen University & Research). Core part of her work is the research collaboration with the fishing industry towards ecologically sustainable and economically viable fisheries. Between 2014 and 2019, she was the Dutch member of the Advisory Committee of the International Council for the Exploration of the Seas (ICES).

Germán Taveira is an MSc student at the Eastern Regional University Center of the University of the Republic of Uruguay (Udelar), working on GIS analysis, remote sensing, and landscape manetrics.

Rudi Voss is a senior researcher at the Center for Ocean and Society (Kiel University) and the German Centre for Integrative Biodiversity Research, Leipzig. In his work, he focuses on ecological-economic fisheries management, including aspects of multispecies interactions, climate change, and stakeholder involvement.

Christian Wagner-Ahlfs is a coordinator for transdisciplinary research at the Center for Ocean and Society and the research platform Kiel Marine Science KMS (both Kiel University). He is responsible for promoting the cooperation of KMS researchers with experts from politics, civil society, and business. The goal is participatory development of research questions.

Jacqueline Winters is the Inuit Research Coordinator (IRC) in Rigolet, Labrador for the Sustainable Nunatsiavut Futures project. She is originally from Makkovik. Jacqueline has a 2-year-old son and was on maternity leave awaiting the arrival of her twin daughters during her contribution to this volume. She is very interested in conducting research that contributes to a sustainable future in Nunatsiavut for her children and other Inuit in Labrador. Jacqueline is passionate about preserving the animals and land in Nunatsiavut for future generations.

John Winters is an Inuit Research Coordinator with the Sustainable Nunatsiavut Futures project. He is located in Hopedale, Labrador (Nunatsiavut). He is experienced in ocean-related work in Nunatsiavut such as involvement in the Indigenous and Northern Affairs Canada's Climate Change Preparedness in the North Program (CCPN) and the Imappivut (Our Oceans) project which covers marine planning, and mapping and interviewing community members within Hopedale, to protect Labrador's interests in coastal marine areas

Melanie Zurba is an associate professor at the School for Resource and Environmental Studies (SRES) at Dalhousie University, Kjipuktuk/Halifax (Canada). Her research focuses on projects developed and implemented in collaboration with Indigenous communities, particularly co-management of species and protected areas, shared forest tenure agreements, land use and occupancy mapping studies, food sovereignty, health promotion, and wellbeing.

PREFACE

The biodiversity crisis and the climate crisis are intrinsically linked. This is threatening the foundations of our social systems. Under a scenario that considers biodiversity loss and climatic changes, an ecological transition towards sustainability implies that human activities must go hand in hand with protecting and restoring nature. Changing the ways in which we consume, work, and live together will require action by scientists, citizens, and communities, placing special focus on socio-ecological relationships and wellbeing. We now need the incorporation of knowledge at multiple scales – from the local to the global. As the need for climate action and ecosystem restoration becomes more urgent, the ocean is gaining attention as a potential part of the solution. Reconnecting societies with nature, ocean-based solutions have been emphasised as measures to protect, manage, and restore marine ecosystems to address societal challenges properly and adaptively provide human wellbeing and biodiversity benefits. However, many of these initiatives remain controversial due to uncertainties around governance systems, potential ecological impacts, and impacts on human populations, to just name a few.

To tackle these challenges requires active cooperation between scientists researching the world's oceans and the societal actors who live off, by, and with marine ecosystems. Against this background, this book asks key questions at the forefront of contemporary transdisciplinary marine research: How can we provide approaches that integrate marine biodiversity and social systems in an appropriate relationship? What methodologies are most suitable to engage stakeholders in participatory processes providing new knowledge and tools for co-designing solutions with balanced socio-ecological embeddedness? How do we best integrate "scientific" with "lay" and "local" knowledge, and how are diverse knowledges a core value in engagement activities? How can we reconcile socio-economic activities and the often-divergent values attached to them to provide ethical principles for fair and equitable policy decisions? Drawing on the expertise of researchers from across the fields of marine natural and social sciences, this book makes a cutting-edge contribution to the question of how we, as both scientists and societies, can return to a sustainable co-existence with the ocean and use the tools of transdisciplinarity to bring together the diverse forms of knowledge needed to achieve this important task.

Transdisciplinarity in marine research

The urgency of addressing socio-ecological problems such as climate change has created the need to develop transdisciplinary research approaches to solving today's pressing environmental challenges. The concept of transdisciplinarity was coined in the 1970s (Jahn et al., 2012) and subsequently theorized in the discipline of ecological economics (Patterson and Glavovic, 2013; Jahn et al., 2012). It is today widely recognized for its suitability to address sustainable development and management of environmental resources (Klein et al., 2001; Scholz et al., 2015, Stock et al., 2011). However, its popularisation did not come about until the late 1990s and early 2000s (Nöel-Knapp et al., 2019; Jahn et al., 2012) with post-normal science approaches (Funtowicz and Ravetz, 1990, 1991, 1993) and the development of knowledge production based on practical real-world problems (so-called Mode 2) (Gibbons et al., 1994; Nicolescu, 2010; Rigolot, 2020; Scholz and Steiner, 2015). Transdisciplinarity gained momentum in the latter part of the 20th century in response to concerns about the mismatch between the specialization and compartmentalization of knowledge and the uncertain reality of a changing world. Addressing "wicked" problems such as poverty, social injustice, and inequalities in the use of and access to ecosystem goods and services in the context of a globalized economy and an environmental crisis, theoretical responses are no longer sufficient. The various moral and ethical angles involved in the inherent complexity of reality become constitutive of socio-ecological problems. Therefore, a research episteme that absorbs ethical principles has to take into account the social concerns that must be ensured in participatory research. As societal collaboration to address complex environmental problems in a changing ocean becomes paramount, scientists must produce advice by shifting their role and responsibility into the sphere of transdisciplinary action (Rudd, 2014).

The socio-ecological crises around the globe resulting from harmful interactions between social, environmental, and health factors have shown that the majority of our societal and economic systems do not go hand in hand with care for people and nature. In the context of this convergence of different global crises (climate change, biodiversity, energy, care, pandemics), it is becoming ever clearer that human beings are eco-dependent. Wellbeing should be at the heart of socio-economic development rather than growth. This shifts the idea of nature towards the economic rationality of conservation, which implies the inclusion of not just monetary, but also non-material values. Moreover, the engagement with stakeholders from all societal groups is needed to bring into play those underlying driving forces such as norms and values about our relationship as humans with nature and the ocean that becomes visible in the governance systems we set up.

Transdisciplinary exchange encourages a rethinking of conceptual scientific approaches to broaden the perspective and include all disciplinary viewpoints in an integrated way of addressing a research topic at the interface of social interactions, political and cultural structures, and socio-economic activities. This epistemological shift moves the focus of research and introduces reconceptualisations of socio-ecological reality that offer new perspectives for environmental conservation. As society is an active part of natural cycles and transformations socio-ecological embeddedness is approached as an analytical methodology, as a constituent part of the problem but also of the solution. Consequently, societal action is integrated as part of the research questions, in the research process, and in the results as socially embedded solutions. In this task of bridging science and society, the role of science in society is questioned, as well as the role of society in science. How can we produce a science-society knowledge system? How do we achieve transdisciplinarity by taking into account different theoretical and methodological perspectives through the involvement of society? How to ensure an equitable knowledge pool in marine science? What are the valid criteria to guide the focus of research in the framework of this socio-scientific symbiosis, and from where – bottom-up or top-down? How can we establish a balanced approach?

Exploring the state of the art of transdisciplinary marine research

Given that local and traditional knowledge entails political issues of environmental rights and access to resources, while science might be traversed by hierarchies and power relations in policymaking, how are social claims entailed by local and traditional knowledge integrated into marine research? Considering that rights and duties are constitutive of social life, notions of justice and ethical and moral values are also implicated in the environment. Since values are context-specific and culturally embedded in social systems, involving society in the research process means integrating cultural diversity, social groups, and gender perspectives. Science is called to shift from a focus exclusively on Western scientific knowledge production to considering traditional or indigenous knowledge, and local knowledge more generally, as a part of knowledge production. Furthermore, contrasting abstract knowledge with empirical and practical application of knowledge through case studies becomes a necessary step in research.

Integrating the incommensurability of the 'imponderables of social life' such as climate change effects into the uncertainty of contemporary socio-ecological phenomena allows for more flexible and realistic scientific results. Adaptive scientific responses to contemporary socio-ecological challenges aim to overcome the rigidity of Western scientific models to open a path toward post-normal science.

New methodological approaches in between modelling and participatory research integrating quantitative and qualitative data are assessed across the contributions of this book.

The rapprochement between science and society through the integration of social and cultural diversity is a part of the process of the democratisation of knowledge that calls for a new ontological approach. However, this approach opens up new questions, highlights scientific gaps, and brings to the table an understanding of the transdisciplinary limitations that this book explores.

How can we effectively produce knowledge for a socio-ecological transition towards sustainability with a transformative goal? Traditional and local knowledge must be integrated not only as objects of analysis but also in the co-production of knowledge and co-design of research at the forefront of current public and political debates on the environment. Furthermore, the integration of knowledge from different sources implies the identification of social actors. While some scholars define them as stakeholders, others define them as marine citizens, which gives them political content as social agents of transformation. While the participation of stakeholders in decision-making processes is highlighted as an ethical consideration that ensures the morality of the procedure by enhancing social consensus on environmental policies, the actual weight of stakeholders in final decision-making remains unclear. The question arises as to whether transdisciplinarity acts as a tool of procedural justice or a democratic framework for action, or whether it is an end that truly integrates the concerns and knowledge of social actors. Beyond being socio-political processes, participatory research initiatives usually take into account the concerns of social actors in policymaking.

Moving towards participatory sustainable ocean management involves clearly defining social actors and their real capacity for power by measuring their inclusion and potential as social activists or socio-environmental transformative agents. Depending on their definition, the direction of society's involvement can be mapped: Who are the stakeholders and marine citizens, how are they selected, and to what extent do they have access to power in decision-making and claiming of rights? Which social group do they represent, and is their viewpoint really representative? The unclear definition of the role of society in transdisciplinary research leads to misunderstanding the task of society in science as mere providers of knowledge on environmental challenges or recipients of educational and environmental awareness-raising initiatives, thus masking claims of rights and conflicts of interest behind participatory processes.

Based on these considerations, further questions arise. What societal values that can and should be integrated into environmental decision-making and policy? The creation of values depends on the positioning within a socio-ecological relationship, whether hierarchical or reciprocal, and thus on the definition of values through the relational position with respect to others in a feedback process. The deontological approach from the perspective of cultural diversity cannot be understood outside of the fields of political economy and political ecology.

Bridging the gap between science and society

From a transdisciplinary perspective, the blurring of disciplinary boundaries can call into question the role of each discipline's knowledge in relation to that of others, the value of stakeholder knowledge when institutionalized by science, the masking of power asymmetries in decision-making processes, and the absorption of dissenting voices or social claims in the light of competing interests. Or, conversely, it can shed light on how ecological considerations intertwine with social justice, rights claims, ethics, and political and economic systems in a just and balanced socio-ecological relationship.

If transdisciplinarity absorbs the interdisciplinary perspective by integrating society as a part of a cooperative science, cooperation must be designed to encompass dimensions of scale and methodological perspectives of different approaches that do not always coincide. Ecosystems generally do not share the same boundaries as social communities that can be defined sociologically, geographically but also ecologically. Nor do all actors share the same interests in resource use, moving between exploitation and conservation to satisfy particular social purposes or lifestyles. Livelihoods, leisure, culture, and wellbeing are at stake in the management of the seas and oceans.

By bridging the gap between science and society, this book explores how disciplines can cooperate in advancing knowledge together with the social actors involved. It highlights the barriers that need to be resolved to facilitate dialogue, overcoming disciplinary hierarchies, and the rigidity of perspectives based on scientific traditions that lead to misunderstanding. At the same time, it highlights common ideas for the transition towards socio-ecological sustainability. It also advocates the close cooperation of professionals from different disciplines rather than the production of a synthesis of different disciplinary approaches from a few dominant disciplines, as this may risk distorting transdisciplinary approaches by not adequately addressing the different spheres that make up social-ecological relationships. Cooperation between disciplines implies cooperation between different professionals. Lessons learned from long-term projects addressing 'wicked problems' from the perspectives of different disciplines, practitioners, or stakeholders shed light on new directions in which transdisciplinary approaches need to move. It highlights reservations between disciplines in valuing local and traditional knowledge, borrowing research methods and techniques from other disciplines without understanding how they work, and difficulties in integrating social data, usually qualitative, into management and governance systems. Finally, this book points out key aspects to consider in future research agendas for overcoming these difficulties and developing new transdisciplinary approaches in the co-production of culturally sensitive knowledge, as well as in the assessment of social-ecological interactions that need to be incorporated into management and governance systems in order to effectively deliver a transition to sustainable oceans and seas.

This book is divided into five parts. The contributions in Part I discuss theoretical approaches to understanding how transdisciplinarity can be appropriately

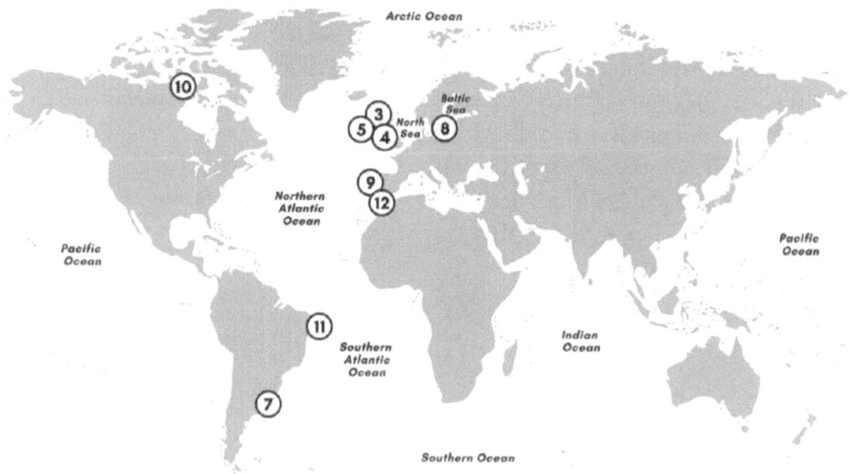

FIGURE 0.1 The geographical scope of this book's transdisciplinary case studies (chapter numbers in circles).

Source: © Vera Köpsel

formulated to provide a common knowledge pool in the framework of socio-ecologically based sustainable management. In Part II, methodology and perspectives at the forefront of research point to participatory systems of gathering societal views, needs, and concerns to assess valid knowledge and social processes embedded in and constitutive of the environment to enhance equity by taking into account all different social groups in ensuring transparency, democracy, and fair decision-making. Part III addresses transdisciplinary practices through geographically wide-ranging case studies that move into the spotlight on ecological and cultural diversity through the lens of diverse social systems, indigenous knowledge principles, and Western societal ways of life (see Figure 0.1). Part IV provides us with examples of long-term projects with a long experience in participatory processes with fishers to integrate their knowledge in management. Although focused on fisheries, which has a long history of stakeholder engagement initiatives compared with other marine activities, the results can be applied in other sectors and serve as the basis to identify new directions for transdisciplinary research. Finally, Part V provides the closing remarks with some conclusions and recommendations. It highlights key aspects to be considered in future research agendas to develop transdisciplinary approaches.

The authors contributing to this book have backgrounds in biology, socio-anthropology, geography, economy, and political science. This diversity of academic fields brings together a blend of qualitative and quantitative mixed methods stemming from different disciplinary traditions that come into play when approaching socio-ecological assessments from a multi-actor, multi-scale, and transdisciplinary approach. The composition of chapters enables us to point out the weaknesses and strengths of transdisciplinarity drawing on a theoretical basis rooted in existing literature, methodological experiences enabling participatory processes, and empirical case studies showcasing transdisciplinary approaches in practice.

"Transdisciplinary Marine Research – Bridging Science and Society" sets on the table the potentials, achievements, misunderstandings, gaps, limitations, and barriers of this common pool of knowledge resources to face today's societal and environmental challenges in marine science – reflections that enable us to identify gaps and solutions for moving effectively forward in the direction of sustainably managing our seas and oceans.

References

Funtowicz, S.O., Ravetz, J.R. (1990). *Uncertainty and Quality in Science for Policy*. Kluwer Academic Publishers, The Netherlands

Funtowicz, S.O., Ravetz, J.R. (1991). A New Scientific Methodology for Global Environmental Issues. In: Costanza, R. (Ed.), *Ecological Economics: The Science and Management of Sustainability*. Columbia University Press, New York.

Funtowicz, S.O., Ravetz, J.R. (1993). Science for the Post-Normal Age. *Futures* 25 (7):735–755.

Gibbons, M., Limoges, C., Nowotny, H. et al. (1994). *The New Production of Knowledge*. Sage, London.

Jahn, T., Bergmann, M., Keil, F. (2012). Transdisciplinarity: Between Mainstreaming and Marginalization. *Ecological Economics* 79: 1–10.10.1016/j.ecolecon.2012.04.017

Klein, J., Grossenbacher-Mansuy, W., Häberli, R., et al. (Eds.) (2001). *Transdisciplinarity: Joint Problem Solving among Science, Technology and Society. An Effective Way for Managing Complexity*. Birkhäuser Verlag, Basel, Boston, Berlin.

Nicolescu, B. (2010). Methodology of Transdisciplinarity – Levels of Reality, Logic of the Included Middle and Complexity. *Transdisciplinary Journal of Engineering & Science* 1. DOI: 10.22545/2010/0009

Noël-Knapp, C., Reid, R.S., Fernández-Giménez, M.E., Klein, J.A., Galvin, K.A. (2019). Placing Transdisciplinarity in Context: A Review of Approaches to Connect Scholars, Society and Action. *Sustainability* 11: 4899. DOI: 10.3390/su11184899

Patterson, M., Glavovic, B. (2013). From Frontier Economics to an Ecological Economics of the Oceans and Coasts. *Sustainability Science* 8: 11–24. DOI: 10.1007/s11625-012-0168-2

Rigolot, C. (2020). Transdisciplinarity as a Discipline and a Way of Being: Complementarities and Creative Tensions. *Humanities & Social Sciences Communications* 7: 100. DOI: 10.1057/s41599-020-00598-5

Rudd, M.A. (2014). Scientists' Perspectives on Global Ocean Research Priorities. *Frontiers in Marine Science* 36 (1): 1–20. DOI: 10.3389/fmars.2014.00036

Scholz, R.W., Steiner, G. (2015). The Real Type and Ideal Type of Transdisciplinary Processes: Part II – What Constraints and Obstacles Do We Meet in Practice? *Sustainability Science* 10 (4): 653–671

Stock, P., Burton, R.J.F. (2011). Defining Terms for Integrated (Multi-Inter-Trans-Disciplinary) Sustainability Research. *Sustainability* 3: 1090–1113. DOI: 10.3390/su3081090

Sílvia Gómez and Vera Köpsel

ACKNOWLEDGEMENTS

The idea for this book emerged from the research insights and experiences in the framework of Horizon Europe's PANDORA project (Grant Agreement No. 773713). PANDORA aimed at providing science-based advice to improve the blue economy in close collaboration with stakeholders, NGOs, and policy-makers.

From the outset, the project design envisaged the generation of new knowledge in co-creation with stakeholders through different participatory processes that have been carried out in the different case study regions. In the course of these participatory processes, it was possible to establish a dialogue between the social scientific and natural scientific partners in PANDORA. New avenues of communication opened up that allowed reflection on the limits and potentials of transdisciplinary approaches and their capacity to enrich policy processes and improve the bridging between science and society.

All of these aspects have transcended the expectations of the project itself. The proof of this is the edited book you have in your hands. In it, we have sought to respond to the challenges at the forefront of marine research by addressing the problems of contemporary society in relation to the environment. In doing so, we hope to make a step further towards the sustainability of the seas and oceans, to improve social and natural life in a fair and equitable world with respect for the environment, and cultural and social diversity.

We thank all project partners in PANDORA whose support for this transdisciplinary look has been key. We much thank the authors contributing their valuable perspectives to this book – without you this book would not exist. Finally, we would like to thank the editors of the environment and sustainability team at Routledge, especially Hannah Ferguson, for trusting in our book idea and helping us to make it a reality.

PART I

Theoretical and conceptual approaches

In order to successfully develop transdisciplinary methods for researching the marine and co-creating sustainable management practices, it is vital to first take a theoretical look at what transdisciplinarity is, how it is conceptualised throughout the marine sciences, and how these conceptualisations influence the ways in which we interact with each other and our environment. The four chapters in this section to exactly that: they move into focus the underlying assumptions and definitions held by different actors about transdisciplinarity, but also about what the ocean is and how it should be managed. In this context, Grünhagen and colleagues (Chapter 1) lay the foundation for this book by analysing the multifaceted picture of transdisciplinarity in the diverse disciplines of marine research. Based on an extensive literature review, they illustrate how differently the term is defined across the fields of scientific methodology, governance, ecosystem services, fisheries and management, hazards and resilience, and geosciences – bringing awareness to the need to stay in close dialogue about what we mean when speaking about transdisciplinarity and what methodological implications result from this meaning.

When engaging with marine transdisciplinarity, it is important to consider that not only the term is defined differently by different disciplines and actors, but that the ocean itself is subject to various meanings, values, and conceptualisations. In Chapter 2, Brum Bulanti and colleagues discuss how 'humanising' the oceans plays a key role in building transdisciplinary processes and co-producing knowledge. This 'humanisation' refers to moving the ocean out of its role as an uninhabited, vast, unknown space and placing it into the light of the human activities that it shapes and is shaped by. This focus on the human dimension of the marine enables us to not overlook the communities living by and off the sea, the values and beliefs that shape their actions, and thus moving closer towards equitable and sustainable management of our oceans.

DOI: 10.4324/9781003311171-1

In a similar vein, Collins (Chapter 3) shifts our attention to the spatial meanings that we give to the sea. By looking through the lens of spatial planning, she brings to light contemporary representations of marine spaces and the role that media images, documentaries, fictions, and stories play in them. This lens can help understand why marine spaces with similar physical characteristics can be associated with very different meanings based on local contexts, historical developments, and personal experiences with these spaces. This has relevance for sustainable ocean management especially in cases where such meanings and values collide, and has much power to shape people's attitudes and actions in relation to the sea. Laying open and comprehending these diverging meanings can aid greatly in finding common ground in marine management and developing solutions based on consensus rather than dissent.

The relationships that we have with the ocean play a key role not only in sustainably managing it but also in building the participatory processes that lead up to successful management. In Chapter 4, Buchan discusses how a democratisation of the human-ocean relationship through marine citizenship is an important cornerstone of transdisciplinary research. Taking on a post-normal science (PNS) perspective, she discusses the relationship between marine citizenship and ocean knowledge. Integrating marine citizens and their knowledge into the scientific peer community enriches the ways in which we, as researchers, can know and understand the marine – and is thus a valuable component in the science-society-policy mosaic that comes into play in decision-making about and for our oceans.

1

THE MULTIFACETED PICTURE OF TRANSDISCIPLINARITY IN MARINE RESEARCH

Caroline Grünhagen, Heike Schwermer, Christian Wagner-Ahlfs, Rudi Voss, Felix Gross, and Marie-Catherine Riekhof

Introduction

Marine Oceans and coastal waters are usually shared by heterogeneous groups, such as commercial fisheries (e.g. exploitation of fisheries resources), tourism (e.g. holiday and recreation), or industry (e.g. generation of renewable energy). The resulting diverse requirements for space and resources often cause problems and conflicts between user and interest groups, so-called stakeholders.[1] Different stakeholders hold diverging perceptions of the systems they are a part of, their interactions and dynamics (Gray et al., 2012; Stier et al., 2015; Aminpour et al., 2020), and the knowledge associated with it (Schwermer et al., 2021a, 2021b). They face different economic interdependencies (Lopes et al., 2017; Schupp et al., 2021; Stelzenmüller et al., 2022) and have different cultural identities (Sterling et al., 2017). Oceans and coastal waters consequently transform into very complex systems. The implementation of "right" management measures that are aligned with the many requirements and desires of stakeholders becomes a difficult task. In such a situation, the implementation of sustainability goals is limited (Burns and Stöhr, 2011; Adams et al., 2003).

These multi-layered, complex, and interlocked problems, also known as "wicked problems", pose increasing challenges to management, society and science (Jentoft and Chuenpagdee, 2009; DeFries and Nagendra, 2017; Jones and Seara, 2020; Hare, 2020). As first described by Rittel and Webber (1973), wicked problems describe a complex and tricky problem that is both symptom and cause of other problems. It is difficult to capture and tackle the problem in its entirety, partly because of the many stakeholders involved and their varying perceptions, knowledge, and interests (Rittel and Webber, 1973; Jentoft and Chuenpagdee, 2009; Hare, 2020). This complexity is further enlarged by the fact that climate change poses additional challenges to oceans and coastal seas (Möllmann et al., 2021;

DOI: 10.4324/9781003311171-2

IPCC, 2022). To find a way forward that not only promotes the health of marine ecosystems but also secures the livelihoods of coastal communities, transdisciplinary approaches are a possible solution.

The Swiss psychologist Jean Piaget is considered responsible for coining the term "transdisciplinarity" in the 1970s. He defines it as a higher level of interdisciplinary relations, placing them in a more complex system. To date, transdisciplinarity research is a growing field in academia, but there is still no uniform definition. It is often contrasted to disciplinary, multi- and inter-disciplinary research: while disciplinary research involves only one discipline, multi- and inter-disciplinary research includes several disciplines (see Figure 1.1). However, the difference of interdisciplinary research is the collaboration of all disciplines to achieve the goals set within a given project. Furthermore, transdisciplinary research involves stakeholder groups from various backgrounds.

We aim to better understand the research landscape in marine research related to "transdisciplinarity". Particularly, we examine whether and how publications in marine research can be clustered and how these groups can be set in relation to one another. The resulting picture should help to better understand how "transdisciplinarity" is used in the context of marine research. To do so, we performed a systematic literature review to identify publications on marine topics using the term "transdisciplinarity". We conducted a cluster analysis and interpreted the detected clusters based on the significant words as well as their spatial positioning in the overall cluster.

We identified 9,228 publications using the term "transdisciplinarity". Among them, 388 include terms from the marine realm and 211 of them could be identified as relevant. Based on a detrended correspondence analysis and an agglomerative hierarchical cluster analysis (programme R), six research clusters could be detected: (i) "Scientific methodology", (ii) "Governance", (iii) "Ecosystem Services", (iv) "Fisheries and Management", (v) "Hazards and Resilience" and (vi) "Geosciences". We display the research clusters in a two-dimensional space (see Figure 1.3 later in the chapter), that is, the *y*-axis shows the application of the concepts of disciplinary, interdisciplinary, and transdisciplinary approaches, while the *x*-axis refers to the topics examined relating to biotic and abiotic factors (e.g. fishery vs. hazards). "Scientific methodology" shows that many publications deal with transdisciplinary methodology: who is involved in research, how is research conducted, and how are the results managed. "Ecosystem services", "Hazards and Resilience" and "Fisheries and Management" indicate the strong relation between humans and nature, while "Governance" focuses on regulating the interaction of humans and nature. Especially the first three clusters range across the inter- to the trans-disciplinary realm while "Governance" stretches more towards the disciplinary realm. The cluster "Geosciences" was particularly surprising. It turned out that transdisciplinarity and interdisciplinarity are used synonymously, such that this cluster contrasts with the other five clusters in terms of the conceptualization and application of the term transdisciplinary. This is a clear indication of the different uses of this concept, requiring a close examination of its origin and the need to clarify concepts when working in teams with different (disciplinary) backgrounds.

Our cluster analysis reveals different traditions in the fields related to the emphasis put on disciplinary, multi-, inter- and trans-disciplinary elements within transdisciplinary approaches.

Literature review: transdisciplinarity – a paradigm shift in science

Etymologically, transdisciplinarity means "beyond disciplines", which can be understood in two ways: either as positioning beyond single scientific disciplines or as motion, with knowledge production moving away from the traditional academic way of disciplinary working (Vilsmaier, 2021). These possible translations reflect two different understandings of transdisciplinarity – one focusing on the relation between traditional scientific disciplines, and the other on epistemology and the question of how knowledge is created. Nicolescu (2010) even called this a "war of definitions". Having a closer look at this academic debate and the origin of the various schools (e.g. Nicolescuian School), it becomes obvious that despite all differences there is one common issue: the discussion about the self-perception of scientists and scientific methodology as part of society. This discussion inevitably leads to the question of which role science plays in solving problems in general and "wicked problems" in specific (Rittel and Webber, 1973; Hare, 2020).

Within the academic discourse, the term transdisciplinarity first came up in the 1970s when the Swiss psychologist Jean Piaget asked for improving the cooperation between different scientific disciplines (Lawrence et al., 2022) (Figure 1.1). Like Piaget in the 1970s, Mittelstraß (2007) suggested broadening the concept of interdisciplinarity to enable science to tackle problems. The intention was not to replace disciplines, rather to have them cooperate to overcome traditional thinking (Figure 1.1). The topic received increased interest with the "First World Congress of Transdisciplinarity" in 1994, followed by the "Manifesto of Transdisciplinarity" (Nicolescu, 2002, 2010). Lawrence et al. (2022) analysed the immense amount of literature published since then and conclude that many publications belong to a "Nicolescuian School". It is characterized by their demand for a "Unity of Knowledge": instead of separating science into hundreds of new disciplines, we should return to an understanding of science as one principle. The "Charter of Transdisciplinarity", adopted by the first "World Congress of Transdisciplinarity" in 1994, even asked for a dialogue between different knowledge cultures, academic and nonacademic disciplines (Nicolescu, 2002).

Another way of understanding transdisciplinarity is summarized by Lawrence et al. (2022) as "Social Engagement Transdisciplinarity" (the Zurich School). In this context, Gibbons et al. (1994) introduced the "Mode 2 Knowledge Production". While "Mode 1" refers to the traditional occidental science and scientific discovery, "Mode 2" is defined as a paradigm shift towards user-oriented research relating to societal needs. This debate about the relation between science and society grew in a context of intense public and political discussion about environmental

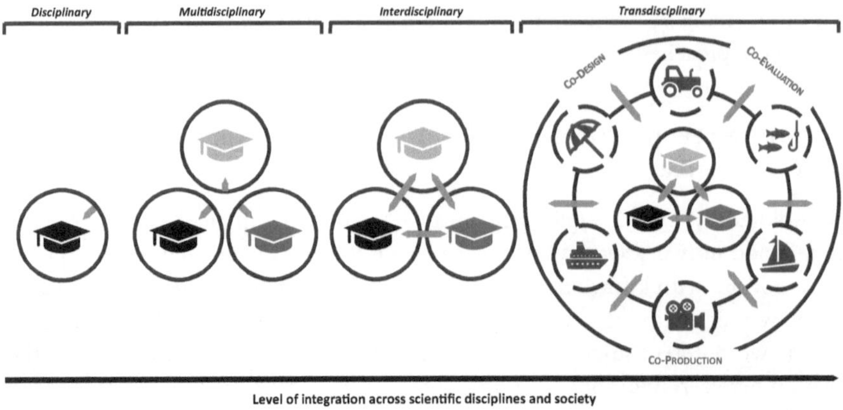

FIGURE 1.1 Conceptual distinction between disciplinary, multidisciplinary, interdisciplinary, and transdisciplinary research related to actors involved. While disciplinary research involves only one discipline, multi- and inter-disciplinary research includes several disciplines. However, the difference of interdisciplinary research is the collaboration of all disciplines (displayed as bi-directional arrows) to achieve the goals set within a given project. Transdisciplinary research further involves stakeholder groups from various backgrounds, for example, tourism, agriculture, or fisheries. In this regard, co-design, co-evaluation, and co-production are key aspects.

problems and social challenges as wicked problems (Gibbons et al., 1994). Central questions emerged regarding the shaping of a sustainable future and that of a society with a focus on bridging the gap between science and society, making knowledge available for necessary decision-making in society, and supporting participatory decision-making (Pohl and Hirsch Hadorn, 2006). With this understanding, science is not isolated from societal issues but is part of its learning process.

Today, almost 50 years after the first mention of the term "transdisciplinarity", there is still no unified concept. However, Lawrence et al. (2022) identify seven key elements: (i) focus on the theoretical unit of knowledge, (ii) inclusion of multi- and inter-disciplinarity, (iii) involvement of societal actors, (iv) focus on real-world problems, (v) working in a transformative manner, (vi) orientation towards a common good, and (vii) reflexivity.

Even if the Nicolescuian and the Zurich School seem to differ in some way, there is a huge overlap. The understanding of transdisciplinarity can even be simplified to three characteristics (WBGU, 2011; Defila and Di Giulio, 2018; Pohl et al., 2017): (i) the goal of transdisciplinarity is to create new knowledge by integrating existing knowledge from various sources (e.g. experience, observation, trainee), perspectives, and cultures. (ii) This knowledge, known as transformative knowledge, supports society to transform towards a more sustainable future. (iii) Science combines abstract knowledge with case specific knowledge and thus becomes part of a social learning process.

From the classical academic point of view, scientists working in research institutions are interacting with stakeholders outside academia (Wagner-Ahlfs et al., 2021) – there is a "we" (academia) and "they" (society). The democratization of knowledge creates a need for a new self-reflection: who is taking which role in a transdisciplinary research process? In this regard, various existing concepts reflect on the different levels of participation (Arnstein, 1969; Stauffacher et al., 2008; Shirk et al., 2012). Arnstein (1969), in his reflections on citizen participation as key element of democracy, draws a "ladder of citizen participation" ranging from manipulation (= no citizen power at all) to informing and partnership to citizen control (= maximum citizen power). Stauffacher et al. (2008) focus on societal decision processes and name different intensities of involvement: information, consultation, cooperation, collaboration and empowerment. Shirk et al. (2012) give three simplified levels how societal actors can be involved in the creation of academic knowledge from a scientific perspective: "Contribution" is a low level of interaction because the project is planned, run and evaluated by scientists, whereby stakeholders only contribute data or information. "Collaboration" is a higher level of interaction, as the stakeholders are involved in planning and analysis. "Co-creation" can be considered the gold standard of transdisciplinary research as stakeholders are involved from the beginning, designing the project as well as formulating the research questions. Thus, stakeholder involvement in scientific projects not only helps to make research more socially relevant and robust but also ensures that it is more easily understood and accepted by a wide range of stakeholders, strengthening the relation between science and society (Köpsel et al., 2021).[2]

Material and methods

To explore the concept of transdisciplinarity in the marine realm,[3] we conducted a systematic literature review adopted after the approach of Abson et al. (2014) and Drupp et al. (2020), that is, a quantitative full-text analysis of scientific publications using a cluster analysis. We used a four-step procedure to identify and analyse relevant publications: (i) data searching, (ii) data screening, (iii) data download, and (iv) data analysis, consisting of three different sub-steps (i.e. conceptual vocabulary, detrended correspondence analysis, and cluster analysis) (Figure 1.2A).

In a first step, we derived relevant publications from the scientific database *Web of Science* (WoS, *www.webofscience.com*) using the search string displayed in Table 1.1 (Figure 1.2A, *data searching*):

The keywords that compose search "Term 1" relate to the marine realm and were discussed in several rounds between all authors, that is, the search string was iteratively developed. As our aim is to better understand how the term "transdisciplinarity" is used in publications related to marine topics, our search "Term 2" included the keyword "transdisciplinarity" and known misspellings of this keyword. To assess whether the search string was well-developed, we first compiled a list of publications that are relevant to the study field according to the authors' knowledge and thus should appear in the WoS hit list. Second, we took a sample of

A.

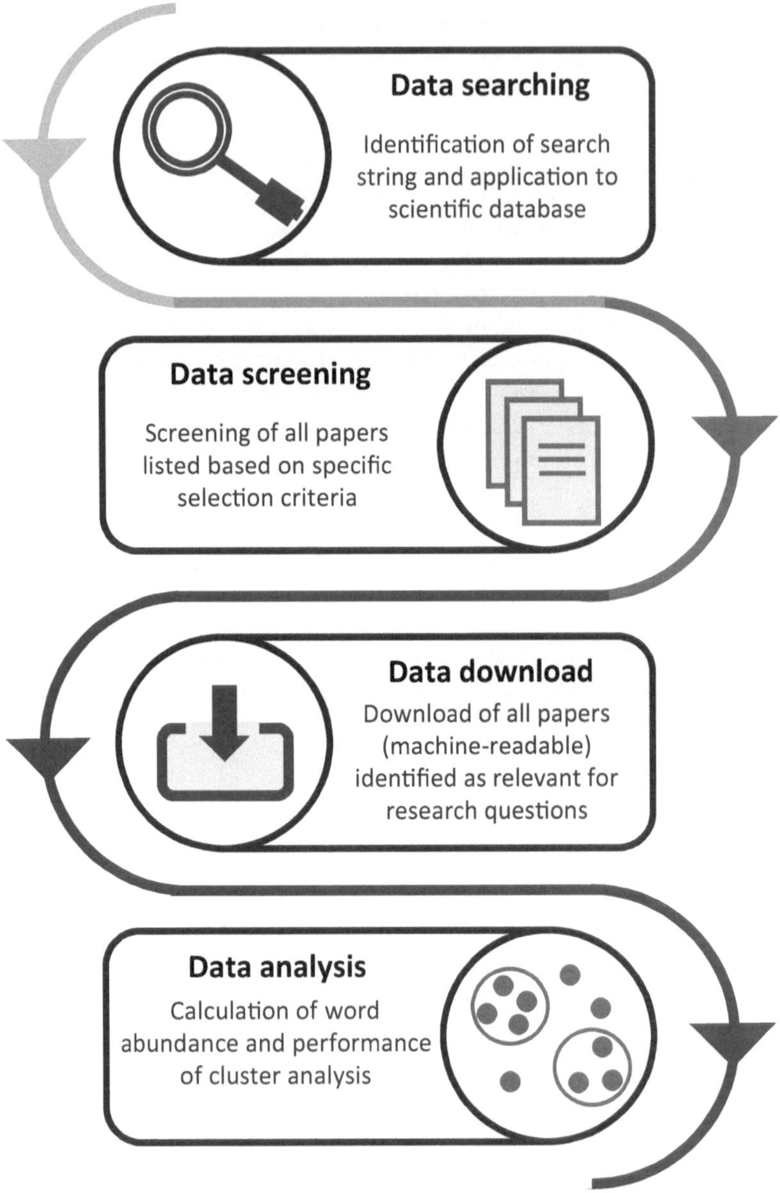

FIGURE 1.2 (A) Four-step methodological approach and stepwise selection of relevant papers (data searching and data screening). Literature was filtered by a predefined search string (*data searching*) and checked for relevance based on five different categories (*data screening*). Relevant papers have been downloaded (*data download*) and further analysed (*data analysis*). (B) A systematic literature review has been conducted focusing on the conceptualization of "transdisciplinarity" in the marine realm. A keyword-based identification strategy has been applied to Web of Science (*data searching, data screening*).

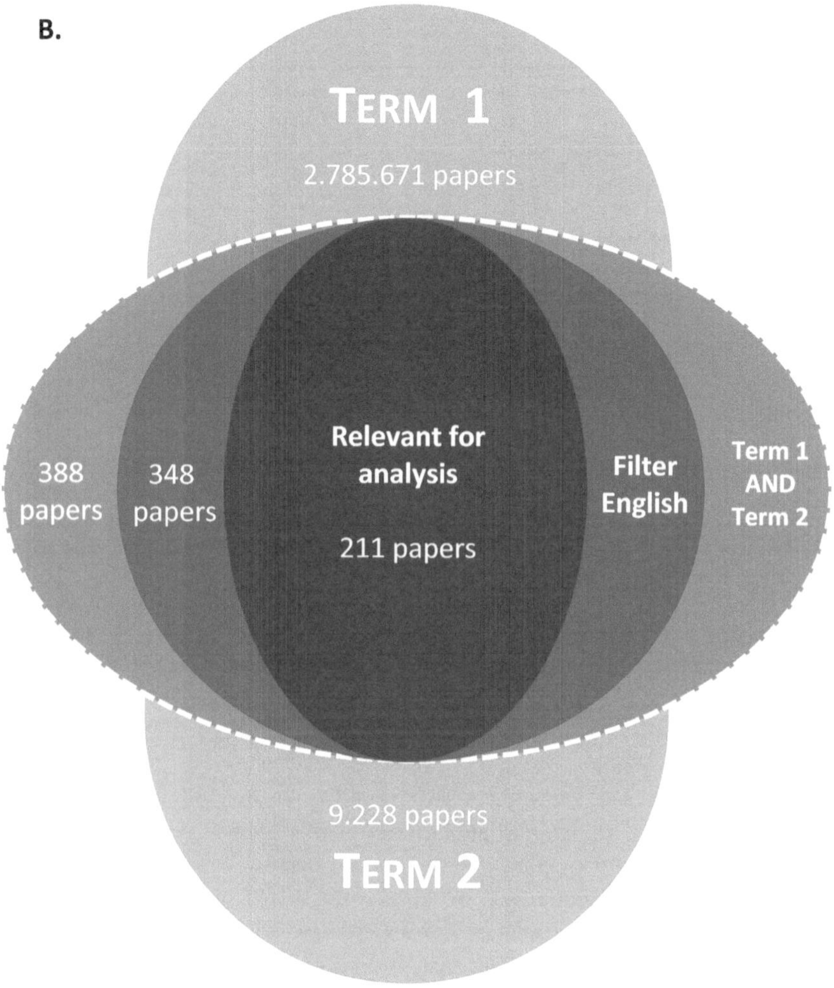

FIGURE 1.2 (Continued)

10% of the publications identified by our search string and checked their relevance to our topic by reviewing title, abstract and keywords. After minor adjustments, we arrived at our final search string identified in Table 1.1, resulting in a total of 388 papers (Figure 1.2B). Applying a language filter focusing only on English publications reduced the number of publications to 348 (Figure 1.2B).

In the second step, all identified publications were divided among the authors of this paper and reviewed in terms of their relevance (Figure 1.2A, *data screening*). For this, a deductive approach was applied, using the following five criteria: (i) marine focus, (ii) transdisciplinarity, (iii) English language, (iv) availability, and (v) quality (Appendix A.1). If a paper was not clearly identified as relevant by the reviewing author, we conducted a group discussion among all authors.

TABLE 1.1 Identified search string

Term 1	*Term 2*
marine OR maritime OR ocean OR oceanic OR oceanograph* OR seabed OR sea OR seawater OR seaside OR seafloor OR seacoast OR seashore OR seafront OR coast OR coastal OR coastline OR coastland OR shore OR shoreline OR shoreside OR shorefront OR bay OR saltwater OR "salt water" OR pelagic OR "high seas" OR demersal OR fish* OR aquaculture OR angling OR "offshore wind*" OR "offshore drilling*" OR shipping OR "blue sector" OR "blue growth" OR mariculture OR "EU water framework directive" OR beach	transdi$ciplinary OR transdi$ciplinarity OR transdi$cipline

Note: An advanced keyword search has been applied to publications title, abstract and authors keywords using all databases of *Web of Science* for the most comprehensive results.

In this context, we included 211 papers that were published between 1974 and 2022 (Appendix A.2) (Figure 1.2B). Third, all relevant publications were downloaded in a machine-readable format as pdf-files for further analysis of full-texts (Figure 1.2A, *data download*).

Fourth, we evaluated how the term "transdisciplinarity" is used within the marine realm. Doing so, relevant publications were analysed on the basis of their full-texts employing a detrended correspondence analysis. This method is usually applied to ecological community data but has been widely tested to identify principle gradients in research landscapes of review papers before (e.g. Abson et al., 2014; Drupp et al., 2020). The data analysis consisted of three sub-steps: (i) selection of conceptual vocabulary from relevant full-texts as the basis for the analysis; (ii) identification of principle gradients in the research landscape using a detrended correspondence analysis; (iii) performance of an agglomerative hierarchical cluster analysis (Figure 1.2A, *data analysis*).

To obtain the conceptual vocabulary, we applied automated and manual filters. At least two authors needed to agree to keep or remove words for further analysis. Using the software R, we then performed a cluster analysis on the basis of the conceptual vocabulary (for details, see Appendix – Additional information to Material and Methods).

Overall, a total of 556 significant words were found within the transdisciplinary marine research environment (Appendix A.7). For each cluster, we selected 15 significant words that we consider to be most representative (see the list of significant word in Table 1.2), based on which the gradients of the research landscape – represented by the axes – have been determined.

TABLE 1.2 Six research clusters in the transdisciplinary, marine landscape

Cluster	*# of papers*	*Significant conceptual words*
Scientific Methodology	22	perspectives, communication, collaboration, engagement, experiences, transdisciplinarity, expectations*, non-academic, interdisciplinarity, stakeholders, academic, participants, societal, recommendations*, guide
Governance	30	marine, biodiversity*, monitoring, sciences, network, science-policy, exchange, ecosystem-based, international, cross-disciplinary, experts, implementing, ecosystems, action, governance
Ecosystem Services	77	ecosystem, habitat, services, conservation, conceptual*, modelling, process, ecological, indicators, consultation, reserve*, quality, recreation, agricultural, trade-offs
Fisheries and Management	22	economic, fisheries, small-scale, fishers, poverty, cooperative, compliance, artisanal, by-catch, management, co-management, policies, socioeconomic, quotas, interviewed
Hazards & Resilience	24	change, climate, planning, public, coast, risk, protection, adaptation, vulnerability, hazards, damage, citizens, resilience*, exposure*, decision-makers
Geosciences	36	morphology, volcanic, isotope, deposition, composition, periods, plate, shoreline, dune, hydrodynamic, satellite, formation, sedimentary, geology, jurassic

Note: For each cluster, the number of relevant publications and the most representative[4] significant conceptual words (N=15) are listed (words marked with "*" have not been plotted in order to improve the readability of Figure 1.3).

Cluster names are chosen manually on the basis of significant words or even consist of them. On top of the cluster analysis, we further checked for connectivity among the research clusters by spanning ellipses around the plotted words of each cluster.

Results

Based on the detrended correspondence analysis of 211 relevant publications, six research clusters could be identified within marine science using the term transdisciplinary. The respective clusters are visualized by means of an ordination plot (Figure 1.3), with axes scaled by DCA units.[5]

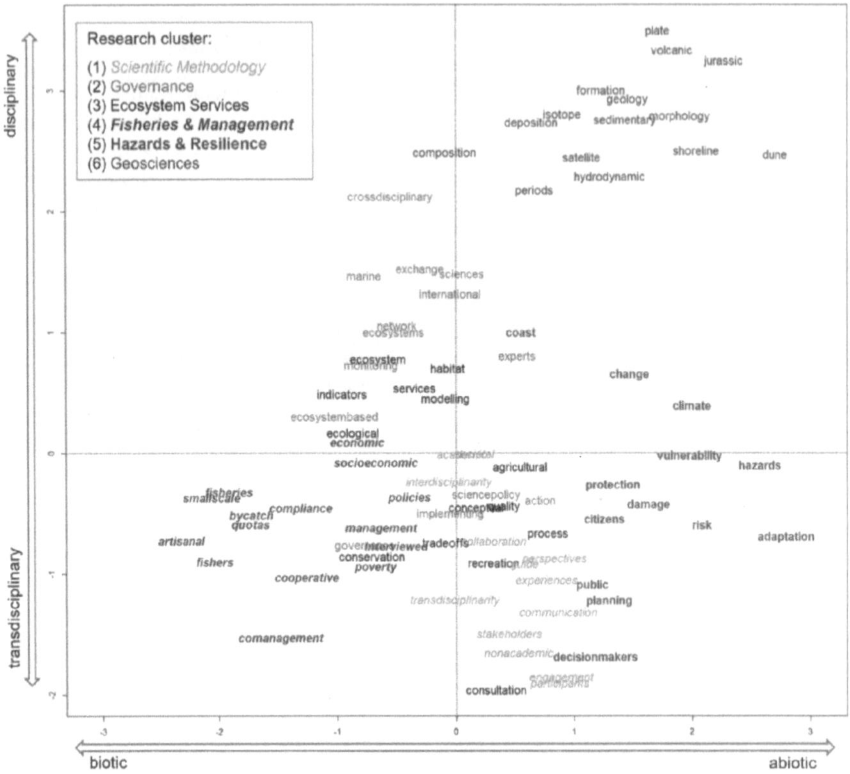

FIGURE 1.3 The multifaceted picture of transdisciplinarity in the marine realm. Six research clusters could be identified on the basis of a full-text analysis of 211 relevant publications. For each cluster, the most significant words are displayed (see Table 1.2 for a list of plotted words).

The ordination plot (Figure 1.3) shows the multifaceted picture of transdisciplinarity in marine-related research, ranging over six DCA units from −3 to +3. We interpret the horizontal axis as the research environments that deal with either biotic or abiotic factors (from left to right). The vertical axis displays the level of integration across scientific disciplines and societies, similar to Figure 1.1. Starting at the top (+3.5), research focuses strongly on its own discipline. Along the axis, the degree of knowledge integration increases continuously, as indicated by the terms "crossdisciplinary" (+2) and "interdisciplinarity" (−0.25), with the highest degree "transdisciplinarity" at the lower end (−2). Thus, we interpret this gradient as the level of knowledge integration, ranging from specific disciplinary academic knowledge to the integration and active involvement of non-academic knowledge. Both gradients allow us to classify and analyse relevant publications into clusters based on their significant conceptual words. Each cluster represents publications sharing similar conceptual vocabulary, whereas dissimilar vocabulary, and thus clusters, is placed far apart. However, the identified research clusters may overlap in the conceptual vocabulary of the respective publications.

Next, we provide detailed thematic descriptions of the six research clusters including their spatial arrangements along the gradients, placing them in a larger context. Further, we analyse how the research clusters are (dis-)connected in terms of their conceptual vocabulary.

Methodological approaches applied in transdisciplinary research

The significant conceptual vocabulary of cluster 1 "Scientific Methodology" can be derived from the area of methodology: (i) who is involved in the research, (ii) how is the research conducted, and (iii) how are the results managed? "Transdisciplinarity" is identified as one of the most significant words, indicating that the methodological approaches are primarily used in a transdisciplinary context. Interestingly, the term "interdisciplinarity" is identified as significant as well, defining the scope of application range.

One thematic area of this cluster focuses on the level of "stakeholder" involvement in transdisciplinary research. The term "society" in connection with the word pair "non-academic" versus "academic" hints to the idea of exchange between different groups, and between science and society. Along those lines, stakeholders outside of academia may act as "participants", exchanging "expectations" and "experiences". Generally, active "communication" is a key element of transdisciplinarity, although it is important to specify what kind of communication is considered: (i) one-way communication or (ii) bi-directional communication. The different kinds of communication become even clearer by considering another significant term: "collaboration". Thus, stakeholders can be actively involved in the research, not only by contributing some information but also by having a more intense communication, either by co-designing the research questions or by co-analysing and co-interpreting the results.

Another research theme focuses on the development of "recommendations" for society or political decisions. While the concept of objective knowledge generation dominates in the classical understanding of science, interaction with society plays an essential role in transdisciplinary research according to our understanding. There is "engagement" and the aim to include different "perspectives", possibly to "guide" policy-makers. Overall, we note that scientific methodology of transdisciplinarity is applied when different research areas overlap as the cluster's location at the intersection of biotic and abiotic environments reveals.

Goals, measures, and stakeholder involvement in ocean governance

The "Governance" cluster is captured as ocean governance addressing the following key questions: (i) where does ocean governance take place, (ii) what are the goals behind ocean governance, (iii) who is involved in the conceptualization and implementation of ocean governance, and (iv) which measures are needed to achieve these defined goals?

In general, "ocean" "governance" is about managing the use of the oceans "worldwide" such that an "international" approach seems to matter. One focus is on "marine" "ecosystems", for example, preserving and enhancing their "diversity" (i.e. "biodiversity"). Central to this is "ecosystem-based" management, often in a "spatial" context. To ensure a successful implementation of the goals set, it is necessary to involve different "sciences" (also called "crossdisciplinary"), as well as other "experts". With regard to the "science-policy" interface, the key role of ocean governance is on the "exchanges" of experiences and good practices and thus the joint creation of knowledge with the aim of informing decision-making. In this regard, various "actions" and "initiatives" can be defined and implemented to achieve these goals. These include, for example, "forums" or "platform" that contribute to strengthening knowledge about the ocean system and associated marine ecosystems but can also support the development of a common "vision". Here, "monitoring" and "observation" are central in order to accompany the implementation of measures aiming for clean, healthy and productive oceans.

Orienting along the vertical gradient in Figure 1.3, the cluster "Governance" mainly covers the area from cross-disciplinary to interdisciplinary research. The clusters thematic focus (horizontal axis) is mainly on biotic ecosystems of the ocean.

Trade-offs in managing ecosystems

The cluster "Ecosystem Services" encompasses mainly literature from a broad field of ecology. It deals with (i) different marine ecosystems and methodological approaches to asses them, (ii) its use values, (iii) nature conservation, and (iv) its inherent close connectivity to other disciplines and society.

The location of the cluster in the ordination plot (Figure 1.3) reveals the complexity of ecological research due to a large shared vocabulary with other clusters. It focuses on "ecosystems" and "habitats" having direct implications on society. To assess the state of an ecosystem, ecological "indicators" are commonly used to identify drivers of change or the habitat structure. Above, dynamic behaviours of a complex, adaptive system can be investigated by "conceptual" or quantitative "modelling". Explaining conditions within landscape units, conceptual models illustrate the connections of environmental stressors, management actions and resulting effects on ecology and society, involving transdisciplinary research (Fiksel et al., 2014). In addition, further insights into the system can be provided due to "consultation" between stakeholders and scientists.

The interaction between humans and nature is demonstrated by values that ecosystems provide in form of ecosystem "services". Significant words such as "recreation" and "quality" refer to cultural services, indicating the relevance of social aspects in transdisciplinary research. As this cluster overlaps with the governance cluster, the quantification of ecosystem services seems to play a crucial role in the conservation and management of nature in the long run, benefiting both nature and humans. In the marine context, conservation issues are directly related to management approaches and can range, for instance, from marine protected areas to tidal river management. Thereby, this thematic area of environmental protection ("conservation", "reserve") is strongly linked to human well-being.

The complexity of ecological systems and their use leads to "trade-offs" and synergies between different sectors. Another research focus is on "agricultural" activities, being of particular interest in landscape ecology. Generally, the "ecosystem services" cluster is an important research area dealing with spill-overs of different biotic and abiotic systems. Nonetheless, the clusters' wide expansion along the vertical axis, including inter-, and trans-disciplinary research (Figure 1.3), suggests that publications within the group might have a different understanding of the involvement of participants outside academia.

Bycatch and overexploitation as key problems in fisheries management

Another cluster can be summarized as "Fisheries & Management". As the title implies, this research landscape deals with "fisheries" and encompasses the following thematic fields: (i) disciplines that study fishery-related issues, (ii) stakeholders, (iii) methodological approaches, and (iv) governance. Figure 1.3 shows that the cluster deals exclusively with biotic factors as marine resources such as fish reflect the basis of this research area. The methodological scope along the vertical gradient is mostly bounded to transdisciplinary approaches, but also extends into interdisciplinary research. Reasoning is the collaboration between economists and ecologists, who make use of ecological-"economic" models to analyse fish stocks and their development under certain management schemes.

Significant terms such as "bycatch" and "poverty" reduction mirror active research fields of the past years. The complete list of significant words allows for detailed insights into the research area (see Appendix A.7). Accordingly, "overfishing", "overexploitation", "illegal", and "collapse" are still challenging issues, being opposed to term "sustainability". In this context, "socioeconomic" considerations are necessary to show the resulting consequences on society, for example, people whose livelihoods depend on fishing. Different stakeholders such as "artisanal" and "small-scale" "fishermen" could be identified. The involvement of these actors in transdisciplinary fisheries research includes methodological concepts from interviews (i.e. actors are "interviewed") to "co-management". Co-management, that is, the inclusion of non-scientific actors in management decisions, can take several forms and is an official goal of the European Common Fisheries Policy. Another thematic focus within the "Fisheries and Management" cluster is related to governance issues or "management" regulations such as "quotas". Transdisciplinary research addresses "policy" questions, a topic of special interest to non-scientific actors, including "compliance" as one important component which requires actors' acceptance for a successful and sustainable management. In this way, the management aims to create a fair balance between the interests of various actors involved.

Risk = hazard × exposure × vulnerability

The cluster "Hazards and Resilience" focuses in the context of transdisciplinary research on the following sub-fields: (i) the relationship of risk in a changing system, (ii) the effects of hazards on the society, (iii) the reaction of nature and society to hazards, and lastly (iv) regional aspects. The cluster covers a large part of the fourth quadrant of the ordination plot, slightly merging with the first one (Figure 1.3). Thus, publications of this cluster focus on abiotic environments like climate. Interdisciplinary approaches form the basis of this cluster, being extended by methodological concepts from transdisciplinary research.

The "Hazards and Resilience" cluster can be described by the so-called risk equation whose components correspond to significant conceptual words, namely: "risk" = "hazard" × "exposure" × "vulnerability". These terms are essential and often used in hazard and resilience studies. Especially the term "risk", being the product of the other factors, plays an important role as transdisciplinary research in this field often aims to reduce it. Another theme within this research cluster seems to relate to the effects on and from society to hazards and potential damage. Here, the most significant terms are concentrating around "citizens", "planning", "public" and "resilience". Transdisciplinarity is paramount for many planning studies on resilience, which includes citizens, academia and public authorities. In this domain, especially in the decision-making process, transdisciplinarity is already a standing term and an applied approach.

Further typical characteristics of this cluster are reactions of nature and society to external forcing, putting the cluster in a larger framework. The term "change" is of great importance as it might relate to climate change and a "warming" climate. Over the past decades, the field of climate change has become a major research area itself. This is in contrast to concepts such as "adaptation" and "protection", which represent the response of societies to changes in the Earth's system. Overall, all societies that are affected by marine hazards are located at the "coast". Hence, a strong connection of marine transdisciplinary research is expected to target both, the coastal area as a research focus and its interaction with "citizens".

Ocean and marine systems from a morphological perspective

The cluster "Geosciences" investigates (i) the structure and age of the Earth in general and rocks of the solid Earth in particular, as well as (ii) the processes of their formation and dynamics within Earth's system. The transdisciplinary research within this area is focusing on "geology", for example, "morphology" or "volcanic", encompassing the "formation" and "composition" of a geological system.

The "Geosciences" cluster is limited to the upper part of the first quadrant, describing a very disciplinary space that focuses on abiotic factors and coastal landscape formations. In contrast to the other five clusters, this cluster does not appear to include non-academic stakeholders when considering significant words in addition to scientific disciplines (e.g. "sedimentary"). Furthermore, Figure 1.3 shows that this cluster has a particular vocabulary, being very distinct from the other clusters and showing no overlap. Even though the term "transdisciplinarity" is used by publications belonging to this cluster, no methodological approaches relating to the transdisciplinary realm (see "Scientific Methodology") are applied.

Connectivity among research clusters

The (dis-)connectivity of the six research clusters in terms of conceptual vocabulary is displayed in Figure 1.4. Based on the cluster analysis given earlier, we found that one cluster focuses on methodological approaches in transdisciplinary research, while the remaining five clusters correspond to research fields of a variety

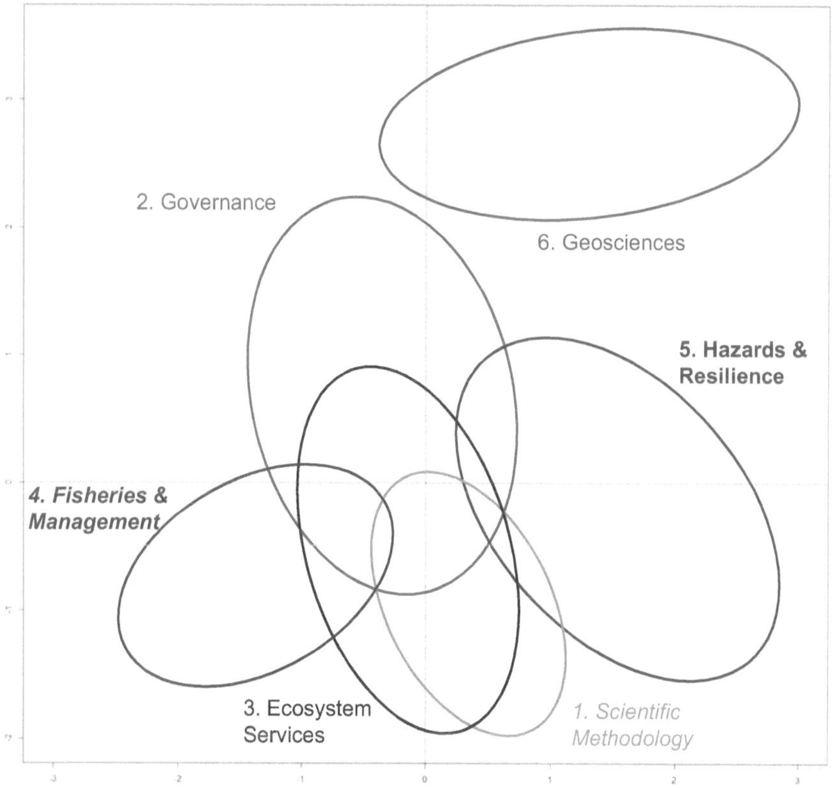

FIGURE 1.4 Connectivity between the six research clusters of transdisciplinarity in the marine realm. Ellipses are drawn around the significant words of the respective cluster (same scale as in Figure 1.3) to display the connectivity among research fields.

of disciplines which conduct transdisciplinary research, according to their understanding of the term.

We observe that "Geosciences" is very disconnected from the other research clusters. Although all publications deal with transdisciplinary approaches, the clusters "Geosciences" and "Scientific Methodology" do not overlap in terms of similar vocabulary. In comparison, the remaining research clusters are more closely linked to each other, whereby the "Governance" cluster plays a very central thematic role (large shared vocabulary with other clusters) across the marine research landscape. To be highlighted, "Fisheries and Management", "Governance", and "Hazard and Resilience" overlap with (i.e. are connected to) "Scientific Methodology", indicating that research in the marine field is subject to different transdisciplinary approaches.

Discussion

Our results demonstrate that there is neither a common understanding nor application of transdisciplinarity in marine research across and within disciplines. One of the underlying reasons is shown in the different terminology of "stakeholders" as they are assigned different expectations and tasks in the transdisciplinary landscape. Various underlying components are discussed next, helping to create an aligned understanding of transdisciplinarity.

The performed analysis showed that "Geoscienes" does not overlap with the other research cluster in terms of vocabulary. The disconnection of the "Geosciences" from the "Scientific Methodology" cluster reveals the fact of divergent understandings of transdisciplinarity. Among all significant words within the "Geosciences" cluster (Appendix A.7), none of them deal with the definition or application of transdisciplinary research according to our understanding (Figure 1.1). The reason for this gets obvious by taking a closer look at publications representing this cluster. For instance, Regier et al. (1974) argue that transdisciplinary researchers can address their research issues independently of stakeholders from different fields. This understanding is in line with several studies from the early 2000s, conceptualizing transdisciplinarity as the intersection of different disciplines (e.g. Karl, 2002; Becker and Grupe, 2012; Seabloom et al., 2012), contrasting our understanding of the concept (Figure 1.1). Even though the perception of transdisciplinary has evolved over time in this research cluster, some geoscientists still understand this concept in a manner of interdisciplinary cooperation or the use of different methods to collect data (e.g. Aucelli et al., 2021). We note that most researchers have had, and some still have, different understandings of transdisciplinarity. This emphasizes the necessity to consciously address one's own understanding of this concept to develop a common terminology for the collaboration between science and society. In turn, this can promote the exchange and success of transdisciplinary projects.

The evolution of the changing conceptual use of transdisciplinarity is also emerging in the literature of the "Ecosystem Services" cluster. While Paterson et al. (2010) observed the interchangeability of the terms "multi-, inter-, trans-disciplinarity", some more recent publications focus on the perception of knowledge integration in transdisciplinary research. Similar to Shirk et al. (2012) describing collaboration, co-creation and co-design as different levels of participation, Mobjörk (2010) uses a slightly different terminology with "consultative" or "participatory" transdisciplinarity. The former describes an approach in which non-academic stakeholders merely gather and collect data for scientists or respond to their research, while society is not directly involved in knowledge production (e.g. Torell et al., 2012). The latter approach values societal and scientific knowledge equally, considering a collective knowledge production process (e.g. Berninsone et al., 2018; Gurney et al., 2019). The publications cited earlier indicate that both concepts of communication exchange are used in recent literature and are thus implemented in practice. A changing use of the term "transdisciplinarity" is discernible, but does not apply to all publications. Accordingly, the term evolves independently of temporal changes.

In the context of sustainability research, even an understanding of transdisciplinary research as "transformation research" arose where science takes an active role in society with the aim of developing society further (Lawrence et al., 2022). However, this transformative level of transdisciplinarity is not found across all research clusters. Only in the area of "Fisheries & Management" a focus on "sustainability" research is indicated as significant (Appendix A.7). This evolved role of transdisciplinarity is not surprising as fisheries-related research already has a very strong understanding of transdisciplinarity in terms of active stakeholder engagement, for example, co-management (e.g. Esther et al., 2021).

Conceptualization and involvement of relevant actors

Stakeholders (in the broad conceptualization) are a central pillar of transdisciplinary research. However, the conceptualization of the term in general and in particular of the actors involved in each specific case sometimes reveals major variations. These differences are evident in the general inclusion of stakeholders but also in the definition of the stakeholder itself, ranging from very broad terms like expert (e.g. Johnson, 2021) to the specific identification of relevant stakeholders like fishers (e.g. Esther et al., 2021). On this basis, the question of who is named as a stakeholder in the various clusters and included in the field of transdisciplinary research will be explored in more detail further.

While from the "Geosciences" cluster, stakeholders or stakeholder groups are not among the significant conceptual words, "Ecosystem Services" focuses at least on one methodological approach (i.e. "consultation"), indicating the engagement or participation of stakeholders (e.g. Celliers et al., 2021). However, a more detailed investigation of the clusters "Scientific Methodology", "Governance", and "Fisheries and Management" reveals a distinctly different picture. In fact, this includes scientists from different disciplines, including several concepts such as "disciplinary," "cross-disciplinary," or "interdisciplinary," as well as stakeholders outside of science who are involved or affected in the different areas (e.g. "fisheries"). This very broad picture is closely related to Freeman's (2010) definition of stakeholder, according to which stakeholders are described as any group or individual who, for example, can influence or is affected by the achievement of a defined goals. While Tiller et al. (2015) are criticizing this concept as too broad, this reflects the situation given in transdisciplinary research. To further illuminate Freeman's (2010) definition and the diversity we have identified in the marine transdisciplinary context, we will exemplify the conceptualization of the term stakeholder in relation to the "Governance" and "Fisheries and Management" clusters.

Based on the results of the cluster analysis, it can be seen that in the context of governance as a research field, "experts" are considered to be the key stakeholders, in addition to "science". In this regard, a precise conceptual distinction needs to be made in the use of the term "expert". On the one hand, this term is used exclusively in a scientific context, namely "expert from science" (Hoerterer et al., 2020). However, according to Rudd et al. (2018), experts are a group of actors across sectors and disciplines, which places the definitions of expert in a broader context including stakeholders outside academia. In contrast, Johnson (2021) and Levesque et al. (2021) use the term "technical expert" to explicitly refer to the function of an "expert", namely consultation with stakeholders, such as decision-makers or community stakeholders regarding for example, the display and interpretation of data. Schmidt et al. (2014) take a completely different approach and use the term in a geographical sense (i.e. "regional experts"). Even if by definition an expert is the one we make it in the individual case (this includes the different disciplines inside and outside academia). This digression shows the diverse uses and frames the diversity of the term applied in marine transdisciplinary research.

In the context of fisheries, stakeholders are conceptualized as actors outside academia like "fishers", equivalently "fisher-men" (-"women"). In the underlying publications of the cluster "Fisheries and Management", the term "fishers" refers to either fishers represented by fishing cooperatives (Bojórquez-Tapia et al., 2021) or fishers themselves (Esther et al., 2021). Transdisciplinary research focuses on stakeholders from small-scale or artisanal fisheries. In addition, recreational fishers such as anglers, subsistence, and indigenous fisher are considered as well. Probably, successful transdisciplinary research requires close contact and the building of trust which might be less complicated in small-scale/artisanal fleets with their connection to the local socio-economic network than in larger companies.

Through our methodological approach,[6] we have demonstrated the existence of different understandings of transdisciplinary concepts. Although we can observe a gradient from disciplinary to transdisciplinary research along the vertical scale, it is interesting to note that many words explaining methodological transdisciplinary approaches are located close to the field of interdisciplinary research. Reasoning might be that interdisciplinarity is a part of the transdisciplinarity definition as shown in Figure 1.1. Nonetheless, no strong gradation of the transdisciplinary knowledge integration level is apparent in the ordination plot. For example, "consultation" is allocated at the lowest intercept of the y-axis. According to our understanding, consultation represents a lower degree of knowledge integration and thus should be located further.

Conclusion

We quantitatively examined the usage of the concept "transdisciplinarity" in the marine realm on the basis of relevant identified papers published between 1974 and beginning of 2022. Our cluster analysis revealed six larger research fields using the terminology transdisciplinarity. We found that across and even within research clusters, no common understanding of the concept exists. Especially geosciences seem to often consider transdisciplinarity synonymously to interdisciplinarity. In addition, the degree of knowledge integration of non-academic stakeholders plays a decisive role in whether and to what extent research is understood as transdisciplinary.

Through our conceptual distinction of multi-, inter- and trans-disciplinarity, we have created awareness for the various existing concepts. Especially our reflection on the use of transdisciplinarity in the (marine) scientific community contributes to a better understanding of how differently the term has been used so far. In this context, it is necessary to further sharpen the understanding of transdisciplinary methods. Here, a conscious examination of one's own understanding of transdisciplinarity is the first step and can thus promote cooperation between science and society due to the exchange of various conceptualizations in a second step.

Data availability statement

The data presented in this study are available on request from the corresponding author. The data are not publicly available due to restrictions, that is, privacy and ethics.

Appendix

For the traceability and reproducibility of our data, all information regarding the step data screening and data analysis are presented in detail in the appendix. In addition, this comprehensive document contains a list of all publications considered relevant for the cluster analysis: http://ssrn.com/abstract=4132482

Funding

CG was funded by the Federal Ministry of Education and Research, RETAKE (no. 03F0895K). HS was funded by the Federal Ministry of Education and Research balt_ADAPT (no. 03F0863D). CWA was funded by the European Commission reSEArch-EU (no. 101017454). MCR acknowledges additional funding through the project SpaCeParti (no. 03F0914A). RV acknowledges funding from European Union's Horizon 2020 Blue Growth programme under Grant Agreement No. 101000318 "SEAwise".

Notes

1 We use the term stakeholder for user groups outside the university system. But – as discussed in the chapter "Conceptualization and involvement of relevant actors" – the use of term stakeholder varies greatly in the reviewed literature.
2 Additional information to the literature review can be found in the appendix.
3 Our definition of the marine realm encompasses the marine system itself and sectors that are directly linked to it. Zones from the open ocean to the continental shelf, including adjacent coastal areas, form the framework. Thus, research studies that deal directly with these systems as well as sectors that make use of them are considered, for example, fisheries, offshore wind and shipping. Studies that address the interaction of terrestrial and marine systems, for example, via inputs of nutrients or sewage are included. Publications dealing exclusively with water systems such as freshwater or inland waters are excluded.
4 According to our understanding of representative words, we chose significant words that reflect the different facets of the respective research field and those that are connected to transdisciplinarity. If similar terms such as "fishing", "fisheries" and "fish" were among the significant words, only one of them was depicted. Same holds true for redundant words, for example, we selected the word "hazards" but ignored terms such as "storms", "flooding" and "erosion".
5 Ordination describes a multivariate technique where a multidimensional dataset (here significant words by publications) is projected onto a lower dimensional space to understand the intrinsic pattern of the underlying data (Pielou, 1984). In ecology, ordination is used on community data to describe the relation between species composition and environmental gradients. Hence, similar species are plotted close to each other, whereas dissimilar species are plotted further apart. A difference of three to four DCA units already indicate a complete turnover in species composition. Figuratively speaking in ecological terms, this corresponds to the transition of a forest to an adjacent meadow.
6 The discussion of our applied method can be found in the appendix.

References

Abson, D. J., Von Wehrden, H., Baumgärtner, S., Fischer, J., Hanspach, J., Härdtle, W., Heinrichs, H., Klein, A. M. & Walmsley, D. (2014). Ecosystem Services as a Boundary Object for Sustainability. *Ecological Economics, 103*, 29–37. https://doi.org/10.1016/j.ecolecon.2014.04.012

Adams, W. M., Brockington, D., Dyson, J., & Vira, B. (2003). Managing Tragedies: Understanding Conflict Over Common Pool Resources. *Science, 302*(5652), 1915–1916. https://doi.org/10.1126/science.1087771

Aminpour, P., Gray, S. A., Jetter, A. J., Introne, J. E., Singer, A., & Arlinghaus, R. (2020). Wisdom of Stakeholder Crowds in Complex Social – Ecological Systems. *Nature Sustainability, 3*(3), 191–199. https://doi.org/10.1038/s41893-019-0467-z

Arnstein, S. R. (1969). A Ladder of Citizen Participation. *Journal of the American Institute of Planners, 35*(4), 216–224. https://doi.org/10.1080/01944366908977225

Aucelli, P. P., Mattei, G., Caporizzo, C., Cinque, A., Amato, L., Stefanile, M., & Pappone, G. (2021). Multi-Proxy Analysis of Relative Sea-Level and Paleoshoreline Changes During the Last 2300 Years in the Campi Flegrei Caldera, Southern Italy. *Quaternary International, 602*, 110–130. https://doi.org/10.1016/j.quaint.2021.03.039

Becker, C., & Grupe, G. (2012). Archaeometry Meets Archaeozoology. Viking Haithabu and Medieval Schleswig Reconsidered. *Archaeological and Anthropological Sciences, 4*, 241–262. https://doi.org/10. 1007/s12520-012-0098-z

Berninsone, L. G., Newton, A., & Icely, J. (2018). A Co-designed, Transdisciplinary Adaptive Management Framework for Artisanal Fisheries of Pehuen Co and Monte Hermoso

(Argentina). *Ocean & Coastal Management, 152*, 37–47. https://doi.org/10.1016/j.ocecoaman.2017.11.002

Bojórquez-Tapia, L. A., Ponce-Díaz, G., Pedroza-Páez, D., Díaz-de-León, A. J., & Arreguín-Sánchez, F. (2021). Application of Exploratory Modeling in Support of Transdisciplinary Inquiry: Regulation of Fishing Bycatch of Loggerhead Sea Turtles in Gulf of Ulloa, Mexico. *Frontiers in Marine Science, 8*. https://doi.org/10.3389/fmars.2021.643347

Burns, T. R., & Stöhr, C. (2011). Power, Knowledge, and Conflict in the Shaping of Commons Governance. The Case of EU Baltic Fisheries. *International Journal of the Commons, 5*(2), 233. https://doi.org/10.18352/ijc.26

Celliers, L., Scott, D., Ngcoya, M., & Taljaard, S. (2021). Negotiation of Knowledge for Coastal Management? Reflections From a Transdisciplinary Experiment in South Africa. *Humanities and Social Sciences Communications, 8*(1). https://doi.org/10.1057/s41599-021-00887-7

Defila, R., & Di Giulio, A. (Eds.) (2018). *Transdisziplinär und transformativ forschen.: Eine Methodensammlung*. Wiesbaden, Germany, Springer VS. https://doi.org/10.1007/978-3-658-21530-9

DeFries, R., & Nagendra, H. (2017). Ecosystem Management as a Wicked Problem. *Science, 356*(6335), 265–270.

Drupp, M. A., Baumgärtner, S., Meyer, M., Quaas, M. F., & Von Wehrden, H. (2020). Between Ostrom and Nordhaus: The Research Landscape of Sustainability Economics. *Ecological Economics, 172*, 106620. https://doi.org/10.1016/j.ecolecon.2020.106620

Esther, M. C. C., Ángel, C. M. M., Gabriela, M. M., Ileana, E., Andrés Miguel, C. M., & Luis, M. C. (2021). Analysis of the Gulf of California Cannonball Jellyfish Fishery as a Complex System. *Ocean & Coastal Management, 207*, 105610. https://doi.org/10.1016/j.ocecoaman.2021.105610

Fiksel, J., Bruins, R., Gatchett, A., Gilliland, A., & ten Brink, M. (2014). The Triple Value Model: A Systems Approach to Sustainable Solutions. *Clean Technologies and Environmental Policy*. Advance online publication. https://doi.org/10.1007/s10098-013-0696-1

Freeman, R. E. (2010). *Strategic Management: A Stakeholder Approach*. Cambridge, Cambridge University Press. https://doi.org/10.1017/CBO9781139192675

Gibbons, M., Limoges, C., Nowotny, H., Schwartzman, S., Scott, P., & Trow, M. (1994). *The New Production of Knowledge: The Dynamics of Science and Research in Contemporary Societies*. London, SAGE Publications

Gray, S., Chan, A., Clark, D., & Jordan, R. (2012). Modeling the Integration of Stakeholder Knowledge in Social – Ecological Decision-Making: Benefits and Limitations to Knowledge Diversity. *Ecological Modelling, 229*, 88–96. https://doi.org/10.1016/j.ecolmodel.2011.09.011

Gurney, G. G., Darling, E. S., Jupiter, S. D., Mangubhai, S., McClanahan, T. R., Lestari, P., Pardede, S., Campbell, S. J., Fox, M., Naisilisili, W., Muthiga, N. A., D'agata, S., Holmes, K. E., Rossi, N. A. (2019). Implementing a Social-Ecological Systems Framework for Conservation Monitoring: Lessons From a Multi-Country Coral Reef Program. *Biological Conservation, 240*, 108298. https://doi.org/10.1016/j.biocon.2019.108298

Hare, J. A. (2020). Ten Lessons From the Frontlines of Science in Support of Fisheries Management. *ICES Journal of Marine Science, 77*(3), 870–877. https://doi.org/10.1093/icesjms/fsaa025

Hoerterer, C., Schupp, M. F., Benkens, A., Nickiewicz, D., Krause, G., & Buck, B. H. (2020). Stakeholder Perspectives on Opportunities and Challenges in Achieving Sustainable Growth of the Blue Economy in a Changing Climate. *Frontiers in Marine Science, 6*. https://doi.org/10.3389/fmars.2019.00795

IPCC (2022). *Climate Change 2022. Impacts, Adaptation and Vulnerability: IPCC WGII Sixth Assessment Report*. Retrieved from https://report.ipcc.ch/ar6wg2/pdf/IPCC_AR6_WGII_FinalDraft_FullReport.pdf.

Jentoft, S., & Chuenpagdee, R. (2009). Fisheries and Coastal Governance as a Wicked Problem. *Marine Policy, 33*(4), 553–560. https://doi.org/10.1016/j.marpol.2008.12.002

Johnson, D. R. (2021). Integrated Risk Assessment and Management Methods Are Necessary for Effective Implementation of Natural Hazards Policy. *Risk Analysis, 41*(7), 1240–1247. https://doi.org/10.1111/risa.13268

Jones, K., & Seara, T. (2020). Integrating Stakeholders' Perceptions into Decision-Making for Ecosystem-Based Fisheries Management. *Coastal Management, 48*(4), 275–288. https://doi.org/10.1080/08920753.2020.1773211

Karl, D. M. (2002). Nutrient Dynamics in the Deep Blue Sea. *TRENDS in Microbiology, 10*(9), 410–418.

Köpsel, V., de Moura Kiipper, G., & Peck, M. A. (2021). Stakeholder Engagement vs. Social Distancing – How Does the Covid-19 Pandemic Affect Participatory Research in EU Marine Science Projects? *Maritime Studies, 20*(2), 189–205. https://doi.org/10.1007/s40152-021-00223-4

Lawrence, M. G., Williams, S., Nanz, P., & Renn, O. (2022). Characteristics, Potentials, and Challenges of Transdisciplinary Research. *One Earth, 5*(1), 44–61. https://doi.org/10.1016/j.oneear.2021.12.010

Levesque, V. R., Wake, C., & Peterson, J. M. (2021). Facilitating Use of Climate Information for Adaptation Actions in Small Coastal Communities. *Elementa: Science of the Anthropocene, 9*(1). https://doi.org/10.1525/elementa.2020.20.00048

Lopes, P., Mendes, L., Fonseca, V., & Villasante, S. (2017). Tourism as a Driver of Conflicts and Changes in Fisheries Value Chains in Marine Protected Areas. *Journal of Environmental Management, 200*, 123–134. https://doi.org/10.1016/j.jenvman.2017.05.080

Mittelstraß, J. (2007). *Methodische Transdisziplinarität: Mit der Anmerkung eines Naturwissenschaftlers*. LIFIS online. ISSN 1864–6972

Mobjörk, M. (2010). Consulting versus Participatory Transdisciplinarity: A Refined Classification of Transdisciplinary Research. *Futures, 42*, 866–873. https://doi.org/10.1016/j.futures.2010.03.003

Möllmann, C., Cormon, X., Funk, S., Otto, S. A., Schmidt, J. O., Schwermer, H., Sguotti, C., Voss, R., & Quaas, M. (2021). Tipping Point Realized in Cod Fishery. *Scientific Reports, 11*(1), 14259. https://doi.org/10.1038/s41598-021-93843-z

Nicolescu, B. (2010). Methodology of Transdisciplinarity – Levels of Reality, Logic of the Included Middle and Complexity. *Transdisciplinary Journal of Engineering & Science, 1*(1). https://doi.org/10.22545/2010/0009

Nicolescu, B. (2002). *Manifesto of Transdisciplinarity*. State University of New York (SUNY) Press.

Paterson, B., Isaacs, M., Hara, M., Jarre, A., & Moloney, C. L. (2010). Transdisciplinary Co-Operation for an Ecosystem Approach to Fisheries: A Case Study from the South African Sardine Fishery. *Marine Policy, 34*, 782–794. https://doi.org.10.1016/j.marpol.2010.01.019

Pielou, E. C. (1984). *The Interpretation of Ecological Data: A Primer on Classification and Ordination*. John Wiley & Sons.

Pohl, C., & Hirsch Hadorn, G. (2006). *Gestaltungsprinzipien für die transdisziplinäre Forschung. Ein Beitrag des td-net*. München, Deutschland, Oekom Verlag.

Pohl, C., Truffer, B., & Hirsch Hadorn, G. (2017). Addressing Wicked Problems through Transdisciplinary Research. In Frodeman, R. (Ed.). *The Oxford Handbook of Interdisciplinarity* (2nd ed.). Oxford University Press. https://doi.org/10.14512/9783962388621

Regier, H. A., Bishop, P. L., & Rapport, D. J. (1974). Planned Transdisciplinary Approaches: Renewable Resources and the Natural Environment, Particularly Fisheries. *Journal of the Fisheries Board of Canada, 31*(10), 1683–1703.

Rittel, H. W. J., & Webber, M. M. (1973). Dilemmas in a General Theory of Planning. *Policy Sciences*, *4*(2), 155–169. https://doi.org/10.1007/BF01405730

Rudd, M. A., Moore, A. F. P., Rochberg, D., Bianchi-Fossati, L., Brown, M. A., D'Onofrio, D., ... & Worley, A. N. (2018). Climate Research Priorities for Policy-Makers, Practitioners, and Scientists in Georgia, USA. *Environmental Management*, *62*(2), 190–209. https://doi.org/10.1007/s00267-018-1051-4

Schmidt, A., Striegnitz, M., & Kuhn, K. (2014). Integrating Regional Perceptions into Climate Change Adaptation: A Transdisciplinary Case Study From Germany's North Sea Coast. *Regional Environmental Change*, (14), 2105–2114.

Schwermer, H., Aminpour, P., Reza, C., Funk, S., Möllmann, C., & Gray, S. (2021a). Modeling and Understanding Social – Ecological Knowledge Diversity. *Conservation Science and Practice*, *3*. https://doi.org/10.1111/csp2.396

Schwermer, H., Blöcker, A. M., Möllmann, C., & Döring, M. (2021b). The 'cod-multiple': Modes of Existence of Fish, Science and People. *Sustainability*, *13*(12229). https://doi.org/10.3390/su132112229

Schupp, M. F., Kafas, A., Buck, B. H., Krause, G., Onyango, V., Stelzenmüller, V., . . . & Scott, B. E. (2021). Fishing Within Offshore Wind Farms in the North Sea: Stakeholder Perspectives for Multi-Use From Scotland and Germany. *Journal of Environmental Management*, *279*, 111762. https://doi.org/10.1016/j.jenvman.2020.111762

Seabloom, E. W., Ruggiero, P., Hacker, S. D., Mull, J., & Zarnetske, P. (2012). Invasive Grasses, Climate Change, and Exposure to Storm-Wave Overtopping in Coastal Dune Ecosystems. *Global Chance Bilogy*, *19*, 824–832. https://doi. 10.1111/gcb.1207

Shirk, J. L., Ballard, H. L., Wilderman, C. C., Phillips, T., Wiggins, A., Jordan, R., . . . & Bonney, R. (2012). Public Participation in Scientific Research: A Framework for Deliberate Design. *Ecology and Society*, *17*(2). https://doi.org/10.5751/ES-04705-170229

Stauffacher, M., Flüeler, T., Krütli, P., & Scholz, R. W. (2008). Analytic and Dynamic Approach to Collaboration: A Transdisciplinary Case Study on Sustainable Landscape Development in a Swiss Prealpine Region. *Systemic Practice and Action Research*, *21*(6), 409–422. https://doi.org/10.1007/s11213-008-9107-7

Stelzenmüller, V., Letschert, J., Gimpel, A., Kraan, C., Probst, W. N., Degraer, S., & Döring, R. (2022). From Plate to Plug: The Impact of Offshore Renewables on European Fisheries and the Role of Marine Spatial Planning. *Renewable and Sustainable Energy Reviews*, *158*, 112108. https://doi.org/10.1016/j.rser.2022.112108

Sterling, E. J., Filardi, C., Toomey, A., Sigouin, A., Betley, E., Gazit, N., . . . Jupiter, S. D. (2017). Biocultural Approaches to Well-Being and Sustainability Indicators Across Scales. *Nature Ecology & Evolution*, *1*(12), 1798–1806. https://doi.org/10.1038/s41559-017-0349-6

Stier, A. C., Samhouri, J. F., Gray, S., Martone, R. G., Mach, M. E., Halpern, B. S., . . . Levin, P. S. (2015). Integrating Expert Perceptions into Food Web Conservation and Management. *Conservation Letters*, *10*(1), 67–76. https://doi.org/10.1111/conl.12245

Tiller, R. G., Mork, J., Liu, Y., Borgersen, Å. L., & Richards, R. (2015). To Adapt or Not Adapt: Assessing the Adaptive Capacity of Artisanal Fishers in the Trondheimsfjord (Norway) to Jellyfish (*Periphylla periphylla*) Bloom and Purse Seiners. *Marine and Coastal Fisheries*, 7(1), 260–273. https://doi.org/10.1080/19425120.2015.1037873

Torell, E., Redding, C. A., Blaney, C. L., Hernandez, E., Sison, O., Dyegula, J., & Robadue, D. D. (2012). Population, Health, and Environment Situational Analysis for the Saadani National Park Area, Tanzania. *Ocean & Coastal Management*, *66*, 1–11. https://doi.org/10.1016/j.ocecoaman.2012.05.005

Vilsmaier, U. (2021). Transdisziplinarität. In Schmohl, T. & Philipp, T. (Eds.), *Handbuch transdisziplinäre Didaktik*. Bielefeld, Transcript Verlag

Wagner-Ahlfs, C., Brennecke, D., Cutajar, J. A., Jiménez, J. I., Koerth, J., Kuljis, M. B., Letortu, P., Markiewicz, M., Morawska-Jancelewicz, J., Morvan, J., Quere, Y., Rioja, C., Vassallo, M. T., Vučković, M., & Zukowska, A. (2021). *Strategies of Stakeholder Engagement. Analysis of Approaches and Strategies within the SEA-EU Alliance.* Kiel, Germany

WBGU (2011). *World in Transition. A Social Contract for Sustainability* (2nd ed.) Berlin, Germany

2

EMPTY OCEANS – HUMANIZING OCEAN AND SEASCAPES FOR BUILDING TRANSDISCIPLINARY KNOWLEDGE AND PRACTICE

Laura Brum Bulanti, Ximena Lagos, and Laura Marrero Beramendi

Introduction

The contemporary environmental crisis and climate emergency calls for a radical change at different levels of our life experience, in the ways we inhabit the planet, and how we relate to others. It also challenges the ways in which we do science, demanding deep epistemological and paradigmatic changes. To properly understand and manage these complex socio-ecological problems of our time, it is necessary to work on a dialogical and diverse knowledge with a multi-scale integrative perspective.

For the past two decades, oceans and coasts have received increased scientific, political, and public attention. They have been explicitly included in the Sustainable Development Goals (SDGs) set up by de United Nations in 2015 (Arbo et al., 2018; ONU, 2018). This is encompassed by new development models like the blue economy and blue growth (Eikeset et al., 2018). These models are moving the frontiers of capitalism to new territories and ecosystems and are fraught with uncertainties and criticism (Barbesgaard, 2018; Ertör and Hadjimichael, 2020), especially considering the ineffectiveness of their predecessor, the 'green economy'. The concept of 'green economy' intends to minimize environmental degradation and poverty (Brand, 2012; Wilson, 2013), being climate change its most visible failure. As a reaction to those models which are sustained in modernity ideals of faith in progress and endless economic growth, some alternative theories have emerged. Such is the case of the degrowth movement (Escobar, 2015), the blue degrowth movement (its marine version) (Ertör and Hadjimichael, 2020), or the Latin-American post development critique (Escobar, 2015) as alternatives to overcome the human and ecological capitalist crisis (Escobar, 2012).

Two major transformations must occur to rethink marine science in times of the environmental crisis. The first one is ontological, and it consists of discussing

DOI: 10.4324/9781003311171-3

how Western Eurocentric societies build scientific knowledge, representations, and relations regarding oceans, seas, and coasts,[1] which deprives them of any human connotation and preserves them as a blue desert.[2] The second transformation requires building new epistemologies to enable other ways to produce theories, methods and practices from diverse disciplines and different forms of knowledge, while considering the gaps and omissions in marine science and management. Transdisciplinary work is the most appropriate framework for an integrative, critical, and situated practice, especially considering the increasingly urgent need for social sustainability.

The building of the void

To understand the socio-environmental problems of our time, particularly in oceans and seas, a critical and historical analysis is required in terms of how Western society has built its knowledge and relations to these aquatic environments. This knowledge has been achieved through an ontology of exclusion, configuring a space without evidence of human experience. As John Mack describes: a space without history, a wild territory (Mack, 2011). Before the Renaissance, Western-Eurocentric representations of the seas and oceans depicted them as chaotic, dangerous, terrifying places, inhabited by the unknown, materialized in fantastic and repulsive creatures (Corbin, 1994) like old cartographic beasts (Pinzón Ríos and Lois, 2015). Those representations drew the attention and control efforts to the land, leaving oceans and seas as dehumanized spaces, and coasts as liminal and inhospitable places. The wild, the "barbarous", the uncontrollable was associated with ocean spaces. By contrast, land spaces (the continents) had the right conditions to be controlled by the laws of God and humans. Schmitt suggests that the land-sea opposition and the human-nature disconnection are key elements to understanding this particular configuration of Western society. Humanity's terrestrial condition is present in sacred texts, which link our origins and living and sensitive experience to the land. On the other hand, he recognizes myths and ideas from classical thinking and philosophy that connect our origins to the sea, which were further developed by Darwinism and scientific thinking (Schmitt, 1997). In both cases, however, land and water are represented as binary opposites and not as complementary elements, sparking the dispute of which element has predominated in our existence (Steinberg and Peters, 2015).

During the 18th and 19th centuries, oceans and seas were essentialized as places of wilderness, sanctuaries no longer terrified or feared, but to be explored, mapped, and written by science. The Modernist episteme, with its confidence in progress and technology, was oriented to produce knowledge to control and govern new domains (especially those including nature, resources, and all the creatures considered inferior) as European mercantilism expanded to the rest of the world. During this exploring and expansion process, oceans and seas remained as an empty space, excluding humans and cultures form its domains (Steinberg, 2001).

The consolidation of the scientific method and science as a field of expert and truthful knowledge about the world laid the constitutive foundations of modernity, gradually unveiling the mysteries of the oceans and the lands beyond. Meanwhile, confidence in mankind and its methods increased in a context of colonial expansion, together with the industrial and scientific revolutions. The military and commercial conquest of the oceans by the European nations was also a technological and cognitive one, thanks to the many explorations and systematic chronicles of voyages and expeditions. The knowledge produced by modernity is rooted in the separation of nature and culture from a mechanistic viewpoint, where nature is governed by science and society is governed by politics, clearly delimiting two separate fields (Latour, 2007). This generates a universal dualism and the nature-culture disjunction which constitutes the basis of modern Western ontology (Hviding, 2001).

This disjunction and binarism are an integral part of the modern scientific paradigm, allowing us to understand some biases in the production of knowledge and in the representations of oceans and seas, with various consequences:

- The exclusion of humans meant neglecting seas and oceans from the social sciences and humanities. Studies and approaches from these disciplines have been relegated (Braudel, 1949; Erlandson, 2001).
- The dehumanized seas are allegedly neutral. Oceans and seas are associated with nature (as opposed to culture). However, this is not seen from a gynocentric (predominantly female) perspective of nature, but rather from the androcentric (predominantly male) one which prevails in the Western world, as shown by discursive studies on these environments (Milstein and Dickinson, 2012). These typically Western binary and opposite categories of analysis relate the feminine to the land and the masculine to the sea, hindering the study of relational dimensions in gender construction processes and subordinating the feminine dimension (Lee Davis and Nadel-Klein, 1992), excluding it from the object of study (oceans and sea) and its management (Diamond et al., 2003). Various disciplines seek to engender these territories, highlighting the contributions of women and other gender identities in the construction of these territories and their role in maritime communities (Davis and Nadel-Klein, 1992; Flatman, 2003; Ransley, 2005; Colville et al., 2015).
- The naturalness and neutrality of oceans and seas has invisibilized oppressed groups: enslaved populations have long been commercialized and/or exploited through international networks. These sociocultural groups have intrinsically integrated maritime communities and the commercial and political expansion of the colonial and postcolonial world. These groups (e.g. African slaves; Filipino seafarers) are gradually being incorporated into the study of oceans as social spaces (Baderoon, 2009; McKay, 2014).
- The notion of free navigation (no obstacles to movement) and the physical difficulties to establish borders in aquatic environments have contributed to constructing the seas as spaces free of asymmetries, control or domination. Little interest has been shown in the power exerted over navigation routes, and

the technological and military control of seas during colonial expansion as well as in the current geopolitical configuration of the world (DeLoughrey, 2019).

These knowledge gaps have consequences in terms of the disciplines which are traditionally dedicated to studying the seas and oceans, favouring the natural and exact sciences. Classical oceanography is being increasingly criticized (Moura, 2019), and new perspectives are emerging from the field of human sciences: blue cultural studies, blue humanities, hydro-criticism, and hydro-colonialism (Mentz, 2009; Alaimo, 2019; Hofmeyr, 2019). New approaches have emerged from the field of social sciences, such as marine social science, together with decolonial perspectives, like socio-environmental oceanography and social oceanography (Narchi et al., 2018; Bennet, 2019; dos Santos et al., 2019), claiming for a space to produce knowledge in/about ocean territories in all their complexity, improving the understanding of ocean socio-ecological dynamics and their problems.

Oceans of power: re-inhabiting the oceans from a critical perspective

Re-inhabiting the oceans implies overcoming the difficulties that modern science imposes on addressing the problems of an increasingly complex, interconnected and global world. These difficulties are a product of the Cartesian, dualistic, reductionist, simplifying and fragmenting approach that characterizes positivist science. All of it has been criticized, claiming for a new paradigm and a science of complexity (Morin, 2009). Overcoming disciplinary approaches beyond addition (multidiscipline) or integration (interdiscipline) opens the space for transdiscipline[3] as a conceptual and methodological proposal, which, albeit polysemic and permanently reviewed and discussed (Morin, 1994; Aronson, 2003; Thompson Klein, 2004; Max Neef, 2005), opens up to a new rationality.

Latin America has been a germinative place of currents and critical thought perspectives amidst historical social inequalities resulting from colonialism. Mainly during the 1960s and 1970s, in a context of popular movements and in the search for profound changes in the prevailing modern Eurocentric scientific rationality, different theories emerged: liberation theology, dependency theory, participatory action research, and popular education, which became relevant schools for the development of ideas, creativity and practices (Escobar, 2012: 232). After an impasse of a few decades of dictatorial governments, the recovery of democracy reignited debates and theoretical production in diverse burgeoning fields: decolonial critical thinking, post-development, political ecology and ecofeminism (Escobar, 2012; Mignolo, 2000; Arrigada and Zambra, 2019). Latin American political ecology resignified and revalued nature, not as an economic and exploitative purpose, but in its ontological dimension and in terms of the place it occupied due to the power relations that configured the continent's territory from an extractivist logic. However, southern theories did not originally pay attention to gender power relations. These were introduced by ecofeminism and feminist political ecology, which enriched decolonial thinking by reflecting on power relations between humans and nature, men and women, how they were mediated by inequalities in the exploitation of resources, and the subordination of ones over others (Arriagda and Zambra, 2019). This differentiation helps to understand other dimensions of oppression, such as race and social class, intersected with gender: an intersectional approach. Latin American critical feminism postulates that these categories, which are far from being totalizing, have not interacted with different ontologies and precolonial cultures and must be used with caution (Marcos, 2013). The intersectional approach allows us to identify differences in how the ocean, its resources and spatiality are accessed and used, without disregarding the multiple representations around this territory.

The currents of thought and practice mentioned earlier articulate with a situated, critical, decolonial transdiscipline open to other ontologies that coexist in the continent, giving space to non-academic knowledge, which we define as "substantive transdiscipline".[4]

That substantive condition is related to pragmatism: it emphasizes the praxis over formal structures or models, and it is substantiated in a concrete and situated problem. It is proposed as a project of "hybridization" or in-disciplining, overcoming the colonial biases that perpetuate inequalities and asymmetries inherited and reproduced by Europeanized academic systems (Walsh and Castro Gomez, 2002; Restrepo and Escobar, 2015). Substantive transdisciplinarity is grounded in a long Latin American academic tradition of collaborative practice and social commitment based on participatory action research (PAR) (Fals Borda, 1999) and, particularly, on the theory of practice for transformation and the integration of popular knowledge in co-production processes, overcoming asymmetrical subject-object relations. It is also inspired by Boaventura de Souza Santos' *Epistemology of the South*, aiming to subvert the focus of analysis towards the processes that make domination and exclusion (typical of androcentric colonial capitalism and modern rationality) through a metonymic reason (the reason that contracts, reduces and simplifies the present). It develops a sociology of absences, capable of understanding how these cuts are produced, how elements of reality are invisibilized, and how the discourses that build the dominant (or hegemonic[5]) vision of the world operate. In this approach, science must be used as a resistance (or counter-hegemonic) tool and abandon its positivist claim of monoculture of knowledge and hierarchical rigor (as the only true knowledge about the world) to open up to an ecology of knowledge. This requires a dialogue with other forms of knowledge (popular, marginal, indigenous, etc.); that is, to accept other (non-linear) temporalities, to recognize the mechanisms that produce hierarchy and difference, to develop a multi-scale perspective, and to overcome the productivity logic (Santos, 2006). These epistemological and theoretical movements share a common element: the Latin American universities' outreach programmes, which date back to the reforms of the first half of the 20th century. Outreach programmes are a tradition which reinforces the connection between teaching, knowledge production and attending to the needs and problems of society, in political-academic projects that differentiate university and outreach models over time and in different socio-historical contexts (Tomasino and Cano, 2016; Cano, 2022).

Case studies: humanizing ocean and coastscapes to promote trans-disciplinarity

The next three case studies of transdisciplinary experiences, related to higher education training, research and public policy assessment, illustrate some scenarios for practice and reflection for substantive transdisciplinarity. They intend to contribute to developing a sociology of absences in oceans and coasts, remarking context-situated aspects and social sustainability needs.

Transdisciplining training: Integrated Coastal Management of the Southern Cone postgraduate programme

In 2007, the master's degree in integrated coastal management of the Southern Cone (ICM) was initiated at the University of the Republic of Uruguay (Conde et al., 2012). It was the result of a long process of diagnosis and inter-institutional regional projects for integrated coastal management in the country (Martínez and Fournier, 1999). The programme was supported by international cooperation funds from Canada and several Latin American countries were actively involved in a North-South process, resulting in a local postgraduate programme with an international profile.

This master's degree seeks to develop training in ICM as a methodological framework to meet the SDGs and Agenda 21, which was agreed at the 1992 Rio Summit. It is based on integrated principles: integrating disciplines, environments (a land-coast-oceans continuum), and institutions and stakeholders linked to coastal issues. Developing the programme implied structural and institutional challenges. First, it involves professors from five schools in distant disciplines (Law, Social Sciences, Natural Sciences, Engineering and Architecture) who participate in teaching, research and outreach activities in the programme and at the Interdisciplinary Center of ICM in the Southern Cone. The programme's outline and curricula were designed with the progressive and sequential incorporation of theoretical and methodological topics related to coastal socio-environments and their management, integrating different disciplines to characterize these socio-ecosystems in their several dimensions (social, physical, and biological) and problems. The theory is encompassed by a workshop-like process to develop skills in a practical way, dealing with real-world problems and using specific techniques and methods. Multi- and inter-disciplinary approaches stand out together with the work with non-academic actors, promoting a strong integration throughout the entire education process (de Álava, 2014). The regular participation of professors from other Latin American universities (Brazil, Argentina, Mexico, Canada, Colombia, and Chile) enriches the teaching process with international expertise and opens up the training to other scales and problems.

The programme targets students from all areas of scientific research, professors and teachers, professionals, and government technicians. Since its first edition, students from very diverse disciplinary and technical fields have participated (e.g. biology, geology, anthropology, law, engineering, social sciences, agronomy, and urbanism, among others). Each cohort attends the theoretical and practical courses as a group, in an in-person modality, promoting teamwork dynamics, fostering dialogue, aiming for a common language, and building a community of practice, which is necessary to establish inter-/transdisciplinary processes. One of the most outstanding aspects of the programme is the training given in Scientific Epistemology and Participatory Action Research (PAR), which are integrated in workshops, working with problems in the territory and articulating theory and praxis (Vienni Baptista and Goñi Mazzitelli, 2021).

ICM is based on the premise of horizontal and vertical articulation and integration of different stakeholders linked to coastal problems, applying standard methodologies to local situations, aiming to connect science, society and politics. Disciplinary integration is one of the main postulates of ICM, but integration is unequal and some disciplines are underrepresented, especially those from the field of social science (except for economics, law and politics). This asymmetry, which stretches across publications, manuals, and methodologies, is defeated progressively in this master's degree, which performs self-diagnoses and students' assessments to adjust and update its curricula. Reinforcing PAR and theory and epistemology of science, while introducing courses in political ecology, archaeology, tourism studies, anthropology and environmental history are part of the programme's evolution and improvements. These actions reinforce critical thinking, promote dialogue between diverse epistemologies and paradigms, incorporate concepts and glossaries from different fields and knowledges, and comprise a variety of methodologies and techniques.

Since its creation, the programme has produced rich knowledge in research (Centro Interdisciplinario MCI Cono Sur, 2011; Brum Bulanti, 2013; Lagos et al., 2017), outreach activities (Echeverría et al., 2013) and projects with other institutions and organizations while keeping a strong transdisciplinary imprint. The ICM model was thus enriched with contributions from disciplines and theories which are invisible to mainstream approaches, regarding the usually neglected socio-historical dimension. This is essential to understand complex problems and to meet the SDGs, especially in sensitive matters as gender, and social and environmental justice, among others.

Small-scale fisheries: women in action

Artisanal fishing in Uruguay is one of the most vulnerable productive sectors, both socioeconomically and culturally. This is explained by the power of market intermediaries, a high bioclimatic variability for the extraction of species, low state protection and support, the lack of social recognition, the stigmatization of the job and poor organizational and agency skills by the fishermen (Migliaro and Santos, 2010).

Since 2013, a PAR experience has been conducted with fishing families in Laguna de Rocha, where fisherwomen have played a leading role. It is one of the most traditional fishing communities in the country, settled since the early 19th century in the southern area of the lagoon. The community is made up of 22 families over five generations, with a strong identity associated with "living *off* and *in* the lagoon" (Lagos, 2016). It is organized in family productive units for the activities of capturing, processing and selling fish, and women participate in the entire fishing value chain (ibid.).

Around the mid-1990s, the first technicians arrived in the territory to help set up an organization of fishermen (Lagos et al., 2019) to represent them during the long process of protecting the area. In addition, different productive projects were conducted, resulting in 20 years of cumulative experience through exchanges with other actors such as researchers, non-governmental organizations (NGOs), and government managers. This improved the quality of life of the fishermen but, at the same time, eroded the organizational structure, which resulted in male leaders eventually abandoning their roles, leaving the association inactive for some years (Lagos et al., 2019; Laporta and Lagos, 2021).

In 2014, a team of two female technicians began to work with a small group of fisherwomen within the framework of an institutional project of the Ministry of Livestock, Agriculture and Fisheries, whose initial objective was to strengthen community ties and their associativity. One of the project's actions took up a culinary initiative that the women had previously designed but could not be developed due to the lack of material conditions (the sanitation network reached the lagoon in 2011). This is how *La Cocina de La Barra* (The Kitchen of La Barra) emerged, being a culinary enterprise inaugurated in 2015 and made up of nine local fisherwomen.

The Kitchen of La Barra (KoLB) is committed to enhancing the value of the fishing culture through its cuisine as a local tradition. It offers a sensory experience of contemplating the landscape, and it revalues the fishing heritage, conceived as the biocultural synthesis of artisanal fishing, species, ecosystems, human work, and social and natural life in interdependence (Laporta and Lagos, 2021). KoLB was a starting point for a deeper process of social transformation. For the women who are part of the project, it represented the beginning of a journey of self-recognition, making visible artisanal fishing at Laguna de Rocha and fisherwoman as drivers of development at the domestic and community levels. Likewise, it is also a place of reunion, belonging, and pride for women (ibid.).

Although the project was not conceived with a gender perspective, this was regarded as an emerging issue to be incorporated since its inception. As Alvarez (Alvarez-Burgos et al., 2017) points out, public policies are assumed from a gender neutrality when, by default, they are not. This apparent neutrality, also present in fishermen's organizations, have consolidated the inequity in gender roles to this day (Alvarez-Burgos, 2020). If artisanal fishing is one of the most invisible sectors, women in fishing are doubly invisible. This is also expressed in the scientific and technical field around artisanal fishing and marine sciences in general. In KoLB, there was a meeting between an all-female technical team and a group of fisherwomen who together recognized that this gender differential was also a decisive factor to the project and the process.

A flexible work strategy was developed on the basis of the PAR methodology. This means that the group of fisherwomen identify and make requests based on their needs and interests related to solving problems of their reality. Those problems are addressed in a constant process of action, learning, reflection, and adaptation for new actions. Another key element of the work strategy has been to respect the emergent group (Pichón Riviére, 1983), and the events or elements of social life that need to be processed collectively. This requires a great flexibility of the technical team, overcoming the typical notions of project productivity, following indicators, and achieving key results which are intrinsically linked to research and development programmes. It was crucial to understand that the project has an underlying social process. This involves expectations, dreams, and hopes from the local population as well as from the technical/academic teams that are put into dialogue and debates, without losing sight of the fact that the focus is on the well-being of local communities. An example of this are the weekly planning meetings between fisherwomen and technicians, together with specific issues of KoLB which are related to the more domestic and daily aspects, the community, and productive nature.

Fisherwomen have historically been subordinated to the domestic space, resulting in the construction of an imaginary of fishing from the male researchers' point of view, depicted mainly as a masculine sphere and underestimating the significant role of women (Alvarez-Burgos et al., 2017; Alvarez-Burgos, 2020). The appearance of women in fishing organizations has to do with these issues: whenever a social reproduction system is threatened, women become part of organizations to protect the entire family and collective system (Bennett, 2005). This, of course, entails a great workload, which materializes in a triple schedule (domestic, productive, and political) to which these women are subject (Laporta and Lagos, 2021: 67). In this sense, the fisherwomen of the KoLB transcend the domestic, community, and productive spheres, addressing them in an integral way within the organizational space.

Coastal space and climate change: planning with a gender approach

Literature on adaptation to climate change argues that local knowledge and women's capacity of agency are essential to implementing adaptation actions within communities (Alston, 2014). However, some authors point out the tension between recognizing the relevance of integrating gender perspectives and those strategies that effectively reduce inequalities to respond to the climate change emergency (Rao et al., 2019). It is also stated that almost all policies to develop and strengthen the adaptive capacity of local communities fail to recognize the gendered nature of everyday realities and experiences (ibid.). Gender issues are completely ignored or incorrectly formulated, placing women as vulnerable subjects without acknowledging their capacity for action, leadership, and negotiation.

The integration of the gender perspective in a planning instrument for climate change adaptation is a process which aims to incorporate actions that respond to social inequalities, contributing to improving the robustness of the socio-ecosystem. It therefore requires integrating fields of knowledge linked to physical and social dimensions. The methodology to identify gender-responsive actions involves developing an analysis that makes it possible to highlight these inequalities. Along process of developing Uruguay's National Coastal Plan (NAP Costas), it became evident that there it was necessary to integrate techniques from environmental sciences with social sciences. A mixture of quantitative and qualitative methodologies of social analysis (disaggregated by sex) was applied to integrate the socioeconomic and demographic variables connected to the access and use of resources in the territory. Simultaneously, qualitative techniques were used to understand the observed phenomenon and, fundamentally, to recognize daily practices of occupation, use of resources, and services in the territory which may be affected by climate change. The statistical data sources, referring to sociodemographic variables disaggregated by sex, presented greater limitations with a more local scope, signposting the limits of census data in terms of scale and representativeness. However, regardless of the quality and diversity of statistical data, a qualitative approach to articulate the collected data remains necessary for a situational perspective. In this way, techniques such as semi-structured interviews, workshops, mapping on the use and perception of space, analyses of statistical and documentary information, and discussion groups were added to the study, depending on the requirements of the case in question. These techniques contributed to studying aspects related to the use of coastal public spaces, the development of recreational, cultural and leisure activities, the main characteristics of mobility in the locality and, in general, the availability of public services of care, healthcare, and gender-based or domestic violence, how accessible they are to women, as well as the perception of safety and enjoyment.

Within the framework of NAP Costas, a gender analysis of territorial scope was implemented in the first phase to identify potential gender-responsive actions by collecting and processing social data disaggregated by sex, analysing documentary information with a gender perspective, identifying and consulting key actors and qualified informants (mainly women) linked to local gender and environmental issues, analysing the available information and corresponding recommendations. These recommendations vary depending on the site, reinforcing the concept that adaptation is contextual and its response is local/situated. However, they can be grouped as follows:

(i) Adaptation infrastructure: gender-sensitive measures such as recreational spaces for women, inclusive infrastructure (in terms of mobility); design oriented to the care of people in situations of dependency (boys/girls or senior citizens)
(ii) Participation and decision-making: gender-responsive measures such as discussing women's representation and facilitating their integration in decision-making processes

(iii) Capacity building and knowledge generation: gender-responsive measures focused on generating and democratizing information on climate change and coastal zones

This experience intends to articulate general frameworks to the local context, to situate these frameworks. It contributes to identifying symbolic and qualitative aspects to address changes in the material dimension of resources and in the exercise of the rights of women. These elements, including symbolic aspects, are particularly relevant for equality and the dismantling of patriarchal norms (Rao and Kelleher, 2005). From the "gender mainstream" approach, and with equality as a main goal (Garcia Prince, 2008), building capacities for adaptation opens a field for action to multiple responses, depending on their physical and social vulnerability. It is also based on the representations and expectations that social actors have both in the territory and the site as a situated environment and as a symbolic representation of gender relations around water, the oceans and coasts.

Conclusion

The examples presented in this chapter help us to expose ways in which to build substantive transdisciplinary processes and co-produce knowledge on seas, oceans, and coasts, both in educational experiences and in applied research. In all cases, this is done by emphasizing the human-social dimension, which is usually unnoticed or merely subsidiary in marine sciences and ecosystem management (Enter-Wada et al., 1998), especially in marine-coastal environments (Visser, 2004).

To this end, two major cross-cutting movements are identified in these experiences. On the one hand, there is disciplinary aperture and dialogue with other academics. At this point, we highlight the relevance in the three cases of integrated work between very diverse scientific disciplines as a key to opening new ontological and epistemological horizons. Integrating disciplines, particularly the opening to other academics who are often alien to seas and coasts (such as social sciences and the humanities) promotes a dialogue between paradigms, epistemologies and specialized vocabularies. This step is key to problematizing and deconstructing common places and quasi-naturalized ideas about human-nature relations, enabling oceans and coasts to transform into landscapes produced by material and symbolic relations. Signposting the most neglected social aspects of oceanography, ICM and land planning, such as gender, race, and social class, among others, is also critical. Integrating social and human sciences in marine-coastal studies is a slow and progressive process which involves hard work, time, and dedication to articulate epistemological frameworks and paradigms. It implies abandoning the disciplinary comfort zone and letting others cross it and transcend it, keeping an open mindset to learn and exercise disciplinary humbleness. In each case, that may depend on the type of process underway (higher education training, participatory research, design of public policies), generating tensions that must be considered from the beginning of the project as part of the process itself. What is clear is that the conditions for transdisciplinarity are not given, they must be built.

The other movement has to do with the dialogue with and between people outside academia. Within this space we can find two types of social actors, which poses different challenges in the process of co-building knowledge and developing joint actions. The first are managers/technicians from the public sector related to marine coastal management, who, although they may be university professionals carrying their disciplinary backpacks, are focused on public action (management, project development, steering). Their aim is to develop concrete actions and they are usually subject to political and economic pressures. This type of social actor demands the construction of spaces for dialogue, valuing field experience and the complexities of management, overcoming academic rhetoric (or, in Fals Borda's words, "doctor's arrogance" (Fals Borda, 1981)) in order to promote transversal dialogues oriented towards connecting action with knowledge. This is even more relevant when new knowledge seeks to be developed jointly with local communities, social organizations, or neighbours. It is key to understand the multiple ways of conceiving and receiving with nature, the multiple relational ontologies (Escobar, 2014) and to build knowledge from diversity, and from the ignorance of the individual towards an ecology of knowledge from a collective standpoint (Santos, 2009). In this sense, academic knowledge, proposed as a substantive transdiscipline, must point to transformative processes of social change and, within it, of society-maritime territorial[6] relationships. This is directly connected to the role Latin American universities play in the democratization of knowledge through outreach programmes (not present in Anglo-Saxon universities), which consist of academic work with a dimension of social commitment (Santos, 2015).

These movements for transdisciplinarity still encounter different obstacles. On the one hand, and most importantly, the command-control paradigm remains strong in academia regarding environmental and marine sciences, where disciplinary visions and deeply rooted technical-professional traditions predominate (Diegues, 2003). This has its correlation at the state, administration and public-policy levels, where a sectorial and fragmented management of socio-ecosystems predominates. Also, to promote basic and applied science, universities and institutions have a structure of fields or areas of knowledge organized in disciplines and sub levels. This rigid structure is used to evaluate, finance, teach, and other functions. Besides, different models related to social commitment in Latin American universities[7] coexist, and some of them question the time and value assigned to critical outreach programmes (Tomasino y Cano, 2016; Cano, 2022) and substantive transdisciplinarity as postulated in this chapter, which are divergent to a productivist conception of science and a difusionist-transferencist extension perspective (ibid.), which enjoys better mechanisms of financing, publishing, evaluation, and career promotion.

Seas, oceans, and coasts in crisis are also the product of systematically forgotten dimensions. The renewed sustainable development agendas and the new promises of blue growth cannot remain isolated from social sustainability (Foladori, 2007).

While disagreements about sustainable development remain, the concept continues to validate and legitimize practices that many times are far from being sustainable. Among these practices, it is essential to point out that thirty years after the Rio 1992 summit, women are still absent. Chapter 17 of Agenda 21 recognized ICM as the rightest approach to achieve coastal, sea, and oceanic sustainability (Vallega, 1993). Women's participation in the management of their territories was highlighted both by Principle 20 at the Rio summit and by action plans present in Agenda 21 (Lee Davis and Nadel-Klein, 1992). However, the gender aspects mentioned in these documents and in other UN publications were not transversalized and ended up being stereotyped (Diegues, 2003). This is a result of the process of compartmentalization of marine sciences, where social aspects, the knowledge of social sciences, and local knowledge were marginalized and made invisible. The positivist and productivist approach of command and control over biomass and ecosystems still prevails in oceanic science and management (ibid.).

Managing oceans as empty social spaces overlooks local populations and women and, despite increasing research on women and their relation with natural resources, there are few studies with a gender perspective in marine sciences and sustainability in general. In ICM and environmental management, a gender mainstream is recognized but merely "in paper", not effectively instrumented (de la Torre Castro et al., 2017). Gender neutrality in management and planning processes on coasts and oceans deepens social inequality, particularly the inequities to which women are subject (ibid.), but also childhoods in the maritime territory. It is necessary to abandon gender blindness and adult-centrism in the approach to territories. An intersectional and situated analysis must be incorporated into ICM and ocean studies to perceive how different gender, ethnic and class dimensions enhance or limit the paths towards the wellbeing of socio-ecological systems.

We believe that the strategies and examples presented, based on theories and methodological frameworks that seek a reflective and critical analysis of our epistemologies, promote working in situated dialogue. Collaborative actions are a way forward to develop a transformative transdiscipline capable of bringing us closer to equity and deep sustainability, committed to the social and ecological dimensions of our life experience in oceans and coastscapes and to the local challenges of a global crisis.

Notes

1 As part of his work, Peter Steinberg (2001) developed the history of the social construction of the oceans and their key role in trade and the territorial power of states.

2 The notion of desert used is related to the concept of void in allusion to the representations of the ocean in pre-modern societies (Steinberg, 2001:49) as an ungovernable, undifferentiated space, with attributes similar to the desert according to Indian societies. This idea of void should not be confused with *emptiness*, since, as the author states, oceans are spaces charged with sociability. The desert is also used as a metaphor, studied by Latin American historiography, and understood as a political construction developed to sustain the idea of tabula rasa, thus legitimizing the development plans in modern states and the capitalist model in the continent during the 19th century. In this sense,

Uriarte speaks of "desert makers" as part of this strategy of producing deserts to deploy the geopolitical control of the State (Uriarte, 2020).

3 We use *transdiscipline* as a noun which represents a theory or perspective that has some application in several disciplines, *transdisciplinarity* as a condition or attribute, and *transdisciplining* as an action.

4 This transdisciplinarity is framed within the theory of praxis and in Latin American critical thinking, differentiating from other critical currents (Hummel et al., 2017).

5 *Hegemony* is a concept used in different fields of social sciences and humanities. The Italian philosopher Antonio Gramsci developed the concept of hegemony as the cultural and political-ideological consensus and leadership of a social sector or class. This notion also implies domination, persuasion, and coercion. On the other hand, *counter-hegemony* implies the production of alternatives of resistance which criticize and question this hegemonic discourse-order (de Moraes, 2010).

6 "Maritorio" (*maritory* in previous publications in English) emerged as a concept in the 1970s in Latin America. It is associated with the idea of territorialized sea in marine areas such as islands, fjords, archipelagos, and inland seas of southern Chile. Currently, the concept has spread in the Southern Cone and refers to thinking *of* and *from* the sea, regarding it as a space inhabited through processes of social appropriation which include culture and identity, and a strategic political sense of life and sovereignty for its inhabitants (Álvarez et al., 2019).

7 Cano (2015) refers to three models to understand university-society relationships in Latin America: Entrepreneurial universities, universities for development, and popular universities, which have a direct impact on the way a university establishes its links with society, defines its outreach programmes, and builds its academic-political project.

References

Alaimo, S. (2019). Introduction: Science studies and the blue humanities. *Configurations*, 27(4), 429–432. https://doi.org/10.1353/con.2019.0028

Alston, M. (2014). Gender mainstreaming and climate change. *Women's Studies International Forum*, (47), 287–294. https://doi.org/10.1016/j.wsif.2013.01.016

Álvarez, R., Ther-Ríos, F., Skewes, J. C., Hidalgo, C., Carabias, D., & García, C. (2019). Reflexiones sobre el concepto de maritorio y su relevancia para los estudios de Chiloé contemporáneo. *Revista Austral de Ciencias Sociales*, 36, 115–126. https://doi.org/10.4206/rev.austral.cienc.soc.2019.n36-06

Álvarez Burgos, M. C. (2020). "No queremos ser pesca acompañante, sino pesca objetivo". Interfaces socioestatales sobre enfoque de género en la pesca artesanal en Chile. *RUNA, Archivo Para Las Ciencias Del Hombre*, 41(2), 67–85. https://doi.org/10.34096/runa.v41i2.8691

Álvarez-Burgos, M. C., Stuardo, G., Collao, D., & Gajardo, C. (2017). La visualización femenina en la pesca artesanal: transformaciones culturales en el sur de Chile. Polis 47–2017.

Arbo, P., Knol, M., Linke, S., & St Martin, K. (2018). The transformation of the oceans and the future of marine social science. *Maritime Studies*, 17(3), 295–304. https://doi.org/10.1007/s40152-018-0117-5

Aronson, P. (2003). La emergencia de la ciencia transdisciplinar. Cinta de Moebio. *Revista de Epistemología de Ciencias Sociales* (18). https://cintademoebio.uchile.cl/index.php/CDM/article/view/26136/27434

Arriagda Oyarzún, E., & Zambra Alvarez, A. (2019). Apuntes iniciales para la construcción de una ecología política feminista de y desde Latinoamérica. *Revista Polis*, 18(54), 14–38. http://dx.doi.org/10.32735/s0718-6568/2019-n54-1399

Baderoon, G. (2009). The African Oceans – tracing the sea as memory of slavery in South African literature and culture. *Research in African Literatures*, 40(4), 89–107. www.jstor.org/stable/40468163

Barbesgaard, M. (2018). Blue growth: Savior or ocean grabbing? *The Journal of Peasant Studies*, 45(1), 130–149. https://doi.org/10.1080/03066150.2017.1377186

Bennett, E. (2005). Gender, fisheries and development. *Marine Policy*, 29(5), 451–459. https://doi.org/10.1016/j.marpol.2004.07.003

Bennett, N. J. (2019). Marine social science for the peopled seas. *Coastal Management*, 47(2), 244–252. https://doi.org/10.1080/08920753.2019.1564958

Brand, U. (2012). Green economy – the next oxymoron? No lessons learned from failures of implementing sustainable development. *GAIA-Ecological Perspectives for Science and Society*, 21(1), 28–32. https://doi.org/10.14512/gaia.21.1.9

Braudel, F. (1949). *La Méditerranée et le monde méditerranéen à l'époque de Philippe II*. Armand Colin, Paris.

Brum Bulanti, L. (2013). Gestión del patrimonio arqueológico en el litoral oeste del departamento de Maldonado (Uruguay). La investigación como práctica integral. *Revista del Museo de La Plata*, 13(87), 417–428. https://publicaciones.fcnym.unlp.edu.ar/rmlp/article/view/2236

Cano Menoni, A. (2015). La extensión universitaria en la transformación de la Universidad Latinoamericana del siglo XXI: disputas y desafíos. En *Los desafíos de la Universidad en America Latina y el Caribe*, pp. 287–380. CLACSO, Buenos Aires.

Cano Menoni, J. A. (2022). University extension in dispute: Neoliberal counterreform and alternatives in Latin American Universities. *Latin American Perspectives*, 49(3), 49–65. https://doi.org/10.1177/0094582X211004911

Centro Interdisciplinario para el MCI del Cono Sur (2011). *Manejo Costero Integrado en Uruguay: ocho ensayos interdisciplinarios*. UDELAR/CIDA, Montevideo.

Colville, Q., Jones, E., & Parker, K. (2015). Gendering the maritime world. *Journal for Maritime Research*, 17(2), 97–101. https://doi.org/10.1080/21533369.2015.1095539

Conde, D., de Álava, D., Gorfinkiel, D., Menafra, R., & Roche, I., (2012). Sustainable coastal management at the public university in Uruguay: A southern cone perspective. In Leal Filho, W. (Ed.) *Sustainable Development at Universities: New Horizons*, pp. 871–885. Peter Lang Scientific Publishers, Frankfurt.

Corbin, A. (1994). *The Lure of the Sea: The Discovery of the Seaside in the Western World, 1750–1840*. University of California Press, Berkeley, CA.

Davis, D. L., & Nadel-Klein, J. (1992). Gender, culture, and the sea: Contemporary theoretical approaches. *Society & Natural Resources*, 5(2), 135–147.

De Álava, D., (2014). La Interacción Interdisciplinaria en la Maestría en ManejoCostero Integrado. En Fernández, V., Repetto, L., Vienni, V., & Von Saden, C. (Eds.) *Seminario En_clave Inter 2014. Educación Superior e Interdisciplina*, pp. 65–71. Espacio Interdisciplinario – UdelaR, Montevideo.

De la Torre-Castro, M., Fröcklin, S., Börjesson, S., & Okupnik, J. (2017). Gender analysis for better coastal management – Increasing our understanding of social-ecological seascapes. *Marine Policy*, 83, 62–74. https://doi.org/10.1016/j.marpol.2017.05.015

DeLoughrey, E. (2019). Toward a critical ocean studies for the Anthropocene. *English Language Notes*, 57(1), 21–36. https://doi.org/10.1215/00138282-7309655

de Moraes, D. (2010). Comunicação, hegemonia e contra-hegemonia: a contribuição teórica de Gramsci. *Revista Debates*, 4(1), 54. https://doi.org/10.22456/1982-5269.12420

Diamond, N. K., Squillante, L., & Hale, L. Z. (2003). Cross currents: Navigating gender and population linkages for integrated coastal management. *Marine Policy*, 27(4), 325–331. https://doi.org/10.1016/S0308-597X(03)00044-7

Diegues, A. C. (2003). A interdisciplinaridade no estudo do mar: o papel das ciências sociais. *Conferência proferida na XV Semana de Oceanografia*. São Paulo, USP.

Dos Santos, C. F., Martins, M. S. L., & de Avellar Mascarello, M. (2019). Oceanografia socioambiental: O que queremos com isso?. *Ambiente & Educação*, 24(2), 42–67. https://doi.org/10.14295/ambeduc.v24i2.9462

Echevarría, L., Gómez, A., Piriz, C., Quintas, C., Tejera, R., & Conde, D. (2013). Capacity building for local coastal managers: A participatory approach for Integrated Coastal and Marine Zones Management in Uruguay. *Revista de Gestão Costeira Integrada*, 13, 445–456. https://doi.org/10.5894/rgci402

Eikeset, A. M., Mazzarella, A. B., Davíðsdóttir, B., Klinger, D. H., Levin, S. A., Rovenskaya, E., & Stenseth, N. C. (2018). What is blue growth? The semantics of "Sustainable Development" of marine environments. *Marine Policy*, 87, 177–179. https://doi.org/10.1016/j.marpol.2017.10.019

Enter-Wada, J; Dale B, Krannich, R., & Brunson, M. (1998). A framework for understanding social science contributions to ecosystem management. *Ecological Applications*, 8(3), 891–904. https://doi.org/10.2307/2641275

Erlandson, J. M. (2001). The archaeology of aquatic adaptations: Paradigms for a new millennium. *Journal of Archaeological Research*, 9(4), 287–350. www.jstor.org/stable/41053178

Ertör, I., & Hadjimichael, M. (2020). Blue degrowth and the politics of the sea: Rethinking the blue economy. *Sustainability Science*, 15(1), 1–10. https://doi.org/10.1007/s11625-019-00772-y

Escobar, A. (2012). Más allá del desarrollo: postdesarrollo y transiciones hacia el pluriverso. *Revista de antropología social*, 21, 23–62.

Escobar, A. (2014). *Sentipensar con la tierra: Nuevas lecturas sobre desarrollo, territorio y diferencia*. UNAULA, Medellín.

Escobar, A. (2015). Decrecimiento, post-desarrollo y transiciones: una conversación preliminar. *Interdisciplina*, 3(7). http://dx.doi.org/10.22201/ceiich.24485705e.2015.7.52392

Fals Borda, O. (1981). La ciencia y el pueblo. En Vío Grossi, F., Gianotten, V. & De Wit, T. (Eds.) *Investigación participativa y praxis rural: nuevos conceptos en educación y desarrollo comunal*, pp. 19–47. Mosca Azul Editores.

Fals Borda, O. (1999). Orígenes universales y retos actuales de la IAP. Análisis político No. 38, IEPRI, Instituto de Estudios Políticos y Relaciones Internacionales. UN, UniversidadNacional de Colombia, Santa Fe de Bogotá, Antioquia, Colombia.

Flatman, J. (2003). Cultural biographies, cognitive landscapes and dirty old bits of boat: "theory" in maritime archaeology. *The International Journal of Nautical Archaeology*, 32(2), 143–157. https://doi.org/10.1111/j.1095-9270.2003.tb01441.x

Foladori, G. (2007). Paradojas de la sustentabilidad: ecológica versus social. *Trayectorias*, 9(24), 20–30.

García Prince, E. (2008). Políticas de igualdad, equidad y gender mainstreaming.¿ De qué estamos hablando? Marco conceptual. Proyecto Regional América Latina Genera, Gestión del Conocimiento para la Equidad de Género en Latinoamérica y El Caribe. San Salvador.

Hofmeyr, I. (2019). Provisional notes on hydrocolonialism. *English Language Notes*, 57(1), 11–20. https://doi.org/10.1215/00138282-7309644

Hummel, D., Jahn, T., Keil, F., Liehr, S., & Stieß, I. (2017). Social ecology as critical, transdisciplinary science – conceptualizing, analyzing and shaping societal relations to nature. *Sustainability*, 9(7), 1050. https://doi.org/10.3390/su9071050

Hviding, E. (2001). Naturaleza, Cultura, Magia, Ciencia. Sobre los metalenguajes de comparación en la Ecología Cultural. En P. Descola & G. Palsson (Coords.), *Naturalaeza y Sociedad. Perspectivas antropológicas*, pp. 192–213. Siglo XXI, México.

Lagos, X. (2016). Cultura de la pesca en Laguna de Rocha. Enfoque cultural para el manejo integral del patrimonio costero. En Gianotti, C., Barreiro, D., & Vienni, B. (Coords), *Patrimonio y Multivocalidad. Teoría, práctica y experiencias en torno a la construcción del conocimiento en Patrimonio*, pp. 135–146. Publicaciones CSIC, Montevideo. https://doi.org/doi: 10.26359/costas.0207

Lagos, X., Dabezies, J. M., Delgado, E., & Cetrulo, R. (2017). Vínculos para la gestión: dinámicas socio-institucionales y perspectivas para el manejo integrado de la pesca artesanal en Laguna de Rocha (Uruguay). Redes. *Revista hispana para el análisis de redes sociales*, 28(1), 47–60. https://doi.org/10.5565/rev/redes.669

Lagos, X., Laporta, C., Álvarez, C., Baptista, M., & Fernández, I. (2019). Asociatividad y fortalecimiento comunitario desde el Manejo Costero Integrado: la experiencia de mujeres pescadoras en el área protegida de Laguna de Rocha (Rocha, Uruguay). *Revista Costas*, 1(2), 111–134.

Laporta, M., & Lagos, X. (2021). Remadoras del cambio: Mujeres pescadoras del Paisaje Protegido Laguna de Rocha (Rocha, Uruguay). Tekoporá. *Revista Latinoamericana De Humanidades Ambientales y Estudios Territoriales*. 3(2), 188–209. https://doi.org/10.36225/tekopora.v3i2.142

Latour, B. (2007). *Nunca fuimos modernos. Ensayo de antropología simétrica*. Ed. Siglo XXI, Buenos Aires.

Lee Davis, D. & Nadel-Klein, J. (1992). Gender, culture, and the sea: Contemporary theoretical approaches. *Society & Natural Resources: An International Journal*, 5(2), 135–147. https://doi.org/10.1080/08941929209380782

Mack, J. (2011). *The Sea. A Cultural History*. Reaktion Books, United Kingdom.

Marcos, S. (2013). Descolonizando al feminismo: la insurrección epistemológica de la diferencia. En Méndez Torres, G., Intzin, J. L., Marcos, S. & Osorio Hernández, C. (Coords.), *Senti-pensar el genero: Perspectivas desde los pueblos originarios*, pp. 145–172. Red Interdisciplinaria de Investigadores de los Pueblos Indios de México, Guadalajara.

Martínez, C. M., & Fournier, R. (1999). EcoPlata: An Uruguayan multi-institutional approach to integrated coastal zone management. *Ocean Coastal Management*, 42, 165–185. https://doi.org/10.1016/S0964-5691(98)00052-0

Max-Neef, M. A. (2005). Foundations of transdisciplinarity. *Ecological Economics*, 53(1), 5–16. https://doi.org/10.1016/j.ecolecon.2005.01.014

McKay, S. C. (2014). Racializing the high seas. En Park, J. & Gleeson, S. (Eds.) *The Nation and Its Peoples: Citizens, Denizens, Migrants*, pp. 155–176. Routledge, London, 312p.

Mentz, S. (2009). Toward a blue cultural studies: The sea, maritime culture, and early modern English literature. *Literature Compass*, 6(5), 997–1013. https://doi.org/10.1111/j.1741-4113.2009.00655.x

Migliaro, A. & Santos, C. (2010). La pesca no es sólo eso: producción, reproducción social y ambiente. Sobre pesca artesanal y variabilidad climática en el Uruguay. *Revista Sociedad Latinoamericana Seminario de Pensamiento Social Latinoamericano (FES Aragón-UNAM)*, 2(3), 1–17.

Mignolo, W. D. (2000). La colonialidad a lo largo y a lo ancho: el hemisferio occidental en el horizonte colonial de la modernidad. En E. Lander (Ed.) *La colonialidad del saber: eurocentrismo y ciencias sociales. Perspectivas latinoamericanas*, pp. 52–82. CLACSO-Unesco, Buenos Aires.

Milstein, T., & Dickinson, E. (2012). Gynocentric greenwashing: The discursive gendering of nature. *Communication, Culture & Critique*, 5(4), 510–532. https://doi.org/10.1111/j.1753-9137.2012.01144.x

Morin, E. (1994). Sur l´interdisciplinarieté. *Bulletin Interactif du Centre International de Recherches et Études transdisciplinaires* n° 2, Juin 1994.

Morin, E. (2009). *Introducción al pensamiento complejo*. Gedisa, España.

Moura, G. G. M. (2019). Construção da crítica à oceanografia clássica: contribuições a partir da oceanografia socioambiental. *Ambiente & Educação*, 24(2), 13–41. https://doi.org/10.14295/ambeduc.v24i2.9728

Narchi, N. E., Cariño, M., Mesa-Jurado, M. A., Espinoza-Tenorio, A., Olivos-Ortiz, A., Early Capistrán, M. M., . . . & Moreira Moura, G. G. (2018). El CoLaboratorio de oceanografía social: espacio plural para la conservación integral de los mares y las sociedades costeras. *Sociedad y ambiente*, (18), 285–301. https://doi.org/10.31840/sya.v0i18.1888

ONU (2018). *La Agenda 2030 y los Objetivos de Desarrollo Sostenible: una oportunidad para América Latina y el Caribe* (LC/G.2681-P/Rev.3), Santiago de Chile

Pichón Riviére, E. (1983). *El proceso grupal*. Editorial Nueva Visión, Buenos Aires.

Pinzón Ríos, G. & Lois, C. (2015). Bestiarios cartográficos. Criaturas del mar en los mapas de América (siglos XVI–XVII). En. G. Pinzón Ríos & F. Trejo Rivera (Coords.), *El mar: percepciones, lectura y contextos Una mirada cultural a los entornos marítimos*, pp. 131–158. UNAM, México.

Ransley, J. (2005). Boats are for boys: Queering maritime archaeology. *World Archaeology*, 37(4), 621–629. https://doi.org/10.1080/00438240500404623

Rao, A., & Kelleher, D. (2005). Is there life after gender mainstreaming?. *Gender & Development*, 13(2), 57–69. https://doi.org/10.1080/13552070512331332287

Rao, N., Lawson, E. T., Raditloaneng, W. N., Solomon, D., & Angula, M. N. (2019). Gendered vulnerabilities to climate change: Insights from the semi-arid regions of Africa and Asia. *Climate and Development*, 11(1), 14–26. https://doi.org/10.1080/17565529.2017.1372266

Restrepo, E., & Escobar, A. (2015). Red de Antropologías del Mundo: intervenciones en la imaginación teórica y política de la práctica antropológica. En *Prácticas otras de conocimiento (s). Entre crisis, entre guerras, Tomo II*, pp. 381–402. CLACSO, Buenos Aires.

Santos, B. D. S. (2006). *Renovar la teoría crítica y reinventar la emancipación social*. Buenos Aires, CLACSO. 110 págs.

Santos, B. D. S. (2009). *Más allá del pensamiento abismal: de las líneas globales a una ecología de saberes. Una epistemología del sur. La reinvención del conocimiento y la emancipación social*. Siglo XXI Editores, CLACSO.

Santos, C. (2015). Sobre la Interdisciplina. En Vienni, B., Cruz, P., Repetto, L., Von Sanden, C., Lorieto, A., Fernández, V. (Coords). *Encuentros sobre Interdisciplina*, pp. 69–78. Espacio Interdisciplinario, UDELAR, Montevideo.

Schmitt, K. (1997). *Land and Sea*. Plutarch Press, Washington DC.

Steinberg, P. E. (2001). The *Social Construction of the Ocean*. Cambridge University Press, Cambridge, 239p.

Steinberg, P., & Peters, K. (2015). Wet ontologies, fluid spaces: Giving depth to volume through oceanic thinking. *Environment and Planning D: Society and Space*, 33(2), 247–264. https://doi.org/10.1068/d14148p

Thompson Klein, J. (2004). Prospects for transdisciplinarity. *Futures*, 36(4), 515–526. https://doi.org/10.1016/j.futures.2003.10.007

Tomasino, H., & Cano, A. (2016). Modelos de extensión universitaria en las universidades latinoamericanas en el siglo XXI: tendencias y controversias. *Universidades*, (67), 7–24. http://beu.extension.unicen.edu.ar/xmlui/handle/123456789/99

Uriarte, J. (2020). *The Desertmakers: Travel, War, and the State in Latin America*. Routledge, New York, 294p.

Vallega, A. (1993). A conceptual approach to integrated coastal management. *Ocean Coastal Management* (21), 149–62. https://doi.org/10.1016/0964-5691(93)90024-S

Vienni Baptista, B., & Goñi Mazzitelli, M. (2021). Aportes para los estudios sobre Interdisciplina y Transdisciplina: Modalidades, estrategias y factores para la integración. *Utopía y Praxis Latinoamericana*, 26(94), 110–127. https://produccioncientificaluz.org/index.php/utopia/article/view/36113

Visser, L. E. (2004). Reflections on transdisciplinarity, integrated coastal development, and governance. En Visser, L (Ed.) *Challenging coasts. Transdisciplinary excursions into integrated coastal zone development*, pp. 23–47. Amsterdam University Press.

Walsh, C., & Castro-Gómez, S. (2002). *Indisciplinar las ciencias sociales: Geopolíticas del conocimiento y colonialidad del poder. Perspectivas desde lo andino*. Editorial Abya Yala, Quito.

Wilson, M. (2013). The green economy: The dangerous path of nature commoditization. *Consilience*, 10, 85–98. https://doi.org/10.7916/consilience.v0i10.3934

3

CO-PRODUCTION OF KNOWLEDGE AS PRODUCTION OF SPACE

How we all give meaning to the sea

Kathryn Collins

Introduction

This chapter explores the importance for marine management of acknowledging that marine spatial meaning is produced by everyone connected to it. In other words, to manage a marine space firm, understanding of the nature of the space being managed is required. In this chapter space is defined as more than physical. Borrowing theories from the urban design, this chapter uses Henri Lefebvre's Production of Space thesis to explore three interrelated spatial moments: physical space, mental (or conceived) space, and lived experience (Lefebvre, 1991).

Public participation within marine management can often be challenging, particularly where value conflicts arise. Ethics and cultural studies help to identify the antecedent values which underlie conflicts. The Production of Space thesis allows for the divergent representations made within marine decision-making scenarios to be seen as opportunities to learn more about marine spaces and how values contribute strongly to the co-production of marine knowledge.

The chapter begins with an explanation of the Production of Space. The urban planning literature is used here and using terrestrial examples aids understanding of this abstract. The importance of understanding lived experience within spatial value formation and spatial imaginaries is also discussed. From here the discussion turns to how underlying ethical positions influence spatial value formation and, using examples from both desk-based and doctoral field research (Collins, 2020, 2022), the chapter concludes with a discussion which firmly locates the co-production of knowledge of marine spaces into spatial theory and considers why thinking in this (socially produced) spatial way is so valuable for marine resource management.

Transdisciplinarity is inherent within this approach. Socially produced spaces draw on experiences and representations from across the academic discipline

DOI: 10.4324/9781003311171-4

spectrum. This chapter advocates for the value of seeking out and including trans-discipline voices within marine management and marine spatial understanding. This method is transdiscipline rather than multidiscipline due to the holistic nature of appreciating all spatial values are equal in order to explore where conflicts and tensions exist and how these can be mediated. The wider we cast the net, the more complete our understanding of the marine spaces we manage will be.

Revisiting 'the marine turn'

Much has been written about the rise of marine spatial planning (MSP), its connections to terrestrial – or 'town and country' – planning, and its establishment as its own independent discipline (Ritchie, 2014). It is not the intention of this chapter to revisit these discussions. What is clear is that MSP focuses largely on strategic-level marine planning and, as part of its establishment as an independent discipline, has largely moved away from reference to the wider 'spatial planning' (town planning, town and country planning, urban planning, terrestrial planning, etc.) conceptual debates. Overall, this is a welcome move. MSP is no longer 'new'; no longer a niche sub-discipline of its terrestrial relative, and the marine environment has distinct planning and management challenges not seen on land.

However, there is still much to learn from spatial theory associated more closely with the domain of the urban theorists and urban planners. This is particularly so when considering the operational marine management consenting regimes which permit, or otherwise, development within marine space. As a spatial planning discipline, these marine consents are determined within a space which is both physical *and* imbued with meaning formed from the activities and values which operate within it (Lefebvre, 1991). Central to this co-production of meaning is the context and the publics which comprise a particular space (Dimendberg, 1998; Madanipour, 2003). Taking marine space as a public space, in virtue of its regulation, access and ownership status (Madanipour, 2003), presents an opportunity to reconceptualise the context within which marine consent decisions are made. As a socially produced public space, marine space can therefore be understood more holistically. This provides a mechanism through with marine managers and researchers can gain a greater understanding, or appreciation, of how stakeholder values, opinions and objections are formed.

In short, to understand marine space, it is useful to conceptually remove the 'marine' and understand first and foremost the nature of the space under examination.

The Production of Space: borrowing from urban planning

When working with the Production of Space thesis, space is not seen as a neutral concept. It is more than a Euclidean, physical area, or a container to be filled with 'stuff'. The space in which planning operates is socially constructed, imbued with meaning and actively produced through human activity. Lefebvre's thesis shows its

Marxist, Hegelian and Schopenhauerian roots in its use of materiality, dialectic and the use of a concrete universal to explain the nature of space. As a concrete universal concept space takes triadic form, in which subjects, objects, and activities Operate to co-produce social space. The use of "three elements and not two" (Lefebvre, 1991, p. 39) negates the relationships of "oppositions, contrasts or antagonisms" (Ibid.) which hold historical significance in philosophical dualisms such as subject and object or human and non-human. For Lefebvre, the use of dualisms, or binaries, has "the magic power to turn obscurity into transparency and to move the 'object' out of the shadows into the light merely by articulating it" (Ibid.). Dualisms have the power to oversimplify situations, and this is seen in everyday life and use to great effect in political discourse: for UK nationals the Brexit 'leave versus remain' arguments presented oversimplified narratives of what an exit from the EU, or otherwise, would mean. This plays on in marine space through post-Brexit fishing rights. What is missing from spatial binaries is the *experience of the relationship* between the two terms. Using the body as an example, lefebvre explains that "the 'heart' as *lived* is strangely different from the heart as *thought* and *perceived*" (Ibid., p. 40). Likewise, the human and non-human – or nature/culture – binary, regularly seen within MSP, in marine impact assessments, and more accessibly within media representations of space such as David Attenborough's Blue Planet series (BBC, 2017), appears absent of the experience which explains the dualism.

Lefebvre's thesis makes explicit the "dialectical relationship which exists within the triad of the perceived, the conceived and the lived" (Lefebvre, 1991, p. 39). The interconnectedness of these three spatial moments is crucial within the Production of Space, and direct reference to the original Lefebvre text is key to understanding this:

> *The perceived-conceived-lived triad (in spatial terms: spatial practice, representations of space, representational space) loses all force if it is treated as an abstract 'model'. If it cannot grasp the concrete (as distinct from the 'immediate'), then its import is severely limited, amounting to no more than that of one ideological mediation among others.*
>
> *(Ibid., p. 40)*

Applying the triad of spatial terms to marine space allows for a deeper understanding of its nature. Conceived marine space, imbued with "ideology, power and knowledge" (Ibid., p. 40) can be witnessed in the charting and delineation of ocean zones and boundaries. Perceived, absolute, marine spaces "structure lived reality" (Ibid.: p. 110) and can be thought of as the immediate and material physicality which is the subject of oceanographic study. Finally, lived experience is the moment which gives meaning to marine spatial encounters; what it feels like to enter this space.

Conceived space is the "dominant space of any society" (Merrifield, 2006, p. 109) and provides structure and arrangement to physical space through charting, describing and modelling. Materialism is important here and "the production of space is deeply embedded and centrally located in the overall political, economic and cultural conditions of a society" (Healey et al., 2000, p. 4). Representations of space are artificially separated from the remaining moments and are *ideological (constructed for a purpose or to win an argument)* rather than *concrete (factual, objective)* presentations (Lefebvre, 1991, p. 40; Elden, 2004; Wilson, 2011). They are used to tell the authors story from their point of view and to convince others to reach the same conclusions regarding the meaning of a space. For example, maps and charts provide more detail than the points and vectors they display. This can be seen within the historical representations of marine space which to contemporary eyes seem at first, unfairly, naïve, and amusing but with deeper thought are seen to have been delineated and choreographed in ways which reflect the views of the societies which created them:

> *During the sixteenth and seventeenth centuries, maps portrayed an ocean cluttered with ships, sea monsters, and rhumb lines, all of which were intended to portray the complex 'reality' of a space rich with natural and social features. By the early eighteenth century, however, the ocean was perceived as a space unworthy of social interest. Cartographers reduced the ocean to an empty, blue expanse, at most punctuated by placeless latitude and longitude coordinates and often – as in Lewis Carroll's parody – as "a perfect and absolute blank".*
>
> *(Steinberg, 1999, p. 410)*

This "choreography" is also seen in contemporary marine mapping. The charts presented in Figures 3.1 and 3.2 represent the same physical space, and indeed the socio-political delineation of the UK EEZ boundary. But they *mediate* space in ideologically different ways to present different conceptualisations of the important features of this space. This becomes evident when considering the source of these images. Figure 3.1 appears to be considerably more detailed than Figure 3.2, but both contain layers of hidden power-relations, historical details, and social constructions within their conceptualisations. The first chart represents the expansion of UK marine space through the political, economic and social desires to increase fishing, oil and gas and, latterly, offshore wind production enacted through a series of statutory instruments through which the 2017 marine boundary has been reached (OGA, 2018). It is showing how the UK EEZ has been expanded to include increasing amounts of marine space as maritime resource sectors have developed. This is showing marine space as a resource provider (Steinberg, 2001) and a space defined through its industrial uses.

The second, simpler chart, originates from a Brexit preparedness study for UK fisheries and represents UK marine space in relation to other EU member states illustrating the other parties within post-Brexit fisheries management considerations (House of Lords, 2016). The socio-political background of the creation of this chart is only fully understood within the context of the report in which it is found. The report acknowledges that the chart is a simplification of mapping data "as the full spectrum of an area's characteristics cannot be represented on a map" (Smith and Brennan, 2012, p. 212).

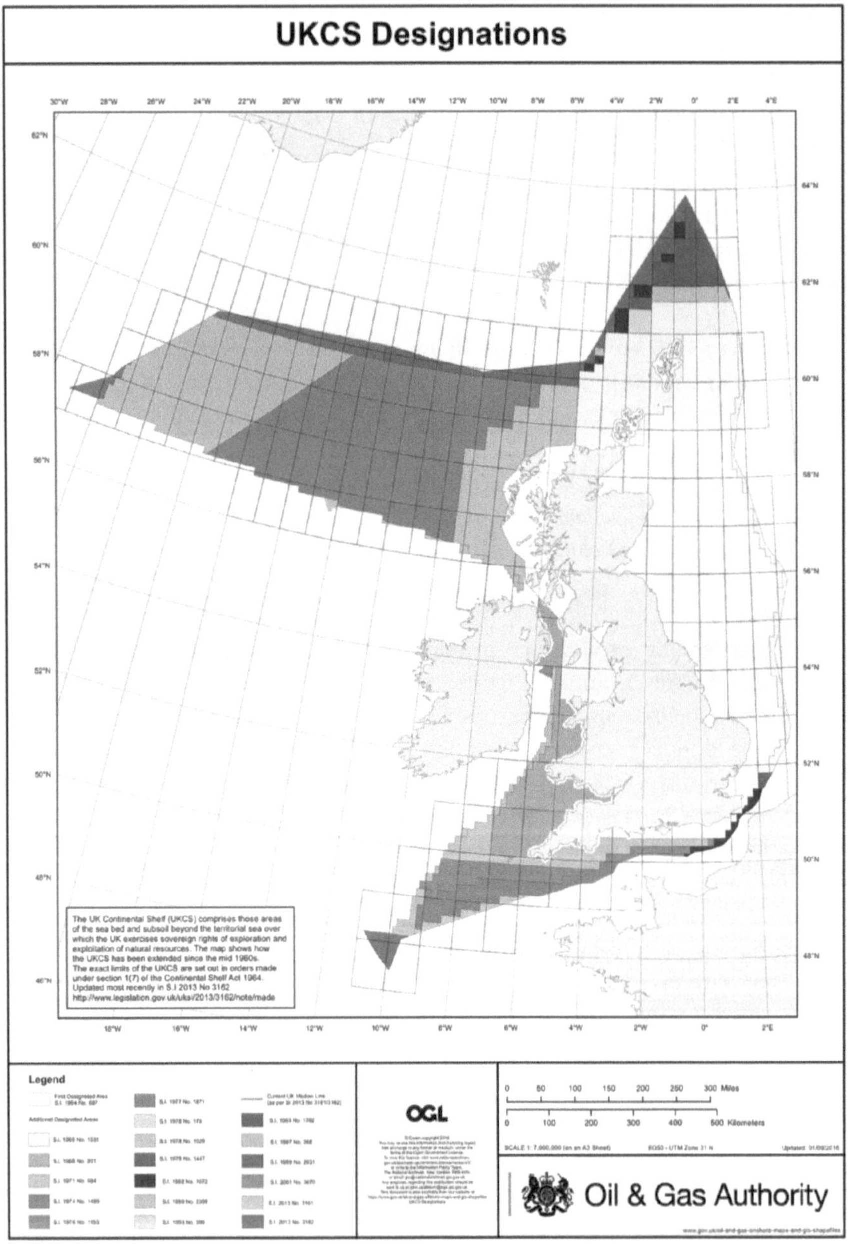

FIGURE 3.1 The Socio-Cultural Historic Formation of the UK EEZ.

Source: (OGA, 2018)

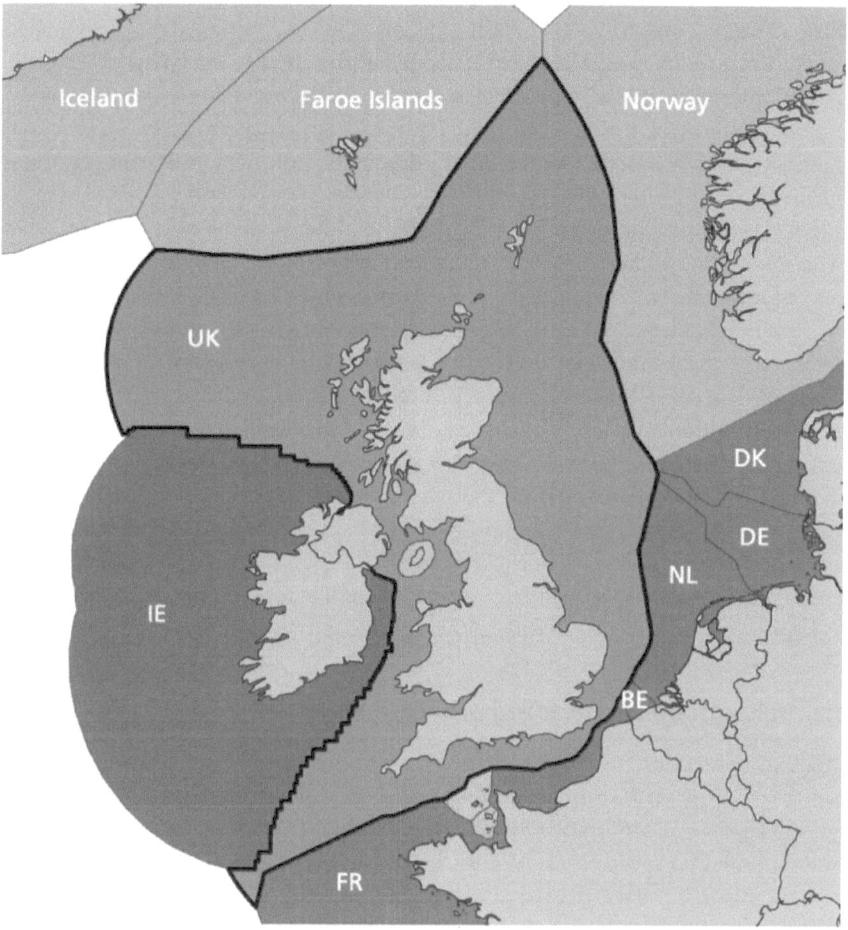

FIGURE 3.2 UK EEZ in Relation to Other EU Member States.

Source: (House of Lords, 2016)[1]

The significance of understanding the context through which charts are created and shared is found in the notion that "how we as social beings use and perceive space cannot be divorced from the wider structuring networks of property relations, the forces of production and the government apparatus itself" (Healey et al., 2000, p. 3; Steinberg, 2001; Elden, 2004).

To summarise the discussion so far, the perceived marine space which is the object of marine scientific study is *mediated* through conceived space. The latter spatial moment operates within marine planning through the charting and ordering of charts and plans, and through the representations used to assess and communicate the impact of marine developments. With perceived marine space now understood as physical, and conceived marine space understood as mental (or conceptual), a discussion of lived experience is needed to complete the triad and this section.

The final triadic moment, lived space, gives meaning to everyday experience. Whilst care is needed not to privilege any of the spatial moments within theoretical or practical thought, considering lived experience separately from the other spatial moments provides useful insight on the limitations of current marine planning practice.

Seen as a social construct, Lefebvre states, in rather impenetrable language, that lived space is a "product of interrelations", a "sphere of the possibility of the existence of multiplicity in the sense of contemporaneous plurality" and "always under construction" (Massey, 2005, p. 9). These are big concepts but can be more simply understood as meaning that lived space is where our experience happens. Lived experience is, then, the experience of being *there*. It is what it feels like to be embodied within a space. The Production of Space thesis requires thinking of space as made up of these three 'types' of space which all impact on each other. In addition, it needs an acceptance of a material reality; a physical world that is 'out there' and whose existence continues without human thought. Without materiality, the power that physical space can exert to enable or limit action is extinguished. Embodiment is also important here. What it feels like for *this body*, with *this conceptual mental landscape*, to inhabit *this space* is personal to every inhabitant.

The value in understanding antecedent values

From the discussion so far, social space can be understood only through grasping the concept of the three moments of space. These moments are in a constant dialectical relationship in which they all add to, therefore materially change, the meaning of a given space. Whilst the physicality of perceived space remained entirely material, conceived space is mediated through both the lived experience of a space and the concepts which are both brought into the space from previous experiences and used to make sense of it. In other words when a space is encountered, either in person or remotely – through media or other representation – it is understood by the inhabitant through their immediate experience, the physical limitations of the space, and also through previous spatial experiences and conceived notions of the space. These notions are formed from individual antecedent moral or ethical values.

To understand the role of antecedent values within the social production of space, this section focuses on conceived space and considers two types of value judgements which form the antecedent ideologies underpinning spatial representations, whether or not they are being knowingly used to ideologically mediate others. The antecedent values discussed in this section are theories of justice and environmental ethics. These are both important within the understanding of spatial relationships. The former helps to understand different beliefs regarding how resources should be distributed, and the latter explores the nature of these non-human resources and what the human relationship with them should be. The urban planning literature is utilised again within these discussions to understand how these values are connected to spatial relationships.

Starting with theories of justice, three main positions are advocated within public space research. These are applied to the access, provision, and maintenance of public goods. In other words, how public resources – including spaces – should be used. This is useful for understanding antecedent values underpinning beliefs about how marine space, and marine resources should be managed.

Distributional justice theories state that the fair – but not quantifiably equal – allocation of goods and services, along with procedural concerns relating to their access, is central to promoting "a sense of citizenship, equality and social justice" (Low, 2013, p. 229). As a response to utilitarian conceptions of ethical fairness, distributional justice theories build on John Rawls's 1971 attempt to "guard against injustice in that the greatest good for the greatest number might, in its most extreme form, represent tyranny by the majority" (McKay et al., 2012, p. 149). In contrast to distributional justice, justice as recognition sees all stakeholders as important and their recognition within access to a public good is key for justice concerns (Young, 1990; Sohlosberg, 2003). In other words, distributional justice is concerned with fair resource allocation, and justice as recognition is concerned with public access to a resource.

Procedural justice – or justice as process – is essentially a compromise position which consolidates ideas regarding the fair allocation of distributional justice with the importance of stakeholder recognition through its demands for "broader and more authentic public participation" (Sohlosberg, 2003, p. 84). For decision-makers, this results in the assertion that "anything less than processes underpinned by an appropriate degree of inclusivity in consultation disregards public moral values, thereby undermining the principles of democracy" (McKay et al., 2012). This presents a deliberative theory which "emphasizes not the decision-making moment but all the processes of opinion formation and public debating that go on before matters come to a vote" (Parkinson, 2012, p. 28). For procedural justice, decision-making processes are greatly improved "by including more diverse experiences and better quality information" (Ibid.), with improvement meaning "more just" decisions. As such, the "right thing to do" – the just thing to do – is the result of "inclusive, democratic encounters, not something that pre-exists them, hiding under rocks waiting to be found by clever truth hunters" (Ibid.). These theories relate directly to environmental ethics, discussed further, and the extent to which a given process can be described as "just" will depend on the ethical position of the evaluator.

Theories of justice are highly relevant to the Ecosystem Based Approaches (EBA) which underpin much of contemporary marine planning and management. The decision whether or not to grant consent for a marine development project is made within the sphere or broader considerations of how best to allocate and manage resources. Under EBA, marine resource allocation and management focuses on goods and service provision for human survival and wellbeing. This, importantly, includes the protection of habitats and species as part of protecting the wider ecosystem. However, the ethical antecedents of this are open to debate. Indeed, as human survival and wellbeing depend on healthy ecosystems, ecosystems themselves benefit from management which supports human flourishing. If marine ecosystems are managed in a way which maintains their health and provides ecosystem services, does the motivation behind resource management decisions matter? The Production of Space thesis allows for these meanings, and the motivations behind them, to be seen as powerful ideological mediations which materially change our understanding of a given space. In other words, the ethical positions behind motivations have power to change the spaces under examination through the way that those spaces are designed. A marine area with a large fish population (perceived space) can be labelled as a primary fishing ground, or a fish conservation area (conceived space). The label used will change the activities permitted in that area and in turn change how that area is used (lived experience).

At the societal level, numerous beliefs exist about how best to allocate resources. Understanding the foundations of these beliefs is crucial for understanding how decisions made by marine regulatory bodies disentangle contradictory public representations and arrive at decisions which are fair, meaningful and accepted as valid. In addition to the multitude of beliefs regarding environmental issues within a society, the complexity of ecosystem goods and services encompass benefits on multiple geographic and human/non-human scales. Whilst marine policy aims to address these value-based conflicts at a strategic level, marine regulation requires engagement with this complexity at the point of conflict. This suggests that "the complex interactions of the highly diverse systems housed with even more complex ecosystems are . . . a cue to up the ante against the simplified answers that are routinely trotted out by well-meaning organizations" (Probyn, 2016, p. 250). To understand the complexity of multiple values within concepts such as "ecosystem services" or "sustainable development", a study of environmental ethics proves insightful. These terms are often accepted uncritically through the rhetoric of environmental agencies (Sagoff, 2010 pp. 394, 399; Probyn, 2016) and their antecedent values left opaque.

Environmental ethics as antecedent values

Environmental ethics adds specificity to wider utilitarian (greatest good to the greatest number), deontological (duty based), consequential (results based), and normative (virtue based) ethical theory. More broadly, environmental ethics addresses normative questions regarding the human/non-human relationship with "the moral status of Nature . . . determined by the contexts within which nonhuman entities are incorporated into human cultural understanding" (King, 2010, p. 352).[2] Moral concerns surrounding issues of intergeneration justice, fairness, respect, and compassion for the non-human world, and the preservation of the intrinsic value of environments are considered through contemplation of this relationship (Sagoff, 2010, p. 392).

A utilitarian environmental ethic appears to complement marine resource decision-making within EBA which supports "utilitarian objectives of sustainable development if not utopian ones . . . [and] appears to place nature on an equal footing with economic interests in debates about development" (Harrison et al., 1999, p. 85). But basing decisions on "the greatest good for the greatest number" remains problematic. Any such preference-satisfaction theory faces challenges regarding "the possibility that a person may be mistaken about what is best for her" (O'Neil et al., 2008, p. 23). To address this, an informed-preference theory can be applied in which education can change personal values and opinions regarding the importance of system or receptor. For example, "On being educated about salt marshes I may subsequently come to value them a great deal, and this education might make a large difference to my well-being" (Ibid., p. 24).

The response given to the challenges of utilitarian ethical models by deontological ethical theorists is to consider the *process* rather than the *outcome* of decision-making (Ibid.; Seip and Wenstop, 2006). For consequentialists, the answer to this challenge is through recognition of the naturalistic fallacy by which subjective values are derived from objective facts about the world. Fish population decline is an objective fact; the necessity to impose management measures to protect further decline is a subjective value. For normative ethicists the virtues of *decision-makers themselves* are the subject of moral enquiry (O'Neil et al., 2008).

To understand the subject of enquiry for environmental ethics, recourse to an 'over-simplified' binary is necessary. Environmental ethicists study the human/non-human relationship and, more fundamentally, the concept of nature itself. In this way, "environmental ethics constitutes critiques of anthropocentrism – some positive, others negative" (Keller, 2010, p. 1) and three main environmental ethical schools can be identified along a somewhat linear spectrum, from anthropocentric, to biocentric to ecocentric. At one end is the value judgement that environmental degradation is seen as undesirable in relation to the harm it will cause to humans (Kortenkamp and Moore, 2001, p. 262). At the other are deep ecological positions which focus on the "rejection of the man-in-environment image in favour of the relational, total-field image" (Wilkinson, 1999, p. 25; Morton, 2010, 2016) and ascribe intrinsic value to ecological features (Kortenkamp and Moore, 2001, p. 262). Countless shades of grey complete the spectrum. Generally, the human/non-human relationship can be expressed as either "humans *and* the environment" or "humans *in, or as part of*, the environment".

The moral consideration behind the desire to protect a particular ecological feature – due to an anthropocentric or ecocentric ethic – determines the reasoning behind the differing moral arguments for public objection towards development projects. Both ecocentric and anthropocentric positions can, and do, lead to positive outcomes in terms of environment protection but "the difference in these two orientations is in the reasons given for supporting conservation" (Gagnon-Thompson and Barton, 1994, p. 149). Understanding these motives is an important element of collaborative planning discussions. For example, public acceptance of the reintroduction of a predator species within an agricultural community is dependent on "the differential access to social power, conflicting ideas about private property, and divergent beliefs about nature" (Bjerke and Kaltenborn, 1999 p. 415).

The language of much of the academic literature regarding environmental ethics paints anthropocentric attitudes as morally inferior to ecocentric positions in that "anthropocentrism is thought to be composed of not only a concern for humankind generally but also an *egocentric* concern" (Karpiak and Baril, 2008, p. 206, emphasis added). It is not fashionable to state a desire for litter-free beaches based on a desire to sunbath in an aesthetically pleasing space. In another example, perceptions towards the reintroduction of wolves in America evidence that "sheep farmers, relative to wild-life managers and research biologists, more often agreed to items like 'too much emphasis has been placed on conservation', and 'I find it hard to get too concerned about environmental issues" (Bjerke and Kaltenborn, 1999, p. 417).

Anthropocentric environmental values are supported by evolutionary sciences and bioethics. In addition, ecocentric ethics can be critiqued in virtue of ethical theorising more generally. All ethical models, including those within the environmental ethics specialism, are axiomatically subject to the anthropocentric nature of value judgements in that "when we make decisions about the environment, we apply our own values; neither animals, nor plants nor mountains are sacred in and by themselves, but may be attributed a status of sacredness from humans" (Seip and Wenstop, 2006, p. 150). This is not to delegitimise deep ecological perspectives, but rather to acknowledge that any, and all, human value judgements, regardless of their motivation, remain deeply embedded in individual and social relations. The meanings individuals attach to the natural environment are both profoundly personal and profoundly social. Applying the Production of Space thesis here demonstrates how ethical values are formed through lived experience, which is dependent on the experiencer. The 'environment' here becomes perceived space, the object of the experience. Values can materially change this material space, however spatial inhabitants cannot speak from this objective position; they can make their best attempt at doing so, but it will be necessarily subjected to embodiment and lived experience.

Central to this ethical discussion is the recognition that decisions made about marine resource allocation or spatial uses subject to the value judgements of both the decision-maker and the spatial inhabitants who have socially produced the space in which the decision is being made. It is therefore important to acknowledge that different environmental ethical positions, and indeed different antecedent ethical theories, affect the values which are attached to the subject of the decision situation.

The language of ethics allows for the articulation that, at its core, marine management, like all spatial management, involves ethical decisions about what to do with space. When this is linked with procedural justice, there is an increased likelihood that stakeholders will accept decisions which would ordinarily contradict with their core interests and values. This latter contention applies a deontological ethical view to decision-making in which the *process* rather than the *outcome* is the subject of ethical concern. Noted here is the deontological position that it is the *process* which is subject to ethical consideration with imperatives attached to ensuring that the process itself is morally right.

Conflicting values within offshore wind planning

The offshore wind literature allows for useful exploration of the challenges faced through failing to acknowledge the value of antecedent beliefs. The early years of renewable energy development in the UK assumed wide public acceptance due to the high level of support expressed in public surveys. Planners operated under five key assumptions regarding this support (Waldo, 2012):

1 A majority of the population is positive towards wind power
2 Opposition is therefore deviant
3 Opponents are ignorant or misinformed
4 The purpose of understanding opposition is to overcome it
5 Trust is a key aspect to achieving this

However, these assumptions are simplistic and do not account for the multiple meanings that individuals attribute to wind technology. Turbines can be seen as representing, for example, " 'stewardship', 'ugly technology', 'responsible energy policy', 'destroying landscape', 'progress' and 'harking back to the past'" (McLachlan, 2009, p. 5343; Stokes et al., 2014). Local communities protesting against – being deviant towards – large-scale developments have been stereotyped as Not In my Back Yard (NIMBY) and been seen as motivated by "individualist self-interest" (Eranti, 2017, p. 286). Negative associations of offshore wind also focus on aesthetic and non-market impacts such as wellbeing, spirituality for local communities and market impacts of tourism, and property value (Dalton et al., 2015 p. 863; Waldo, 2012).

Meaningful local stakeholder engagement which is respectful for local perceptions and values is central to both understanding and challenging these associations. But this is not simply a case of accepting lay – non-expert, public – opinion as fact because "laypeople may not always recognise their inaptitude for making truly informed claim evaluations" (Scharrer et al., 2014, p. 465). Sometimes being too close to a situation, or development, makes it hard to see the bigger picture. Sometimes conspiracy theories based on genuine fear of technology are hard to shift. Acknowledging and mediating between multiple categories of knowledge claims thus becomes a critical judgement for assessors and regulators such that expert and lay knowledge categories can also be supplemented with "institutional knowledge claims" and "expert planner knowledge claims" (Lee, 2017, p. 4). It is important therefore to acknowledge both the origins of wind technology knowledge claims and consider their validity within these categories.

Aesthetic impacts have tended to dominate lay debate regarding offshore renewables. Coastal residents simply do not like their sea-view changed from "blue expanse" to "industrial power plant". What makes for a nice view is extremely subjective; beauty is in the eye of the beholder. In addition to aesthetic and other "proximity dependent impacts", perceived impacts of wind developments can lead to "place-protective action" in which "changes to a place threaten to disrupt emotional attachments and aspects of identity" (Stokes et al., 2014, p. 65). These attachments include consideration of the relationship between individual opinions of the symbolic nature of technology and of the place impacted (McLachlan, 2009, p. 5348). By identifying these, the response to a proposed development is often quite logical. For example, if one views wind turbines as a "blot on the landscape" because one values open and empty countryside, one is likely to object to a local wind farm proposal. Opposition to wind developments can therefore reflect more than direct opposition to the siting of a development in a particular place but concern about this type of development being built in the environment generally.

These debates within offshore wind consenting also highlight additional conflict types which occur between stakeholder groups. These include "local versus global" issues and "green on green" conflicts,[3] with the former also evident in the latter. While local opposition tends to focus on local identity, and impacts to recreation and tourism – as well as emotional attachment to a place – these are situated within wider policy and developmental concerns, operating at national or global scales, such as renewable energy policy, energy security, climate change, and the commodification or industrialisation of nature (Ibid., p. 5345; Ladenberg, 2010; Haggett, 2011). Ensuring that local stakeholder concerns are not dismissed as 'deviant' is crucial here. Instead, those that do object – vocally or otherwise – would more helpfully be seen as opposing development "on the basis of detailed knowledge of their area, the development, and the issue more generally" (Haggett, 2011, p. 504). Likewise dismissing non-local objectors is also unhelpful. Non-local objectors can provide useful perspectives and data for assessment and consent deliberations, for example "tourists are found to be more negative compared to the local residents, because they want to enjoy unspoiled nature" (Ladenberg, 2010, p. 1298; cf. Nimmo et al., 2011). As the title of this chapter suggests, "we can all give meaning to the sea".

Values and the co-production of marine social space

As the discussions outline, marine management operates within a co-produced space in which its physical fluidity is reflected in the dynamic, and often contested, multiple meanings, associations and imaginaries attributed to it. The complexity of this space is further illustrated within contemporary representations of marine space. The influence of media representations of marine and other non-human spaces evidences the known impact of media on public attitudes and behaviour. Increased public engagement with specific conservation projects attributed to Disney animated films such as Finding Nemo and Shark Tale (Jones, 2014, p. 5) have been replicated more recently in relation to the 'Blue Planet Effect' associated with the critically acclaimed Attenborough documentary series from which the effect takes its name (BBC, 2017). There are risks here of isolating marine space concerns to specific, social media status-raising activities, and the over-prioritisation of charismatic species protection associated with both fictional and non-fictional media portrayals.

In addition to contemporary documentaries, imagery and fictions are historical accounts, stories, myths and legends which add significant meaning to marine spaces. Representations of space based on myths and stories imbue an area with both geographical and cultural significance. For example, the Goodwin Sands, a subtidal and intertidal sandbank off the south-east English coast, is believed by some to be the remains of an island sunk by vengeful gods as punishment for its owner, Earl Godwin, failing to make-good on promises to build a church in honour of the saints that saved his life during a treacherous return sea voyage (Collins, 2022). As 'Kent's Atlantis' the Goodwin Sands plays an important role in the cultural identity of residents of the English coastal towns who live within eye-sight of the marine space (Ibid.). These mythical representations appear, at first, to be amusing and of limited productive value, however without their inclusion within social space conceptualisation the 'concrete' nature of the space is limited to the ideological representations which are privileged within the marine planning system. To understand social space, discrimination against 'types' of representations is to be avoided as this delegitimises their productive power. This, in turn, risks delegitimising individuals for whom these representations are meaningful and stakeholder engagement conflict can quickly arise.

But pragmatism is also required. While representations of myths, legends and archival accounts add meaning to social space, they are resource-intensive to research and the nature of these accounts – based on memory and anecdote – have a subjective quality which is challenging to address within positivist-based resource management assessment and decision-making. Put simply, these rich qualitative representations are hard to accommodate within quantitative approaches. Non-physical representations are important for getting closer to the concrete meaning of a social space, but they are complicated, contradictory, and often have limited or no "tangible" evidence.

Through its use of representations of space marine, planning aims to simplify the complexity of marine environments. It aims to manage this complexity through quantifiable evidence-based decision-making. But something appears to be lost within these pragmatic, and necessary, simplifications. The epistemological and ethical questions regarding the human relationship with the non-human – or more-than-human – marine environment remain. Understanding that multiple environmental ethical positions exist within stakeholder communities and publics for a specific development proposal, allows for more meaningful engagement with the representations being voiced. Marine management based on "dominant utilitarian approaches to environmental policy [is] not consistent with the existence of plural and incommensurable values" (O'Neil et al., 2008, p. 91). Being mindful of symbolic logics which differ from the majority view, or seem deviant from presupposed assumptions regarding public perception, is also skill which marine managers need to practice. Care is needed to ensure that consultation material and application assessments themselves are accessible to wide audiences of non-expert stakeholders and that new information provided through public representation is considered with due care and attention. Another way to approach this discussion is through consideration of the multiple value-judgements underpinning symbolic logics, and beliefs about the environment more generally. Where these values clash, or are perceived to clash by certain stakeholders, conflict arises.

Applying a methodology based on the Production of Space thesis helps in this spatial understanding. It can explain why similar marine spaces are attributed wildly different meanings based on their physical characteristics, the conceived notions applied to them, and the experience of engaging with them. While it is acknowledged that not all this data is verifiable, or indeed relevant to marine management or planning decision-making, engaging with these multiple representations allows for the values which underpin local marine spatial values to be acknowledged and maintained. This is important. Representations are a powerful productive force and give meaning to the spaces we occupy, encounter, manage and develop. When thinking about marine management, the Production of Space thesis reminds us that this marine space is socially produced; it is public. Our understanding of this socially produced space is fundamentally easier if apply transdisciplinarity. Including theories, values and voices from across the academic discipline spectrum engenders an approach of openness, honesty and integrity. This is a challenging methodology to apply with resource-stretched marine regulatory frameworks, but at its heart it is democratic and powerful. We all give meaning to the sea, so we can all decide how best to manage it.

Notes

1 Note here how the reports' author considers that scale and legend are unnecessary for this conceptualisation of UK marine space. These details are not needed for the point they are making.
2 King uses the capitalisation of the word 'Nature' along with the non-capitalised 'nature' to distinguish between Nature as a human concept, and nature as the physical assemblage.
3 The concern here is, for example, that "society has gone green (at least in its rhetoric), but what kind of greenness do we want?" (Warren et al., 2005, p. 854).

References

BBC (2017), *Blue Planet II*, BBC One, 29 October 2017

Bjerke, T. & Kaltenborn, B. (1999), The Relationship of Ecocentric and Anthropocentric Motives to Attitudes Toward Large Carnivores, *Journal of Environmental Psychology*, 19, 415–421

Collins, K. (2020), Governance, Decision-Making and Publicness in Marine Space. PhD diss., Newcastle University, Available at: https://theses.ncl.ac.uk/jspui/handle/10443/5059

Collins, K. (2022), Shifting Sands, Layering Meanings: A Tale of the Goodwin Sands, Told through its Social Production, *Shima*, 116: 1, 199–225.

Dalton, G., Allen, G., Beaumont, N. Georgakaki, A., Hacking, N., Hooper, T., Kerr, S., O'Hagan, A., Reilly, K., Ricci, P., Sheng, W. & Stallard, T. (2015), Economic and Socio-Economic Assessment Methods for Ocean Renewable Energy: Public and Private Perspectives, *Renewable and Sustainable Energy Review*, 45, 850–878

Dimendberg, E. (1998), "Henri Lefebvre on Abstract Space", in Light, A. & Smith, J. (eds), *Philosophy and Geography II: The Production of Public Space*, Lanham, Rowman & Littlefield

Elden, S. (2004), *Understanding Henri Lefebvre. Theory and the Possible*, London, Continuum

Eranti, V. (2017), Re-visiting NIMBY: From Conflicting Interests to Conflicting Valuations, *The Sociological Review*, 65(2), 285–301

Gagnon-Thompson, S. & Barton, M. (1994), Ecocentric and Anthropocentric Attitudes Toward the Environment, *Journal of Environmental Psychology*, 14(2), 149–157

Haggett, C. (2011), Understanding Public Responses to Offshore Wind Power, *Energy Policy*, 39, 503–510

Harrison, C., Burgess, J. & Clark, J. (1999), "Capturing Values for Nature: Ecological, Economic and Cultural Perspectives", in Holder, J. & McGillivary, D. (eds), *Locality and identity: Environmental issues in law and society*, Aldershot, Ashgate

Healey, P., Madanipour, A. & Hull, A. (2000), *The Governance of Place: Space and Planning Processes*, Aldershot, Ashgate

House of Lords (2016), European Union Committee 8th Report of Session 2016–17 Brexit: fisheries (HL Paper 78), Available at: https://publications.parliament.uk/pa/ld201617/ldselect/ldeucom/78/78.pdf

Jones, P. (2014), *Governing Marine Protected Areas: Resilience through Diversity*, London, Earthscan from Routledge

Karpiak, C. & Baril, G. (2008), Moral Reasoning and Concern for the Environmental, *Journal of Environmental Psychology*, 28, 203–208

Keller, D. (2010), *Environmental Ethics: The Big Questions*, Chichester, Wiley-Blackwell

King, R. (2010), "How to Construe Nature: Environmental Ethics and the Interpretation of Nature", in Keller, D. (ed.), *Environmental Ethics: The Big Questions*, Chichester, Wiley-Blackwell

Kortenkamp, K. & Moore, C. (2001), Ecocentrism and Anthropocentrism: Moral Reasoning About Ecological Commons Dilemmas, *Journal of Environmental Psychology*, 21, 261–272

Ladenberg, J. (2010), Attitudes towards Offshore Wind Farms – The Role of Beach Visits on Attitudes and Demographic and Attitude Relations, *Energy Policy*, 38, 1297–1304

Lee, M. (2017), Winner of the SLS Annual Conference Best Paper Prize 2016: Knowledge and Landscape in Wind Energy Planning, *Legal Studies*, 37(1), 3–24

Lefebvre, H. (1991), *The Production of Space*, Oxford, Oxford University Press

Low, S. (2013), "Public Space and Diversity: Distributive, procedural and Interactional Justice for Parks" in Young, G. & Stevenson, D. (eds), *The Ashgate Research Companion to Planning and Culture*, Farnham, Ashgate

Madanipour, A. (2003), *Public and Private Spaces of the City*, New York, Routledge

Massey, D. (2005), *For Space*, London, SAGE

McKay, S., Murray, M. & Macintyre, S. (2012), Justice as Fairness in Planning Policy-Making, *International Planning Studies*, 17(2), 147–162

McLachlan, C. (2009), 'You Don't Do a Chemistry Experiment in Your Best China': Symbolic Interpretations of Place and Technology in a Wave Energy Case, *Energy Policy*, 37, 5342–5350

Merrifield, A. (2006), *Henri Lefebvre: A Critical Introduction*, New York, Routledge

Morton, T. (2010), *The Ecological Thought*, Cambridge MA, Harvard University Press

Morton, T. (2016), *Dark Ecology*, New York, Columbia University Press
Nimmo, F., Cappell, R., Huntington, T. & Grant, A. (2011), Does Fish Farming Impact on Tourism in Scotland? *Aquaculture Research*, 42, 132–141
OGA (2018), *OGA-UKCS Designations.* https://assets.publishing.service.gov.uk/government/uploads/system/uploads/attachment_data/file/549620/UKCS_Designations.pdf
O'Neil, J., Holland, A. & Light, A. (2008), *Environmental Values*, Abingdon, Routledge
Parkinson, J. (2012), *Democracy and Public Space: The Physical Sites of Democratic Performance*, Oxford, Oxford University Press
Probyn, E. (2016), *Eating the Ocean*, Durham, Duke University Press
Ritchie, H. (2014), Understanding Emerging Discourses of Marine Spatial Planning in the UK, *Land Use Policy*, 38, 666–675
Sagoff, M. (2010), "Environmental Ethics and Ecological Science", in Keller, D. (ed.), *Environmental Ethics: The Big Questions*, Chichester, Wiley-Blackwell
Scharrer, L., Stadtler, M. & Bromme, R. (2014), You'd Better Ask an Expert: Mitigating the Comprehensibility Effect on Laypeople's Decisions About Science-Based Knowledge Claims, *Applied Cognitive Psychology*, 28, 461–471
Seip, K. & Wenstop, F. (2006), *A Primer of Environmental Decision-making: An Integrative Quantitative Approach*, Dordrecht, Springer
Smith, G. & Brennan, R. (2012), Losing Our Way with Mapping: Thinking Critically about Marine Spatial Planning Is Scotland, *Ocean and Coastal Management*, 69, 210–216
Sohlosberg, D. (2003), "The Justice of Environmental Justice: Reconciling Equity, Recognition, and Participation in a Political Movement", in Light, A. & De-Shalit, A. (eds), *Moral and Political Reasoning in Environmental Practice*, Cambridge, MA, MIT Press
Steinberg, P. (1999), The Maritime Mystique: Sustainable Development, Capital Mobility, and Nostalgia in the World Ocean. *Environment and Planning: D, Society & Space,* 17(4), 403–426.
Steinberg, P. (2001), *The Social Construction of the Ocean.* Cambridge Studies in International Relations, Cambridge, Cambridge University Press
Stokes, C., Beaumont, E., Russell, P. & Greaves, D. (2014), Anticipated Coastal Impacts: What Water-Users Think of Marine Renewables and Why, *Ocean & Coastal Management*, 99, 63–71
Waldo, A. (2012), Offshore Wind Power in Sweden – A Qualitative Analysis of Attitudes With Particular Focus on Opponents, *Energy Policy*, 41, 692–702
Warren, C., Lumsden, C., O'Dowd, S. & Birnie, R. (2005), 'Green On Green': Public Perception of Wind Power in Scotland and Ireland, *Journal of Environmental Planning and Management*, 48(6), 853–875
Wilkinson, D. (1999), "Using Environmental Ethics to Create Ecological Law", in Holder, J. & McGillivary, D. (eds), *Locality and Identity: Environmental Issues in Law and Society*, Aldershot, Ashgate
Wilson, J. (2011), Colonising Space: The New Economic Geography in Theory and Practice, *New Political Theory*, 16(3), 373–397
Young, I. (1990), *Justice and the Politics of Difference*, Princeton, NJ, Princeton University Press

4

TRANSFORMATION THROUGH PARTICIPATION

Democratising the human-ocean relationship

Pamela M. Buchan

Introduction

The consequences of humanity's impact upon the world's physical and biological systems are one of the defining challenges of our generation. The role of the World Ocean in climate change and sustainability issues is coming to the fore, with the COP26 meeting resulting in a 2021 Glasgow Pact that commits to increasing evidence on ocean-based action. The United Nations Decade of Ocean Science for Sustainable Development (2021–2030) outlines ten key challenges, which range from a natural scientific approach of collecting data and developing understanding of ocean systems, through application in practice to reduce pollution and restore ocean ecosystems, and as far reaching as changing humanity's relationship with the ocean, recognising that the human-ocean relationship is a social-ecological system that spans people and the environment via biological, physical, and sociological processes.

Miller et al. (2008) argue that research into social-ecological systems can be limited by the privileging of specific epistemologies according to discipline, and advocate for epistemological pluralism (i.e. interdisciplinarity) to allow a more complete understanding of these complex systems. Epistemological pluralism is embraced by environmental pragmatism, which draws on any tools in the research box that will support the development of effective solutions. Pragmatism is focused on the nature of plural human experience, social institutions, participatory democracy, and the need for continuous reassessment of the moral and ethical principles of society (Parker, 1996), therefore a key concern of environmental pragmatism is "the articulation of practical strategies for bridging gaps between environmental theorists, policy analysts, activists, and the public" (Light and Katz, 1996, p5).

DOI: 10.4324/9781003311171-5

This theoretical context can be plainly understood in real world cases. Take, for example, the problem of marine plastic pollution. There is a requirement to identify and measure the range, scale and dynamics of impacts on water quality and biological life. There is a need also to bring in innovative and technological thinking to filter pollution sources, design plastic alternatives, and clean up environments. The economy must be analysed to understand how it can support or respond to such pressures and changes, and how funding can be delivered to drive innovation. A regulatory framework might be invoked using environmental law or governmental policy to place pressure on plastic producers and plastic consumers to drive a change.

Such a comprehensive approach is well underway on climate and ocean issues, via substantive international agreements, such as the 1992 Rio Declaration on Environment and Development and the 1997 Kyoto Protocol, and United Nations initiatives including the 2015 Sustainable Development Goals. Yet the trajectory of emissions isn't responding to these measures (IPCC, 2014) and post-COVID-19 pandemic data from Oceanic and Atmospheric Research (2021) indicates that levels of carbon dioxide and methane continue to rise in the atmosphere year on year, threatening catastrophic ocean acidification (Findlay and Turley, 2021). As scholars of the ocean, we should therefore reflect on what is missing in this comprehensive approach.

Historically, there has been insufficient research focus on factors within the social component, with much marine research activity focused on gaining ever more confidence in the 'normal' scientific (Kuhn, 1962) factors relating to the ecological system and its perturbations. In this chapter therefore, I argue that a key missing component is people – their knowledges, values, and rights; and lack of recognition of the political aspect of marine citizenship as an important piece of the transition puzzle. The ever-growing wealth of scientific knowledge, defining the problem and the solutions, has not been matched by an equal wealth of knowledge about how hearts and minds might share the sense of urgency and morality needed to support and contribute to the actions proposed by science. In this chapter, I use my interdisciplinary research into marine citizenship to exemplify how addressing today's most urgent issue of ocean health can be supported by the transdisciplinarity of environmental democracy and holistic conceptualisations.

In the next two sections, I give a brief overview of the development of the definition of marine citizenship, followed by an explanation of the post-normal science (PNS) framework. Subsequently, the two core PNS components of plurality of knowledge and the extended peer community are discussed in detail, drawing on empirical data from marine citizenship research (Buchan, 2021). The empirical data is used to emphasise how marine citizens have an implicit understanding of their value as knowledgeable and as legitimate peers for environmental transformation, aligning with the PNS framework. I then bring together my argument with the contemporary global interest in ocean literacy and encourage ocean researchers to consider how their work can promote transdisciplinarity through widening access to the peer community.

Marine citizenship

Marine (or ocean) citizenship as a concept, when first proposed, brought together the principles that the ocean is a common good, that people have individual impacts upon it, and that people relate geographically to the marine environment (Fletcher and Potts, 2007). In keeping with wider environmental education research (Schild, 2016), the proposed solution to fostering ocean citizenship was primarily based upon growing ocean literacy, with little interrogation of the relationship people have with the marine environment as a geographical place. Despite the well-known knowledge or value-action gap (Blake, 1999; Owens, 2000), the limited body of marine citizenship research has been dominated by education and awareness raising as tools for behaviour change, in favour of reducing impacts upon the ocean (Buchan, 2021).

These approaches reflect the prioritisation of scientific knowledge by marine practitioners (e.g. McKinley and Fletcher, 2010; Rees et al., 2013). This prioritisation is normative, with environmental education and decision-making typically dominated by technocratic knowledge and solutions (Robottom, 1991). Though definitions have referenced both rights and responsibilities (McKinley and Fletcher, 2010), the research focus has exclusively considered responsibilities: typically as individual, private, pro-environmental behaviours. Pro-environmental behaviours are promoted through the top-down transfer of scientific knowledge via education and psychological techniques such as social marketing, tailored to people's varying value priorities (Owens, 2000; Walker-Springett et al., 2016). Yet research shows that education is not linearly related to environmental concern and actions, and other factors mediate the relationship (e.g. partisan politics: Hamilton and Safford, 2015; education and climate change concern: Kahan et al., 2012; value-action gap: Blake, 1999; values and self-efficacy: Estrada et al., 2017). It has been argued that scientific environmental controversies require political development and that waiting to develop 'enough' scientific evidence will result in policy paralysis (Sarewitz, 2004).

Here then is an opening for a new way of looking at marine (and by extension, environmental) citizenship as an holistic and relational human experience, grounded not in a one-way receipt of knowledge promoting a desired outward action, but in the geography of place and politics. This invokes classical understandings of citizenship as a relationship between citizen and state, in this case in situ in marine environments. If marine citizenship is to be effective as a policy tool for improving marine environmental health (McKinley and Fletcher, 2012), then this much wider conceptualisation of marine citizenship must be explored. Echoing the wider philosophy of Jelin (2000), here, marine citizenship is defined as "exercising the right to participate in the transformation of the human-ocean relationship for sustainability" (Buchan et al., with editors).

This view of marine citizenship is as dynamic as the ocean itself, and acknowledges people as both individuals and collectives with agency to shape outcomes. Marine citizenship therefore can be considered a transdisciplinary tool that can contribute to the transformation of the human-ocean relationship. However, this potential for marine citizens to act as agents of change can only be realised if societal systems recognise the right to participate in marine decision-making and make it accessible. From a political perspective, this is democratisation of environmental decision-making. From an academic perspective, this means recognising the value of marine citizens as holding useful, diverse knowledge, and as legitimate peers in developing future solutions to environmental problems. In essence, this is PNS.

Post-normal science

PNS provides a useful framework for interdisciplinary and transdisciplinary approaches to developing solutions at the science-society interface. Whilst 'normal' science (Kuhn, 1962) relies on positivist mathematical/scientific-based solutions to defined problems, PNS is identified as relevant to complex 'wicked' problems that are high risk, have high stakes and where there is high uncertainty (Figure 4.1)

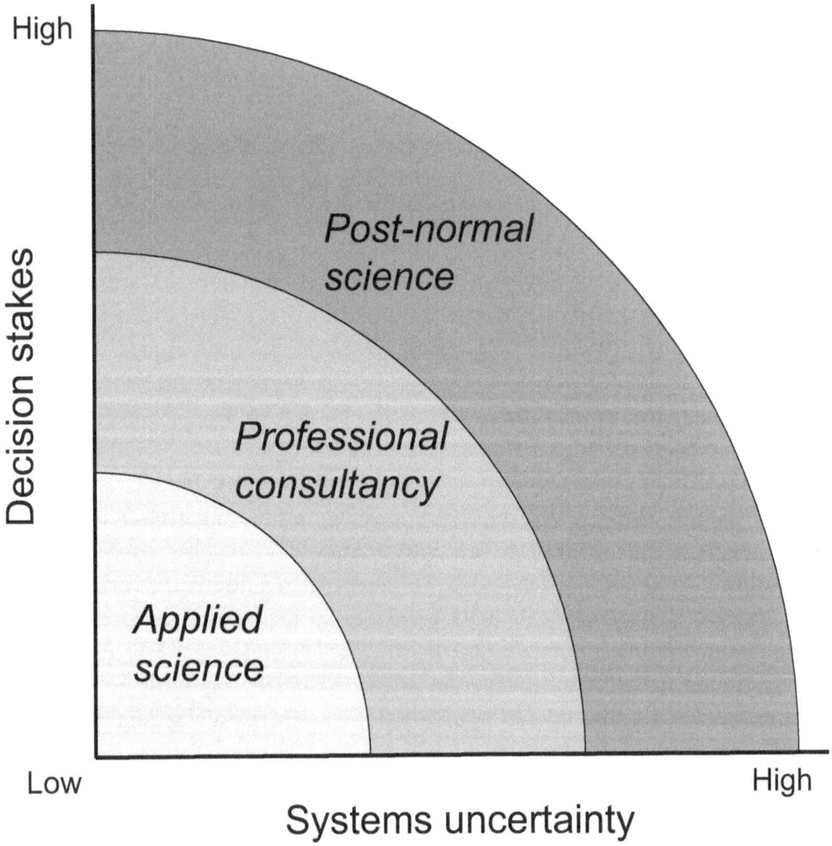

FIGURE 4.1 Diagram of post-normal science.

Source: (Funtowicz and Ravetz, 1993)

(Funtowicz and Ravetz, 1993, 2003). Climate change and transformation of the human-ocean relationship align with these criteria (e.g. Turnpenny, 2012; Jones, 2002): the risk of a 'do nothing' approach is extremely high at local to global scales; a major intervention is required to mitigate and adapt to these consequences which makes these high stakes issues; and whilst we have a great deal of confidence in scientific forecast, the inherent chaos in climate makes certainty impossible, and we have considerably less evidence and confidence in the social responses to these issues.

To manage the complexity, uncertainty, and risk, PNS proposes incorporation of a plurality of knowledges and an extension of the peer community. In practice, this means drawing on knowledge beyond that which is obtained through scientific research; sources such as traditional or indigenous knowledge, local knowledge, and socio-political knowledge. PNS embraces transdisciplinarity by reaching out beyond academics and scholars, towards inclusion of practitioners and communities in the process of identifying problems and developing acceptable solutions. Through this process, scientific knowledge is supplemented with an ethical, moral, and cultural value-base. This is of benefit to participatory decision-making (Lee and Abbot, 2003), and empowers citizens through participatory decision-making processes which are deeper than typical tokenistic consultation (Arnstein, 1969). From a political perspective, this can be understand as a democratisation of science through active citizenship.

Through this post-normal and transdisciplinary lens, this chapter considers, in turn, plurality of knowledge and extended peer communities, in the context of marine citizenship as it is understood by marine citizens themselves. The primary data and analysis which this discussion draws on originates from a wide-ranging interdisciplinary investigation of marine citizenship in the UK setting (Buchan, 2021). The empirical data was gathered from active marine citizens, who were identified through three UK case studies, comprising two marine groups (one local, one regional) and a national citizen science project. A pure mixed methodology was employed, with quantitative and qualitative data obtained through an online survey, and open-ended interview of a purposefully selected sample, representing diversity in key characteristics. Observations of marine citizens participating in a marine citizenship action of their choosing supported researcher understanding of how marine citizenship is understood and practiced by marine citizens (Figure 4.2). (Detailed methodology is described in Buchan, 2021, pp. 90–113.)

The research investigated internal/psychological factors such as basic human values, environmental identity, environmental attitudes, and emotions; and geographical factors such as ocean proximity/visits, marine place attachment, marine place dependency, place identity, and the physical, material, and cultural qualities of the ocean which underpin such relationships. Of specific pertinence to this chapter, marine citizens were asked to share their marine citizenship actions, and their perspectives on the utility of marine citizenship for decision-making and transforming the human-ocean relationship. The study contributed a novel investigation of marine citizenship participatory rights, through extent and experience

FIGURE 4.2 Example marine citizenship activity, releasing juvenile lobsters to replenish wild stocks.

Source: © Pamela M. Buchan

of participation in decision-making by marine citizens, in the context of existing procedural participatory rights in law and regulatory frameworks. The inductive and interdisciplinary approach enabled the emergence of evidence that marine citizens conceive a more sophisticated role for knowledge in marine citizenship than linear, knowledge-deficit models.

The next two sections draw on the empirical findings relating to knowledge, marine citizenship actions, and participation in decision making, in order to provide evidence for the utility of applying transdisciplinarity through a PNS lens. Through this, I argue that marine citizens implicitly take a PNS position on developing environmental solutions and transforming the human-ocean relationship: they view themselves as holding useful local and place-based knowledge; and as legitimate members of the peer community wishing to participate in the transformation process. First, I focus on plurality of knowledge, and then I consider the extended peer community.

A plurality of knowledge

PNS places legitimacy on a plurality of knowledges and perspectives, which necessitates a dialogue to navigate interests, values and conflicts, and the production of an agreed outcome. By drawing on a range of knowledges to determine parameters, the pitfalls of too much sensitivity (in statistics a Type I error) or too much selectivity (Type II error) are navigated, in the 'wicked' problem condition of insufficient quantitative certainty (Funtowicz and Ravetz, 2003).

Using a PNS framework, I extended the marine citizenship research peer community, which had hitherto been limited to marine practitioners and researchers (e.g. through studies such as McKinley and Fletcher, 2010; Rees et al., 2013), by working directly with active marine citizens themselves (Buchan, 2021). I engaged with a plurality of knowledge, by both testing established areas of sustainability research, such as the Schwartz (2012) theory of basic human values, Clayton's (2003) Environmental Identity Index, and theories of place (e.g. Devine-Wright, Price, and Leviston, 2015; Devine-Wright, 2013), and listening to marine citizens using an inductive approach which left space for emergent factors. Through this interdisciplinarity and use of pure mixed methods, the research design enabled marine citizens to challenge the prevailing normative view of marine and environmental citizenship as knowledge-driven, individualistic behaviour change (e.g. Blake, 1999; Hawthorne and Alabaster, 1999; Kollmuss and Agyeman, 2002; Stoll-Kleemann, 2019).

The positivist view of knowledge deficit, in its rawest form, views people as empty vessels who, once filled with information, will behave in the ways intended by those who educate them. In contrast to this, marine citizens demonstrated that they have an intuitive and complex understanding of the role of knowledge in marine citizenship. They described knowledge in three key ways: (i) plurality; (ii) knowledge transfer; (iii) as a citizenship action in its own right.

Plurality of knowledge

Marine citizens acknowledged two key types of knowledge: scientific and local (including place). Scientific knowledge was recognised as valuable in collecting evidence to feed into informed policy-making, implicitly recognising the normative privileging of scientific knowledge, for example, in designating marine protected areas (Pieraccini, 2015). Citizen science, as a way of doing marine citizenship, recognises this relationship between policy and science and provides a vehicle for individuals to support the ocean through the gathering of data.

Local knowledge was particularly recognised in the context of place: "It's probably about living in a place isn't it, and getting to know the people. And getting to know a bit about the shores and what lives on them and what lives in the waters" (interviewee). Within local place knowledge, there were intersections with other factors, such as basic human values. For example, being more knowledgeable about the local wildlife (Figure 4.3) and what is safe to eat or touch, and what should be avoided, reduces fear of nature and promotes sense of security; something that is important for those who prioritise this value. Recognising the landscape, or local topography also promotes security through confidence, and increases place attachment. Local knowledge, however, also related to people, and the social capital of the knowledge held by local residents and collectives such as marine groups and its usefulness for marine management: "I think recognising different groups' expertise and knowledge is really important. Giving them respect," (interviewee).

FIGURE 4.3 Spending time in nature at the coast, to learn about the local wildlife.

Source: © Pamela M. Buchan

Traditional or indigenous local knowledge is commonly recognised as deserving of more respect, particularly in developing nations and the Global South, where the application of Western environmental management approaches is imposed on more traditional regimes (Rahman, 2020). My study showed that in developed nations also, local communities have local and place-related knowledge that is relevant to marine decision-making, and a desire to meaningfully contribute their knowledge alongside more 'legitimised' scientific knowledge.

Knowledge transfer

Knowledge features widely in the environmental citizenship literature as a promoter of action (e.g. Fletcher and Potts, 2007; Guest et al., 2015; Hawthorne and Alabaster, 1999; Heck et al., 2016; McKinley and Fletcher, 2010, 2012; Potts et al., 2012; Rees et al., 2013 *inter alia*). Through the knowledge deficit model, knowledge transfer is one-way from expert to lay person, equipping them with awareness of environmental issues, information about how to change their behaviour to reduce environmental harm, and seeking to shape their understanding and perceptions.

So pervasive is this model that many marine citizens also expressed the need to educate the wider public so that they would participate in marine citizenship. They lamented the lack of knowledge of the wider public: "I'm also amazed at how little is known" (interviewee). Though marine citizens advocated education when engaging with others, they did not describe their own marine citizenship as motivated by knowledge, but rather by their connection with the ocean as a place (Buchan, 2021). It was evident that when knowledge was lacking, this could be a barrier to efficacy of changes in behaviour, whilst being equipped with more extensive knowledge could enable changes in behaviour, where the motivation to do marine citizenship already exists (Buchan, 2021; Buchan et al., 2022).

However, marine citizens highly valued knowledge transfer in a range of additional ways. Local place-based knowledge, as discussed earlier, connected with important feelings such as pride of place – which itself was believed to promote action to preserve the character of the place. (And certainly the wider study evidenced the importance of place attachment for marine citizenship.) Learning develops skills and tools with which to make informed decisions as a marine citizen. Facilitating these sorts of knowledge transfers in the wider public, alongside sharing of environmental values, was a cornerstone of public-facing marine citizenship. Implicitly recognising the power of multiple voices and knowledges, the marine group, as a collective of marine citizens with a diversity of knowledges, was highly praised by marine citizens.

The transfer of scientific and local knowledges was also identified by marine citizens as fundamental for evidence-based and effective decision-making in policy, and research supports this view (Foxwell-Norton, 2013; Lazarus, 2009; Steele, 2001). Communities can contribute to local and national processes through the collection of data from citizen science, and support better decision-making by sharing local knowledge via community participation.

Citizenship action

Marine citizens viewed knowledge and knowledge exchange as fundamental to the practice of marine citizenship as a political action. Informing oneself was cited as an act of citizenship, both for personal decision-making, and as a responsibility to learn about what is happening to the world. Lifelong learning has been cited as a political response to an increasingly technocratic and de-collectivised society (Freire, 1972; Martin, 2003); and promotes inclusive, pluralistic, reflexive, and active citizenship (Johnston, 1999). Through one's own learning, citizenship can move away from fulfilling a set of individualised behaviour changes, to a more critically considered civic participation.

For marine citizens, knowledge was integral to the power of collective action: "the more people get together, and the more people talk about things . . . then they have to listen because it's the power of the people" (interviewee). Through the knowledge exchange taking place within the marine group, the collective was granted more legitimacy externally, as embodying expertise. Group members were individually empowered through learning, sharing of opportunities, and the social experience, leading to increased civic participation.

Through these nuanced and complex understandings of the role of knowledge, marine citizens demonstrated an inherently post-normal view of marine and environmental issues. Whilst acknowledging the power of scientific evidence, marine citizens were not satisfied with a procedure which excluded them. They wanted to bring people together to create change within the environmental social movement; to situate marine and coastal decision-making alongside their local place-based lived experience; and, fundamentally, to be part of the process shaping how humanity uses the ocean.

Extending the peer community for environmental democracy

In PNS, the quality of policy-making is ensured through open dialogue with an extended peer community, defined as all those with "a desire to participate in the resolution of the issue" (Funtowicz and Ravetz, 2003, p. 6). As I have argued earlier, marine citizens have the desire to participate in marine sustainability. When and to what extent should the marine peer community be extended? The answer to that question is clearly subjective, and is itself therefore post-normal, and might be best determined through enhanced democratic and participatory processes engaging with those who wish to participate.

Public participation in decision-making is evidenced as a valuable process across many fields of enquiry. For example, in environmental law, this has been seen to produce better solutions through deliberation of a wider set of voices (Steele, 2001), which are perceived as more acceptable and legitimate (Scharpf, 2003). There is a growing body of research within the science communication, or public engagement with science field, demonstrating the importance of more participatory and deliberative processes for navigating issues at the science-society interface (Smallman, 2014). For many years, there has been debate about the quality of citizen participation, with Arnstein's (1969) classic Ladder of Citizen Participation indicating the relationship between participatory method and how much power is transferred to the citizen.

Participatory processes are in essence an extension of the peer community: scientific experts and knowledge are extended to include multiple actors and perspectives from different stakeholders and the wider public. In the context of marine citizenship, there are two main forms of participation: (i) participation as procedural, via the legislated right to participate in environmental decision-making; and (ii) participation as political, via citizenship activities in the public sphere, aimed at changing the views of fellow citizens and creating grassroots pressure upon decision-makers. Each of these represents an extension of the peer community for marine environmental problem-solving; the first through formal representation on a case-by-case process, and the second through democratisation of the shape of the human-ocean relationship. I will consider each of these in turn.

Procedural participation

The Rio Declaration on Environment and Development, 1992 acknowledges a political and moral responsibility of those with the power to make environmental decisions to include the wider public. In legal terms, the right to participate in environmental decision-making is afforded to the public concerned or affected, which might be interpreted as those with an interest through business or statutory responsibility, or those who live in direct connection with the location of the development. In the United Nations Economic Commission for Europe 1998 Convention on Access to Information, Public Participation in Decision-Making and Access to Justice in Environmental Matters (the Aarhus Convention), environmental non-governmental organisations (NGOs) are given an explicit role in representing public or nature-based interests in environmental management. This model of public procedural participation is not without criticism, for example, for lack of clarity on who are the public concerned/affected, the role of environmental justice NGOs, and the nature of the participation (e.g. no guidance is given about the spectrum of participatory types, from tokenistic consultation to co-creation); for reinforcing existing power imbalances; and for the ability of environmental NGOs to adequately represent wider public interests (Holder, 2006; Lee and Abbot, 2003; Nadal, 2008; Steele, 2001).

There are also uncertainties as to how existing legislature responds to contemporary challenges. For example, in the introduction to this chapter, I highlighted the global scale of ocean degradation and climate change, and that the ocean is understood as a common good. The Kyoto Protocol to the United Nations Framework Convention on Climate Change, 1997 recognises differential financial responsibility between developing nations and developed nations in mitigating climate change, due to uneven historical greenhouse gas contributions from industrialisation. There is therefore both scientific and legal acknowledgement of the global scale of the ocean and climate systems, in which causes and impacts might be under different national or regional legislative jurisdiction, or in areas beyond national jurisdiction (ABNJ). Where existing procedural participatory rights are conferred on the public affected or concerned, it becomes clear that these conditions cannot always be meaningfully defined in the context of global environmental challenges (Peters, 2015). The fluid and continuous nature of the world ocean is a challenge for marine citizenship through procedural means.

Furthermore, whilst marine citizens might have the desire, they do not always have the means to participate as they would wish to (Buchan, 2021). For example, individual behavioural choices are limited by the constraints of a society and economy which place uneven expectations and limitations upon individuals, which intersect with their socio-economic circumstances. I found amongst marine citizens, a wide-scale lack of awareness of their procedural rights to access environmental information, participation in environmental decision-making, and environmental justice.

A lack of knowledge about procedural rights is likely to limit access to those rights: "As a diver, I see a number of things that concern me and other members of the diving club I belong to. But I don't know who to voice these concerns with" (interviewee). Certainly ignorance of law in general is widespread and has implications for compliance (van Rooij, 2020), and it is difficult to see how citizens can effectively hold decision-makers to account if they are unaware of the right to do so.

Whilst formalised means of participation include consultation, citizen science, citizens juries or focus groups which might inform or be considered alongside scientific evidence, the use of local, traditional or place-based knowledge also equips citizens to develop solutions to problems themselves (Funtowicz and Ravetz, 2003). Therefore alongside further democratising environmental decision-making, there is also a case for considering devolution of decision-making, in a way which involved citizens in the places where they live.

Issues about rights, ethics, distribution of resources, and responsibilities are political and need to be debated to reach resolutions. Shove and Walker (2007) argue for explicit illumination of the politics of environmental transition. It is a political process to determine that the contemporary human-ocean relationship is problematic for the future health of nature and humanity, and to consider policy actions to address the challenge. It is also a political process which determines who are the actors who should be managing or influencing the transition. Green politics emerged to give recognition to an ecological and holistic approach to the environment, which is sustainable and equitable, and which recognises grassroots democracy (Capra and Spretnak, 1984). Whilst scientific calls have reached out to policy-makers to better recognise the role of the ocean in climate (e.g. Laffoley et al., 2022) and whole-ecosystem approaches to marine management (e.g. Rees et al., 2020), they have not yet called for a 'blue' politics to place pressure on policy-makers.

Political participation

If one accepts the premise of PNS – that 'wicked' ocean and climate issues cannot be solved through 'normal' scientific endeavour – then one must also accept the need for interrogation of the political and procedural processes that surround environmental decision-making and transformation of the human-ocean relationship for sustainability. The quality of political decision-making depends upon fair representation of different peoples and communities – at its best, democracy is the most extended peer community. Whilst it clearly is not practical to make all environmental decisions in a fully democratic way across the whole globe, the negotiation about who is represented and how is a construct of political action. In democratic states, political actors need the support of the electorate to enact significant policy directions, and so an active citizenry is required to deliver messages from the grassroots.

In my research, marine citizens were very aware of this relationship. They demonstrated a view of marine citizenship that was strongly grounded in civic participation, with higher than average levels of political and civic engagement, and a strong narrative of being champions for the marine environment, through public engagement and political action (Buchan, 2021). Indeed citizenship activities of this sort were much more common than procedural participatory experiences.

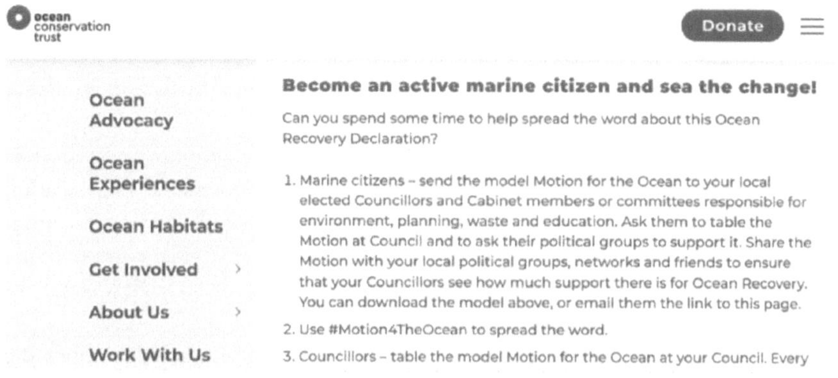

FIGURE 4.4 The 'Motion for the Ocean'. Local government Ocean Recovery Declaration is an example of mobilising marine citizens to exercise their marine citizenship rights to ask for transformation of the human-ocean relationship.

Source: https://oceanconservationtrust.org/project/the-motion-for-the-ocean/

Although the rights of marine citizenship have not formerly been characterised, marine citizens implicitly felt they should be involved in shaping the human-ocean relationship. In some cases, this involvement extended from voluntary activities into professional choices, where occupation and career trajectory was primarily influenced by environmental aspirations. Citizens viewing their occupation as an act of citizenship have been observed before in the environmental sector (Buchan and Yates, 2019).

Marine citizens work to mobilise a grassroots movement, through individual actions to change hearts and minds, and collective actions via place-based marine groups and wider political campaigning (e.g. see the local government Ocean Recovery Declaration, Figure 4.4). Social movements are a collective process (Diani, 1992; Tindall, 2002) and can work across scales from shaping local marine places as communities up to placing pressure of decision-makers to rebalance the benefits for humanity and nature. The potential for marine citizens as activists should not be underestimated by environmental scholars, in favour of simplistic educational approaches to behaviour change (Chawla and Cushing, 2007; Jelin, 2000; Schild, 2016).

Supporting transdisciplinarity

Having laid out how PNS can act as a conceptual framework to support transdisciplinarity through marine citizenship, attention must be given to access and power. Though the sample of marine citizens in my research were quite well-spread, demographically, it was limited to the UK and there was an over-representation of higher levels of education. The sample were most typical of those who volunteer. Volunteering has well-established barriers associated with age, gender, education, income and time (Egerton, 2002; Egerton and Mullan, 2008). Barriers to volunteering and civic participation are often rooted in systemic barriers, reinforcing marginalisation and power imbalances in wider society.

What is clear is that there is a need for reform of public participation processes to improve awareness of procedural rights and increase procedural participation opportunities. As scholars of the ocean and its interface with society, we must consider what we can do within our research and practice that will promote transdisciplinarity and PNS activities. One such activity might be to review what we understand ocean literacy to be.

Ocean literacy has tended to focus on natural scientific knowledge about the ocean. For example, the United States' National Marine Educators Association (2019) defines ocean literacy as understanding the following:

1 The Earth has one big ocean with many features.
2 The ocean and life in the ocean shape the features of the Earth.
3 The ocean is a major influence on weather and climate.
4 The ocean makes Earth habitable.
5 The ocean supports a great diversity of life and ecosystems.
6 The ocean and humans are inextricably interconnected.
7 The ocean is largely unexplored.

However, if the goal of ocean literacy is to promote marine citizenship (e.g. Fletcher and Potts, 2007; Fielding et al., 2019), then there is good reason to consider whether ocean educators might wish to include political and procedural literacy in their activities. As noted, the political context of environmental citizenship has not been well-investigated (Schild, 2016), yet it has been known for some time that multiple types of knowledge are associated with pro-environmental behaviours, and action-based procedural knowledge is more influential than scientific, or declarative, knowledge (Kaiser and Fuhrer, 2003). Marine citizens who are equipped only with marine scientific knowledge may be less able to engage in transformative processes than those who additionally understand their rights and the suite of civic participatory procedures potentially available to them. Additionally, those who do not recognise their own power as citizens to promote changed ocean practices may be limited in the choices and actions they engage in.

For some readers, being a marine researcher or practitioner might be a marine citizenship activity, and this chapter will be food for thought about our own practice and methodologies. Within academic institutions and cultures, how far are we prepared to go as activist researchers to drive transformation? By embracing plurality in practice, new insights will be uncovered, illuminating what methodologies are effective in what circumstances, and how the community of marine problem-solving peers can be extended beyond the limitations imposed by existing structures. The methodologies and case studies in this book will no doubt prompt reflections on putting these ideas into practice.

Conclusion

In this conceptual reflection, I have presented a conceptualisation of marine citizenship which acknowledges its political and participatory potential for promoting transformation of the human-ocean relationship (Schild, 2016). I have summarised the PNS framework (Funtowicz and Ravetz, 2003) and its relevance, both for ocean- and climate-based problems (Jones, 2002), and for transdisciplinarity. Through the PNS lens, the views of marine citizens themselves have been set out to understand the relationship between marine citizenship and knowledge, and the potential for marine citizens to extend the peer community in marine decision-making and policy. I have argued that marine citizens are implicitly plural in their use of knowledge, value scientific and local knowledge, and use knowledge acquisition and exchange for citizenship and transformation. I have discussed how marine citizens can extend the peer community by participating procedurally and politically, and touched on some of the challenges and barriers associated with this in practice.

To meet the challenges of navigating transformation of the complex human-ocean, social-ecological system, there is a need for environmental pragmatism that embraces issues such as morality, democracy and the structure of social institutions (Parker, 1996). These issues are debated within political institutions, which are influenced by existing power imbalances and marginalisation, yet are also responsive to social movements for societal change (Diani, 1992). It is in the interest of ocean researchers who wish to mitigate the ocean's crisis, to be as inclusive as possible and support marine citizens access to processes of change through transdisciplinary research evidence. The empirical data demonstrates that marine citizens view themselves as knowledgeable and legitimate actors but that they are held back by institutional biases about knowledge legitimacy, and awareness of and access to their existing participatory rights.

Transdisciplinarity can be rooted right through problem-solving from the earliest co-production of knowledge, perhaps via citizen science; employing inductive research that amplifies the voices of participants (as in my research presented here); co-designed and co-produced research which directly integrates an extended peer community into a problem-identifying/solving process; through to informing and improving the democratic processes of decision-making. Recognition of the public and political faces of marine citizenship responsibilities, and the rights of marine citizenship as the right to participate, is the first step to developing a transdisciplinary transformation.

Acknowledgements

The empirical findings presented in this chapter originate from research funded by the ESRC. My thanks to Dr Louisa Evans, Dr Margherita Pieraccini, and Prof Stewart Barr.

References

Arnstein, S. R. (1969). A Ladder of Citizen Participation. *Journal of the American Institute of Planners*, 35(4), 216–224. https://doi.org/10.1080/01944366908977225

Blake, J. (1999). Overcoming the 'value-action gap' in Environmental Policy: Tensions between National Policy and Local Experience. *Local Environment*, 4(3), 257–278. https://doi.org/10.1080/13549839908725599

Buchan, P. M. (2021). Investigating Marine Citizenship and Its Role in Creating Good Marine Environmental Health [University of Exeter]. http://hdl.handle.net/10871/126112

Buchan, P. M., Evans, L. S., Pieraccini, M., & Barr, S. W. (2022). Marine Citizenship: The Right to Participate in the Transformation of the Human-Ocean Relationship for Sustainability. *Unpublished manuscript in review*.

Buchan, P. M., & Yates, K. L. (2019). Stakeholder Dynamics, Perceptions and Representation in a Regional Coastal Partnership. *Marine Policy*, 101, 125–136. https://doi.org/10.1016/j.marpol.2018.12.017

Capra, F., & Spretnak, C. (1984). Green Politics. *National Forum*, 64(3), 21–23.

Chawla, L., & Cushing, D. F. (2007). Education for Strategic Environmental Behavior. *Environmental Education Research*, 13(4), 437–452. https://doi.org/10.1080/13504620701581539

Clayton, S. (2003). *Identity and the Natural Environment: The Psychological Significance of Nature*. MIT Press.

Devine-Wright, P. (2013). Think Global, Act Local? The Relevance of Place Attachments and Place Identities in a Climate Changed World. *Global Environmental Change*, 23(1), 61–69. https://doi.org/10.1016/j.gloenvcha.2012.08.003

Devine-Wright, P., Price, J., & Leviston, Z. (2015). My Country or My Planet? Exploring the Influence of Multiple Place Attachments and Ideological Beliefs Upon Climate Change Attitudes and Opinions. *Global Environmental Change*, 30, 68–79. https://doi.org/10.1016/j.gloenvcha.2014.10.012

Diani, M. (1992). The Concept of Social Movement. *The Sociological Review*, 40(1), 1–25.

Egerton, M. (2002). Higher Education and Civic Engagement. *The British Journal of Sociology*, 53(4), 603–620. https://doi.org/10.1080/0007131022000021506

Egerton, M., & Mullan, K. (2008). Being a Pretty Good Citizen: An Analysis and Monetary Valuation of Formal and Informal Voluntary Work by Gender and Educational Attainment. *The British Journal of Sociology*, 59(1), 145–164. https://doi.org/10.1111/j.1468-4446.2007.00186.x

Estrada, M., Schultz, P. W., Silva-Send, N., & Boudrias, M. A. (2017). The Role of Social Influences on Pro-Environment Behaviors in the San Diego Region. *Journal of Urban Health*, 94(2), 170–179. https://doi.org/10.1007/s11524-017-0139-0

Fielding, S., Copley, J. T., & Mills, R. A. (2019). Exploring Our Oceans: Using the Global Classroom to Develop Ocean Literacy. *Frontiers in Marine Science*, 6, 340. https://doi.org/10.3389/fmars.2019.00340

Findlay, H. S., & Turley, C. (2021). Chapter 13 – Ocean Acidification and Climate Change. In T. M. Letcher (Ed.), *Climate Change* (Third Edition) (pp. 251–279). Elsevier. https://doi.org/10.1016/B978-0-12-821575-3.00013-X

Fletcher, S., & Potts, J. (2007). Ocean Citizenship: An Emergent Geographical Concept. *Coastal Management*, 35(4), 511–524. https://doi.org/10.1080/08920750701525818

Foxwell-Norton, K. (2013). Communication, Culture, Community and Country: The Lost Seas of Environmental Policy. *Continuum*, 27(2), 267–282. https://doi.org/10.1080/10304312.2013.766307

Freire, P. (1972). Pedagogy of the oppressed. Penguin.

Funtowicz, S. O., & Ravetz, J. R. (1993). Science for the Post-Normal Age. *Futures*, 25(7), 739–755. https://doi.org/10.1016/0016-3287(93)90022-L

Funtowicz, S. O., & Ravetz, J. R. (2003). Post-normal Science. International Society for Ecological Economics (Ed.), Online Encyclopedia of Ecological Economics at http:// Www. Ecoeco. Org/Publica/Encyc. Htm.

Guest, H., Lotze, H. K., & Wallace, D. (2015). Youth and the Sea: Ocean Literacy in Nova Scotia, Canada. *Marine Policy*, 58, 98–107. https://doi.org/10.1016/j.marpol.2015.04.007

Hamilton, L. C., & Safford, T. G. (2015). Environmental Views from the Coast: Public Concern about Local to Global Marine Issues. *Society & Natural Resources*, 28(1), 57–74. https://doi.org/10.1080/08941920.2014.933926

Hawthorne, M., & Alabaster, T. (1999). Citizen 2000: Development of a Model of Environmental Citizenship. *Global Environmental Change*, 9(1), 25–43. https://doi.org/10.1016/S0959-3780(98)00022-3

Heck, N., Paytan, A., Potts, D. C., & Haddad, B. (2016). Coastal Residents' Literacy about Seawater Desalination and Its Impacts on Marine Ecosystems in California. *Marine Policy*, 68, 178–186. https://doi.org/10.1016/j.marpol.2016.03.004

Holder, J. (2006). *Environmental Assessment: The Regulation of Decision Making*. Oxford University Press.

IPCC. (2014). Climate Change 2014: *Synthesis Report. Contribution of Working Groups I, II and III to the Fifth Assessment Report of the Intergovernmental Panel on Climate Change* [Core Writing Team, R.K. Pachauri and L.A. Meyer (Eds.)].

Jelin, E. (2000). Towards a Global Environmental Citizenship? *Citizenship Studies*, 4(1), 47–63. https://doi.org/10.1080/136210200110021

Johnston, R. (1999). Adult Learning for Citizenship: Towards a Reconstruction of the Social Purpose Tradition. *International Journal of Lifelong Education*, 18(3), 175–190. https://doi.org/10.1080/026013799293775

Jones, P. J. S. (2002). Marine Protected Area Strategies: Issues, Divergences and the Search for Middle Ground. *Reviews in Fish Biology and Fisheries*, 11(3), 197–216. https://doi.org/10.1023/A:1020327007975

Kahan, D. M., Peters, E., Wittlin, M., Slovic, P., Ouellette, L. L., Braman, D., & Mandel, G. (2012). The Polarizing Impact of Science Literacy and Numeracy on Perceived Climate Change Risks. *Nature Climate Change*, 2. https://doi.org/10.1038/nclimate1547

Kaiser, F. G., & Fuhrer, U. (2003). Ecological Behavior's Dependency on Different Forms of Knowledge. *Applied Psychology*, 52(4), 598–613. https://doi.org/10.1111/1464-0597.00153

Kollmuss, A., & Agyeman, J. (2002). Mind the Gap: Why Do People Act Environmentally and What Are the Barriers to Pro-environmental Behavior? *Environmental Education Research*, 8(3), 239–260. https://doi.org/10.1080/13504620220145401

Kuhn, T. S. (1962). *The Structure of Scientific Revolutions Vol.* The University of Chicago Press.

Kyoto Protocol to the United Nations Framework Convention on Climate Change, 1 (1997). https://unfccc.int/resource/docs/convkp/kpeng.pdf

Laffoley, D., Baxter, J. M., Amon, D. J., Claudet, J., Downs, C. A., Earle, S. A., Gjerde, K. M., Hall-Spencer, J. M., Koldewey, H. J., Levin, L. A., Reid, C. P., Roberts, C. M., Sumaila, R. U., Taylor, M. L., Thiele, T., & Woodall, L. C. (2022). The Forgotten Ocean: Why COP26 Must Call for Vastly Greater Ambition and Urgency to Address Ocean Change. *Aquatic Conservation: Marine and Freshwater Ecosystems*, 32(1), 217–228. https://doi.org/10.1002/aqc.3751

Lazarus, R. (2009). *Super Wicked Problems and Climate Change: Restraining the Present to Liberate the Future.* Georgetown Law Faculty Publications and Other Works. https://scholarship.law.georgetown.edu/facpub/159

Lee, M., & Abbot, C. (2003). The Usual Suspects? Public Participation Under the Aarhus Convention. *The Modern Law Review*, 66(1), 80–108. https://doi.org/10.1111/1468-2230.6601004

Light, A., & Katz, E. (Eds.). (1996). *Environmental Pragmatism* (1st ed.). Routledge. https://doi.org/10.4324/9780203714140

Martin, I. (2003). Adult Education, Lifelong Learning and Citizenship: Some IFS and Buts. *International Journal of Lifelong Education*, 22(6), 566–579. https://doi.org/10.1080/0260137032000138130

McKinley, E., & Fletcher, S. (2010). Individual Responsibility for the Oceans? An Evaluation of Marine Citizenship by UK Marine Practitioners. *Ocean & Coastal Management*, 53(7), 379–384. https://doi.org/10.1016/j.ocecoaman.2010.04.012

McKinley, E., & Fletcher, S. (2012). Improving Marine Environmental Health Through Marine Citizenship: A Call for Debate. *Marine Policy*, 36(3), 839–843. https://doi.org/10.1016/j.marpol.2011.11.001

Miller, T., Baird, T., Littlefield, C., Kofinas, G., Iii, C., Stuart, F., & Redman, C. (2008). Epistemological Pluralism: Reorganizing Interdisciplinary Research. *Ecology and Society*, 13(2). https://doi.org/10.5751/ES-02671-130246

Nadal, C. (2008). Pursuing Substantive Environmental Justice: The Aarhus Convention as a 'Pillar'of Empowerment. *Environmental Law Review*, 10(1), 28–45.

National Marine Educators Association. (2019). *Ocean Literacy. National Marine Educators Association*. www.marine-ed.org/ocean-literacy/overview

Oceanic and Atmospheric Research. (2021, April 7). Despite Pandemic Shutdowns, Carbon Dioxide and Methane Surged in 2020–. NOAA Research. https://research.noaa.gov/article/ArtMID/587/ArticleID/2742/Despite-pandemic-shutdowns-carbon-dioxide-and-methane-surged-in-2020

Owens, S. (2000). 'Engaging the public': Information and Deliberation in Environmental Policy. *Environment and Planning A*, 32(7), 1141–1148.

Parker, K. A. (1996). Pragmatism and Environmental Thought. In *Environmental Pragmatism* (1st ed., pp. 21–37). Routledge. https://doi.org/10.4324/9780203714140

Peters, B. (2015). Towards the Europeanization of Participation? Reflecting on the Functions and Beneficiaries of Participatory Rights in EU Environmental Law. In C. Fraenkel-Haeberle, S. Kropp, F. Palermo, & K.-P. Sommermann (Eds.), *Citizen Participation in Multi-level Democracies* (pp. 311–333). Koninklijke Brill NV. https://papers.ssrn.com/abstract=2422378

Pieraccini, M. (2015). Rethinking Participation in Environmental Decision-Making: Epistemologies of Marine Conservation in South-East England. *Journal of Environmental Law*, 27(1), 45–67. https://doi.org/10.1093/jel/equ035

Potts, T., O'Higgins, T., & Hastings, E. (2012). Oceans of Opportunity or Rough Seas? What Does the Future Hold for Developments in European Marine Policy? *Philosophical Transactions of the Royal Society of London A: Mathematical, Physical and Engineering Sciences*, 370(1980), 5682–5700. https://doi.org/10.1098/rsta.2012.0394

Rahman, S. (2020). Environmental Citizenship for Inclusive Sustainable Development: The Case of Kelab Alami in Mukim Tanjung Kupang, Johor, Malaysia. *Journal of the Indian Ocean Region*, 16(1), 100–118. https://doi.org/10.1080/19480881.2020.1704986

Rees, S., Fletcher, S., Glegg, G., Marshall, C., Rodwell, L., Jefferson, R., Campbell, M., Langmead, O., Ashley, M., Bloomfield, H., Brutto, D., Colenutt, A., Conversi, A., Earll, B., Hattam, C., Ingram, S., McKinley, E., Mee, L., Oates, J., . . . Wynn, R. (2013). Priority Questions to Shape the Marine and Coastal Policy Research Agenda in the United Kingdom. *Marine Policy*, 38, 531–537. https://doi.org/10.1016/j.marpol.2012.09.002

Rees, S. E., Sheehan, E. V., Stewart, B. D., Clark, R., Appleby, T., Attrill, M. J., Jones, P. J. S., Johnson, D., Bradshaw, N., Pittman, S., Oates, J., & Solandt, J.-L. (2020). Emerging Themes to Support Ambitious UK Marine Biodiversity Conservation. *Marine Policy*, 117, 103864. https://doi.org/10.1016/j.marpol.2020.103864

Rio Declaration on Environment and Development, Pub. L. No. A/CONF.151/26 (1992). www.un.org/en/development/desa/population/migration/generalassembly/docs/globalcompact/A_CONF.151_26_Vol.I_Declaration.pdf

Robottom, I. (1991). Technocratic Environmental Education: A Critique and Some Alternatives. *Journal of Experiential Education*, 14(1), 20–26. https://doi.org/10.1177/105382599101400103

Sarewitz, D. (2004). How Science makes Environmental Controversies Worse. *Environmental Science & Policy*, 7(5), 385–403. https://doi.org/10.1016/j.envsci.2004.06.001.

Scharpf, F. W. (2003). Problem-solving Effectiveness and Democratic Accountability in the EU. MPIfG Working Paper.

Schild, R. (2016). Environmental Citizenship: What Can Political Theory Contribute to Environmental Education Practice? *The Journal of Environmental Education*, 47(1), 19–34. https://doi.org/10.1080/00958964.2015.1092417

Schwartz, S. (2012). An Overview of the Schwartz Theory of Basic Values. *Online Readings in Psychology and Culture*, 2(1). https://doi.org/10.9707/2307-0919.1116

Shove, E., & Walker, G. (2007). Caution! Transitions Ahead: Politics, Practice, and Sustainable Transition Management. *Environment and Planning A: Economy and Space*, 39(4), 763–770. https://doi.org/10.1068/a39310

Smallman, M. (2014). Public Understanding of Science in Turbulent Times III: Deficit to Dialogue, Champions to Critics. *Public Understanding of Science*, 25(2), 186–197. https://doi.org/10.1177/0963662514549141

Steele, J. (2001). Participation and Deliberation in Environmental Law: Exploring a Problem-Solving Approach. *Oxford Journal of Legal Studies*, 21(3), 415–442. https://doi.org/10.1093/ojls/21.3.415

Stoll-Kleemann, S. (2019). Feasible Options for Behavior Change toward More Effective Ocean Literacy: A Systematic Review. *Frontiers in Marine Science*, 6, UNSP 273. https://doi.org/10.3389/fmars.2019.00273

Tindall, D. B. (2002). Social Networks, Identification and Participation in an Environmental Movement: Low-Medium Cost Activism within the British Columbia Wilderness Preservation Movement. *Canadian Review of Sociology and Anthropology-Revue Canadienne De Sociologie Et D Anthropologie*, 39(4), 413–452.

Turnpenny, J. R. (2012). Lessons From Post-Normal Science for Climate Science-Sceptic Debates. *Wiley Interdisciplinary Reviews-Climate Change*, 3(5), 397–407. https://doi.org/10.1002/wcc.184

United Nations Economic Commission for Europe. (1998). Convention on Access to Information, Public Participation in Decision-making and Access to Justice in Environmental Matters.

van Rooij, B. (2020). *Do People Know the Law? Empirical Evidence About Legal Knowledge and Its Implications for Compliance* (SSRN Scholarly Paper ID 3563442). Social Science Research Network. https://papers.ssrn.com/abstract=3563442

Walker-Springett, K., Jefferson, R., Böck, K., Breckwoldt, A., Comby, E., Cottet, M., Hübner, G., Le, L., Shaw, S., & Wyles, K. (2016). Ways Forward for Aquatic Conservation: Applications of Environmental Psychology to Support Management Objectives. *Journal of Environmental Management*, 166, 525–536. Scopus. https://doi.org/10.1016/j.jenvman.2015.11.002

PART II

Methods and perspectives

In transdisciplinary processes, the *What* matters just as much as the *How*. The first section of this book reflected on how marine transdisciplinarity is defined in the marine sciences, illustrated diverse conceptualisations of the ocean, and highlighted the value of marine citizenship for sustainably managing our seas. The four chapters in this section focus on specific methodological perspectives on marine transdisciplinarity. In Chapter 5, Carrick and colleagues discuss their experiences of using Participatory Action Research (PAR) and Bayesian Belief Networks (BBN) as tools in a participatory process. In the context of a tidal energy project in the Solway Firth (UK), they found that using BBNs in cycles of stakeholder engagement and modelling has the capacity to improve participation and visualise complex system relationships. Importantly, a discussion of the features of BBNs encouraged participants to reflect on their own perspectives and those of others. On the other hand, the complexity of the method must also be viewed with caution as difficulties in understanding and interpreting the BBNs lead to frustration and have the potential to reduce engagement in practice.

On a more theoretical level, McAteer and Flannery discuss in Chapter 6 whether the professionalisation of marine citizen science supports or obstructs processes of knowledge co-production and ocean co-management. The authors point out that citizen science comes with a strong potential to enrich the management of the marine environment and engage non-academic actors in scientific research. However, a professionalisation and institutionalisation of marine citizen science bears the danger of harming the nature of active and open participation as agendas often become top-down in such processes. To safeguard the transformative potential of citizen science in an institutionalised setting, therefore, we should become aware of power relations in and barriers to processes of knowledge co-production and counteract them by setting up project networks, finding collective funding sources, and integrating effective feedback loops into participatory processes.

DOI: 10.4324/9781003311171-6

Not only the methods we use in processes of knowledge co-production influence their success but also the underlying assumptions and definitions that we, as participants in such processes, hold and bring to the table. The case study from Inuit Nunangat (Canada) presented by Petriello and colleagues in Chapter 7 underlines the importance of questioning and deconstructing existing assumptions about knowledge co-production and the terms linked to it. Focusing on traditional environmental knowledge systems, the authors argue that the meanings associated with the concept can differ widely even between the members of one project, and that dialogue about these meanings should be a key pillar of any project in which knowledge is co-produced. From these considerations result the recommendation to leave room in any project for discussing and debating the values, visions, and principles of the co-production process as part of its methodology and not, as often done, as an afterthought of the project.

Deconstructing the implicit meanings and values attached to terms, we often use but seldomly question, has great potential for overcoming misunderstandings, disagreements, and conflicts in any transdisciplinary process. Schaber and colleagues uncover the different meanings that fisheries stakeholders give to the term 'sustainability' via quantitative questionnaires handed out during a participatory workshop in the Western Baltic Sea (Chapter 8). Several fundamental challenges for transdisciplinary approaches arise from their analysis: firstly, the authors show that it can be hard to identify for whom a stakeholder is speaking and whose views they represent – those of a certain group, their own, or both. Secondly, they underline the importance of clearly separating the status quo from normative ideas about the future when discussing goals for sustainable ocean management. Lastly, the stakeholders participating in the workshop highlighted the significance of jointly defining terms such as 'sustainability' at the start of a transdisciplinary project in order to avoid misunderstanding each other throughout the entire process.

5

USING BAYESIAN BELIEF NETWORKS AND PARTICIPATORY ACTION RESEARCH TO IMPROVE STAKEHOLDER ENGAGEMENT

Jayne Carrick, Clare Fitzsimmons, and Tim Gray

Introduction

As indicated in Chapter 4, sustainable ocean management requires joint action by scientists, citizens, and communities to protect and restore nature and fight climate change. Tidal energy has the potential to both contribute to action on climate change and adversely affect the marine and coastal environment, representing a paradox for sustainable ocean management. To resolve this paradox, knowledge and perspectives from scientists, citizens and communities need to be incorporated into the decision-making processes. In this chapter, we illustrate the use of two methods that claim to improve stakeholder engagement and co-production of knowledge: advanced statistical modelling, known as Bayesian Belief Networks (BBNs); and Participatory Action Research (PAR). BBNs are statistical models widely used to aid decision-making in medical and defence settings (Stewart et al., 2014), and, increasingly, to inform environmental decision-making (Uusitalo, 2007; Aguilera et al., 2011). In contrast, PAR is an approach rooted in social sciences. Combining BBNs and PAR therefore represents a truly transdisciplinary approach. We use the case study of a proposed tidal energy scheme in the Solway Firth in the west coast of the UK, to illustrate the novel application of BBNs and PAR to sustainable ocean management. The lessons from this case can be applied to the consenting process for marine renewables and participatory processes in sustainable ocean management more generally.

In this section, we first discuss the potential contribution to, and impacts on, sustainable ocean management of tidal energy. We then explain and justify the use of BBNs and PAR to engage stakeholders in the licencing and consenting process for tidal energy developments. Considering that legitimate decision-making is dependent on procedures rather than outcomes (Ottinger, 2013), the goal is to engage stakeholders in fair decision-making processes, as opposed to achieving specific outcomes such as consent.

DOI: 10.4324/9781003311171-7

Tidal energy

Renewable energy makes a significant contribution to the sustainability agenda by reducing consumption of non-renewable sources of energy and decreasing greenhouse gas emissions from fossil fuels. Increasing renewable electricity generation to phase out fossil fuels is a key part of the UK government's strategy to reach net zero greenhouse gas emissions by 2050 (UK Government, 2021b). However, renewable energy developments have environmental impacts, including the carbon footprint made by the manufacture of their components and the effects on the local environment during their installation and operation.

The UK has the largest tidal energy resource in Europe, which could meet 50% of the UK's electricity demand (UK Government, 2013). However, in 2020, tidal and wave energy combined contributed only 0.004% of the electricity (11 GWh) generated in the UK (UK Government, 2021a). The lack of progress in realising the potential for tidal energy in the UK is attributed to high capital costs, long payback periods (Burrows et al., 2009), novel technologies, many of which are untested at commercial scale, and high electricity grid connection costs. These obstacles are compounded by the lengthy and expensive licensing and consenting processes for renewable energy developments in the UK (González and Lacal-Arántegui, 2016; Mackinnon et al., 2018; and Ocean Energy Forum, 2016), the complexity of the marine and estuarine environment, and the lack of data on the environmental impact of tidal schemes (Quero García et al., 2019).

Tidal energy schemes deploy turbines underwater in a tidal area to capture energy from passing tidal water. Collisions of fish and marine mammals with underwater turbines and impacts on populations of marine fauna during the operation of tidal energy schemes represent obvious risks. Moreover, contentious environmental impacts such as visual intrusion add a further raft of issues to be managed. All this means that a diverse and unique range of stakeholders need to be consulted as part of the licensing and consenting process for each tidal energy proposal. In coastal locations, stakeholders are arguably more diverse than those engaged with an offshore project such as a wind farm that could be dominated by fishers and marine ecologists. Given the diverse range of issues, many of these stakeholders have competing views on issues such as use of shared sea space, and these tensions are often exacerbated by pre-existing hostility and mistrust between stakeholders (Gray et al., 2005). The result is low public confidence and delays in the licencing and consenting processes (Simas et al., 2015).

Such frustrations characterise participation in environmental decision making generally (Carson, 2009). However, attempts to deal with these frustrations often focus on technical assessments, side-lining democratic values (Fiorino, 1990). Participation is frequently framed as a bureaucratic hurdle (Pieraccini, 2015, p. 33) to be overcome rather than an opportunity for meaningful engagement (Rydin et al., 2015). The persistence of prioritising technical assessment over meaningful participation encourages commissioning bodies to adopt models of participation limited to confirmatory consultations on pre-made decisions that disempower participants.

Such procedures fuel distrust and disillusion (Chwalisz, 2017, p. 78) and create a cycle of conflict that hampers the progress of decision-making. This cycle causes further delays, spiralling costs, and derailment which frustrates practitioners and stakeholders alike (Reed, 2008). Although, in general, the public support marine renewable energy (BEIS, 2021; Bailey et al., 2011), specific proposals provoke strong public and stakeholder opposition (Bell et al., 2005). In this research, an attempt is made to break this cycle of conflict by applying two methodologies or approaches to stakeholder engagement –BBNs and PAR.

Bayesian Belief Networks

The consenting and licensing process for tidal energy schemes needs to balance credible assessment of the potential environmental and social impacts with meaningful participation of stakeholders; where meaningful engagement is defined as fair process rather than convincing people to support proposals. Use of better decision-making tools could achieve this balance (Hooper and Austen, 2013). Conventional decision-making tools struggle to deal with the complexity and uncertainty of the natural world, social impacts, and the need to incorporate and reconcile stakeholder views (Elsawah et al., 2015). Advanced systems, such as BBNs, model the non-linear, complex relationships and combine conditional probabilities, and promise to do a better job of addressing different and conflicting interests. BBNs are a type of graphical statistical model, known as directed acyclic graphs (DAGs) comprising networks of nodes, which represent variables, linked by arcs that represent conditional dependencies between the variables (Liao et al., 2017). BBNs model the combined likelihood of events (joint probability distributions), which are represented in the modelling by numbers. The capacity of BBNs to model complexity and uncertainty provides the opportunity to deliver a better outcome (decision) based on more accurate modelling of stakeholder preferences than is provided by conventional decision-making. In addition, BBNs have the capacity to incorporate and combine qualitative and quantitative data, including participant knowledge (beliefs) elicited from a sample of stakeholders (Johnson et al., 2010).

In this study, we focus on three specific features of BBNs that could improve participation in environmental decision making: their ability to produce visual representations; their capacity to incorporate quantitative and qualitative data; and the ability to easily update the models.

Visual representations: As graphical models, BBNs can be displayed as networks of nodes that can be easily understood (Stewart et al., 2014). Visualisations of BBNs can be used to enable participants to see and understand how their individual views have been included and how they relate to others' views. Seeing their views included and working in the model can help stakeholders feel valued. Visualising data in the network can also help participants appreciate the wider context, and interaction with software allows participants to test how changes affect the system, encouraging learning and reflection on their original positions (Barton et al., 2012).

Incorporation of quantitative and qualitative data: Complex empirical quantitative data can be used to generate increasingly accurate models, and accuracy contributes to stakeholder satisfaction. However, if decision-making fails to account for how people feel, the decisions themselves are not readily accepted or effectively implemented. BBNs can incorporate empirical and subjective (belief) data (Chen and Pollino, 2012), including stakeholder knowledge (belief elicitation) (Johnson et al., 2016). BBNs' ability to incorporate qualitative data representing stakeholder preferences, as well as quantitative data, ensures that the broad range of stakeholder interests are included, improving stakeholder satisfaction (Campbell et al., 2012; Renooij, 2001).

Updateable: BBNs can be easily updated (Chen and Pollino, 2012) so that prior beliefs can be revised as more information becomes available (Johnson et al., 2016). By updating the models to demonstrate different scenarios, the distribution and effects of 'belief' on decision networks (Chen and Pollino, 2012) can be tested, evaluated, and revised for prediction, diagnosis, and sensitivity analysis; assessing how responsive specific nodes are to changes elsewhere in the model. The ability to update BBNs facilitates and encourages participants to reflect and appreciate alternative views, thereby reducing potential conflict between stakeholders (Low-Choy et al., 2009).

However, despite the potential for BBNs to be used in participatory decision making, the process of incorporating knowledge BBNs has often been technocratic, top-down, and extractive. Typically, the current process, known as belief elicitation, comprises a single request for data from a limited range of experts. Closed questions are used so responses can be easily incorporated into the model and there are usually no opportunities for the participants to review the model and reflect on their contributions. Previous studies demonstrate the shortcomings of such extractive, one-way engagement processes (Cash et al., 2003; Reid et al., 2016; Sterling et al., 2017; Toomey et al., 2017). To avoid these weaknesses, we drew on PAR to redesign the belief elicitation process, re-focusing the use BBNs on the participatory process and democratic values.

Participatory Action Research

As outlined in Chapter 2, PAR is one of the main research approaches in participatory inquiry (Breu and Peppard, 2003; Reason, 1998), where collaborative participation is focused on social or an environmental change (Kindon et al., 2007; Pain et al., 2012). Originating in grassroots movements to engage people who are usually disempowered (Reason and Bradbury, 2001), PAR is done with (not on) people, to empower them in the production of knowledge and relevant action (Breu and Peppard, 2003).

TABLE 5.1 Stages of a typical PAR process

Phase	*Activities*
Action	Establish relationships and common agenda between all stakeholders Collaboratively scope issues and information Agree on time-frame
Reflection	On research design, ethics, power relations, knowledge construction process, representation and accountability
Action	Build relationships Identify roles, responsibilities and ethics procedures Establish a Memorandum of Understanding Collaboratively design research process and tools Discuss and identify desired action outcomes
Reflection	On research questions, design, working relationships and information requirements
Action	Work together to implement research process and undertake data collection Enable participation of others Collaboratively analyse information generated Begin planning action together
Reflection	On research process Evaluate participation and representation of others Assess need for further research and/or various action options
Action	Plan research-informed action which may include feedback to participants and influential other
Reflection	Evaluate action and process as a whole
Action	Identify options for further participatory research and action with or without academic researchers

Source: (Kindon et al., 2007, p. 15)

To achieve collaborative participation, PAR can be undertaken in alternating phases of action and reflection (Kindon et al., 2007). As shown in Table 5.1, participants are typically engaged in a process of active relationship building and knowledge production, interspersed by phases of reflection, including reflection on the research design data and analysis.

We drew on this description of phases of action and reflection in PAR to facilitate opportunities for reflection, learning, collaboration, and co-production of knowledge. This provided a systematic framework for data collection and analysis based on iterative cycles of engagement and modelling.

This chapter sets out the pros and cons of using these two methods (BBNs- and PAR-inspired cycles) together, explaining how new knowledge was produced and how the process might contribute to socially embedded solutions. The remainder of this chapter is structured as follows. In the next section, we introduce the case study and describe the methods of stakeholder engagement using BBNs and PAR, including the selection of key informants. We explain how BBNs and PAR were employed together and how this approach was applied in the case study.

Then in the 'Discussion' section, we describe the pros and cons of our approach (application of methods). Later, in the 'Conclusion' section, we summarise the findings and identify the lessons for the future use of BBNs and PAR to improve stakeholder engagement in offshore renewables and sustainable ocean management more widely.

Methods

Solway Energy Gateway

Due to its large tidal range, the potential of tidal energy on the Solway Firth has been explored since the 1960s (Aggidis and Feather, 2012; Becker et al., 2017; Burrows et al., 2008, 2009). One of the latest proposals, known as Solway energy Gateway (SEG), comprises a series of underwater turbines across the Solway Firth designed to capture tidal energy as it comes in and reseeds between Bowness-on-Solway in Cumbria (England) and Annan in Dumfries and Galloway (Scotland).

The Solway Firth is home to several environmentally valuable designated sites protected by local, national, and international regulations. It lies on the border between England and Scotland, which have two different regulatory regimes and cultural contexts. The Solway Firth therefore presents a series of extremely sensitive environmental and social issues, and the impact assessments that would be required as part of the consenting process for a tidal energy scheme such as SEG would be extremely complex.

Recognising the environmental sensitivity of the location, the developers (SEG) have proposed a design to reduce environmental impact. In contrast to conventional barrage technology, SEG would not need to impound water to create a substantial 'head' drop to power the turbines (Solway Energy Gateway, 2011). Instead, a novel technology known as Venturi Enhanced Turbine Technology (VETT) would be used, where water passes through reduced diameter ducts, to increase the flow rate. The accelerated water can then either drive a turbine directly or produce a pressure difference which is used to drive a remote turbine (Halcrow Group Ltd et al., 2009). By using the VETT technology, Solway Energy Gateway (2011) claims the proposed development would reduce the environmental impacts of impounding water caused by other tidal range technologies such as barrages.

SEG exemplifies the paradox of tidal energy in sustainable ocean management. The large tidal reach of the Solway Firth offers a significant potential contribution to action on climate change by increasing renewable energy. However, the unique social and environmental features in and around the Solway Firth could be adversely affected during its installation and operation.

Key informants

The complexity of the environment of the Solway Firth and its location on the English/Scottish border introduced a complex and wide-ranging set of, often competing, interests. For example, regulatory authorities on either side of the border hold conflicting views. For this study, our selection of participants, whom we identify as key informants (KIs), aimed to obtain diverse voices, rather than representative voices of the local population. Table 5.2 shows the diversity of participants who took part in this study, including stakeholders from Scotland and England as well as experts and non-experts.

The 16 KIs were identified and engaged via snowball sampling. Four KIs had attended previous meetings associated with the proposed development (KI-B,

TABLE 5.2 Key Informants

Codes	*General description*	*Method of identification*
KI-A	Local resident (England)	Snowball sampling – identified by KI-M
KI-B	On SEG development team and local resident	Initial contact and attendee of previous SEG meetings
KI-C	Works for a government body – conservation and local resident (England)	Snowball sampling – identified by KI-I, KI-O
KI-D	Works for a government body – habitats and landscapes – and local resident (Scotland)	Snowball sampling – identified by attendee of previous SEG meetings
KI-E	Works for a conservation charity (England)	Snowball sampling – identified by attendee of previous SEG meetings
KI-F	Works for a conservation charity and local resident (Scotland)	Snowball sampling – identified by KI-E
KI-G	Works for a Scottish Government – marine planning	Snowball sampling – identified by KI-B
KI-H	Parish Councillor and local resident (England)	Snowball sampling – identified by KI-B
KI-I	Parish Councillor and local resident (England)	Snowball sampling – identified by KI-B
KI-J	On SEG development team	Snowball sampling – identified by attendee of previous SEG meetings
KI-K	Works for a sustainable development charity and local resident (Scotland)	Attendee of previous SEG meetings
KI-L	Works for a sustainable development charity and local resident (Scotland)	Attendee of previous SEG meetings
KI-M	Works for a council funded wetlands conservation body and local resident (England)	Snowball sampling – identified by attendee of previous SEG meetings
KI-N	Works for a government body – marine conservation (England)	Snowball sampling – identified by KI-A
KI-O	Works for a conservation charity and local resident (England)	Snowball sampling – identified by KI-E
KI-P	On SEG development team	Attendee of previous SEG meetings

KI-K, KI-L and KI-P). A further four KIs were identified by attendees of the previous meetings, but who did not themselves attend (KI-D, KI-E, KI-J and KI-M). The remainder were identified by other KIs during the cycles of engagement.

The KIs were assigned a unique code to preserve their anonymity, as listed in Table 5.1. In total, 16 different KIs were engaged over five cycles of engagement, and the maximum number of KIs engaged in a single cycle was 12, in cycle 1. During the study, qualitative data were collected via 32 interactions with the KIs, including one-to-one interviews, a Parish Council meeting, and written contributions. Between the cycles of engagement, 14 individual and 2 combined BBNs were produced.

Using Bayesian Belief Networks and Participatory Action Research together

In contrast to the current use of BBNs, we used PAR-inspired cycles of action and reflection to facilitate the co-production and co-analysis of BBNs with each participant (see Table 5.3). We planned five cycles of engagement with the KIs, where each action would be followed by reflection, using the features of BBNs to encourage participants to review and transform their prior beliefs, creating change. This facilitated an iterative process of knowledge production, where the reflection on knowledge accrued in one cycle informed the next. For example, following feedback from the KIs, the proposed group events planned in cycles 4 and 5 were replaced with other activities. The next section summarises how using BBNs in PAR-inspired cycles of action and reflection worked to co-produce knowledge in the case of SEG.

Co-production of BBNs with key informants

In this section, we explain how BBNs were co-produced with KIs over the cycles of action and reflection. We explain how the KIs engaged with the BBNs in each cycle of engagement and how the data collection process evolved with each cycle.

Action – Cycle 1 interviews were designed to establish relationships with the KIs and to establish baseline data to populate BBN models. During interviews, each KI was asked to identify what issues relating to SEG and the Solway Firth were important to them, why these issues matter, and what the relationships are between these issues. Open discussions were held around maps, photographs, and diagrams to stimulate thoughts about the area (current map), history (historical map) and proposed scheme (plans of the proposed scheme).

Reflection – after the cycle 1 interviews we took time to reflect on the engagement of the KIs in accordance with the features of PAR. For example, a key feature of PAR is the inclusion of local beliefs and values. To achieve this, our KIs included stakeholders who lived locally and contributed expert and

TABLE 5.3 Cycles of action and reflection, based on 'key stages in a typical PAR process'

PAR stages	*Objectives*	*Analysis*	*Method/Sampling point*
Action	Establish baseline stakeholder beliefs on SEG via a process of 'broadening out'. Identify issues (what matters) to demonstrate how participant's views are heard, recorded, and valued. Explore why each issue matters considering their importance, how sure stakeholder feel about their views, and whether there are any co-dependencies. Cultivate trust and respect with individual stakeholders.	Use transcriptions of interviews and data collected from Participatory Diagramming (PD) to identify themes and patterns.	**Cycle 1:** One-to-one unstructured interviews with 5–6 stakeholders. Use PD and/or mapping to facilitate data collection and encourage engagement
Reflection	On the interviews	Compare the results of the interviews with the PAR approach.	
Action	Construct a BBN from each interview*	Create nodes from the issues raised. Parametrise the BBNs with weightings based on the importance and how confident people felt. Use the identified co-dependencies themes and patterns to create relationship arcs.	BBN Modelling (researcher)
Reflection	Integrate the structure of BBNs. Build trust and respect with stakeholders. Identify changes in beliefs and perspectives through interacting with the BBN.	Co-analyse the structure of their own BBN with individual stakeholders discussing any changes in beliefs and perceptions. Review primary data to identify themes and patterns.	**Cycle 2:** One-to-one interviews with stakeholders. Use BBN to facilitate interactive PD.
Action	Revise individual BBNs from each interview	Use primary data to revise individual BBNs.	BBN Modelling (researcher)
Reflection	Participation in integration of the individual BBNs. Cultivate trust and respect between stakeholders. Identify changes in beliefs and perspectives though interacting with the BBNs.	Co-analyse other stakeholder's BBNs. Discuss changes in beliefs and perceptions. Review primary data to identify themes and patterns in how participants feel about the fairness of the process and what they are learning.	**Cycle 3:** One to one interviews (recorded) with stakeholders. Use BBN to facilitate interactive PD.

(*Continued*)

TABLE 5.3 (Continued)

PAR stages	*Objectives*	*Analysis*	*Method/Sampling point*
Action	Synthesise the BBNs	Create nodes from the issues raised. Parametrise the BBNs with weightings based on the importance and how sure people felt. Use the identified co-dependencies to create relationship arcs.	BBN Modelling (researcher)
Reflection	Integrate BBN to explore the potential for collective learning and changes in beliefs and perspectives though interacting with the BBN.	Review primary data to identify themes and patterns in changes in beliefs and perspectives.	**Cycle 4**: One group event (focus group or workshop). Use BBN to facilitate interactive PD.**
Action	Revise structure of BBN based on results of individual and group reflections.	Review primary data to revise synthesised BBNs.	BBN Modelling (researcher)
Reflection	Present final BBN to stakeholder group. Explore how stakeholders felt about it: how the BBN incorporated, valued, and represented their beliefs; how their beliefs changed; and what they learnt through interaction with the BBN.	Obtain feedback during/after the event to record the final positions/beliefs/perceptions held and assess the changes in beliefs from the original positions. Obtain feedback from the participants on whether the process was fair and what they learnt.	**Cycle 5:** Event/exhibition of the findings on either side of the Solway to optimise attendance and access.***

Source: (Kindon et al., 2007, p. 15)

* The modelling was undertaken separately by the researcher, however, the KIs were given opportunities to review and develop the models in subsequent cycles of engagement

** The proposed group event in cycle 4 was replaced with one-to-one interviews in response to feedback from the KIs and to reduce the risk of consultation fatigue and the burden of participation

*** The proposed group events in cycle 5 were replaced with written consultations in response to feedback from the KIs and to reduce the risk of consultation fatigue and the burden of participation.

non-expert knowledge. Reflecting on the cycle 1 interviews, we found that some of the experts interviewed tended to be more defensive about the current consenting process and worried that their data could be used against them in a future application. In contrast, non-expert KIs were less defensive. In response, in cycle 2, we noted if and how the visual representations of BBNs helped both experts and non-experts to be more open and less defensive.

Action – BBN modelling was undertaken with the data captured during the cycle 1 interviews. These first BBN models acted as a baseline to gauge subsequent changes in views and perceptions. An individual BBN was produced for each KI with nodes representing each issue they raised (see Figure 5.1). We coloured each node according to six themes to aid interpretation – living environment, physical environment, conservation, socio-economic, community, and proposed tidal scheme.

Reflection – Cycle 2 interviews enabled the KIs to reflect on their prior beliefs by reviewing visual representations of the BBN models constructed after their cycle 1 interview. Each model displayed nodes that represented the issues they raised and the KIs were asked to reflect on why these issues mattered so that the 'states' of each node could be defined. Each KI could identify irrelevant nodes that should be deleted; any missing issues; where nodes should be added; if any nodes could be merged; or if nodes should be split up to provide more detail. The KIs were also invited to consider the relationships between nodes so that arcs could be added.

Cycle 2 provided the first opportunity to review the visual representation (the beginnings) of a BBN. The KIs easily understood that the nodes represented the

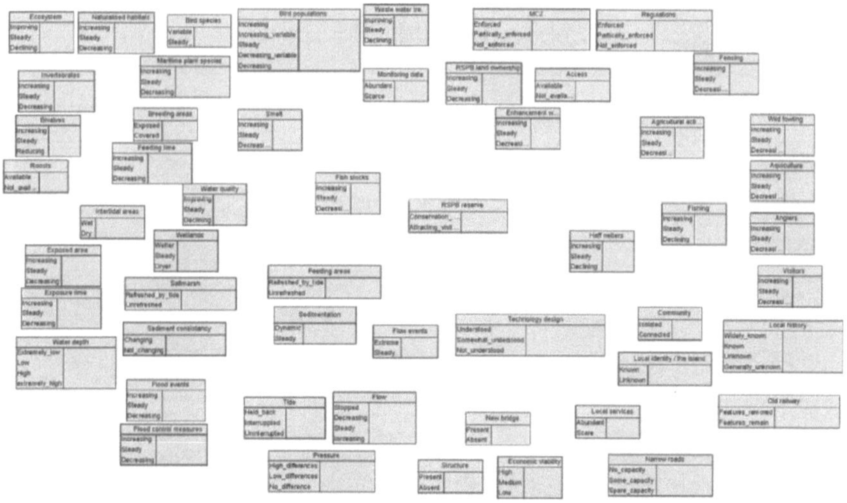

FIGURE 5.1 BBN model developed in cycle 1 with KI-E; nodes represent issues raised during one-to-one interview.

issues they had previously raised and were able to work around the model discussing ideas about connections between the nodes and their 'states'. Identifying the themes by colour worked well: KI-E stated that it provided a "logical order" to work round the nodes.

Considering the previously noted defensiveness of the expert KIs, engagement with the visual representations of the BBNs was variable. We reflected that KIs representing organisations that are formally involved with existing consenting processes for tidal energy schemes, such as KI-G and KI-D, seemed constrained by their knowledge and investment in the existing processes which reduced their openness to an alternative approach.

Action – BBN modelling was used to incorporate the qualitative data captured in the cycle 2 interviews. The structure of each 'individual' model was developed by amending the nodes and adding states to them as directed by the KIs (see Figure 5.2). Arcs were placed between the nodes to represent relationships (conditional dependencies) between the issues (variables).

Reflection – Cycle 3 interviews were conducted in which each KI was again asked to reflect on their own 'individual' model that showed revised nodes, states, and arcs, as discussed in cycles 1 and 2, to check they understood it and that it represented their views. In addition, each KI was asked to reflect on another KI's model to see if it enabled them to understand an alternative perspective.

The types of issues raised in cycles 2 and 3 were similar, but in cycle 3, the KIs were thinking about them more deeply. Generally, KIs concentrated on developing the nodes present in cycle 2; redefining them or adding other associated nodes to

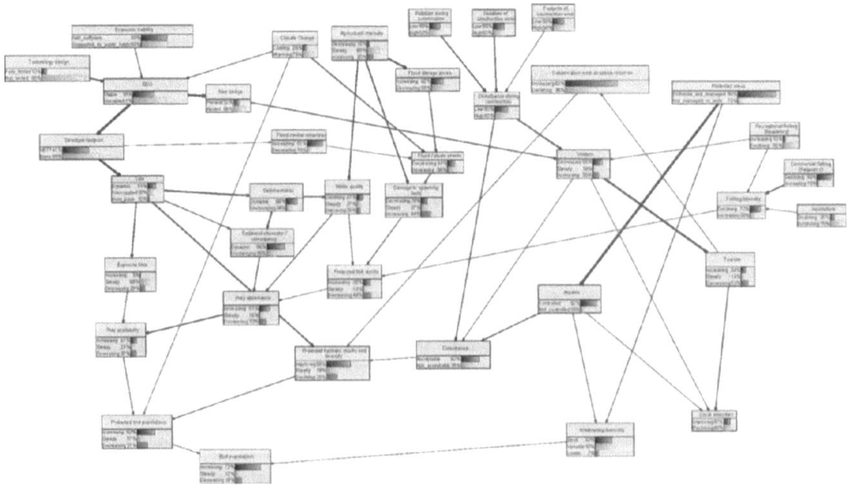

FIGURE 5.2 BBN model developed in cycle 2 with KI-E; nodes represent the issues raised during a one-to-one interview.

clarify meaning. This process was aided by the review of a BBN by another KI (KI-A) who, after looking at KI-B's BBN from cycle 2, went back to their own BBN and considered the difference between visitor numbers (a node in their BBN from cycle 2) and tourism (a node in KI-Bs BBN from cycle 2). In cycle 3, KI-A considered that increasing visitor numbers would have a negative effect on community peace of mind due to the strain on local services. However, in cycle 2, KI-B had predicted that more tourism would have a positive impact on the community by increasing local jobs and pride in the area. To account for this apparent discrepancy, in cycle 3, KI-A considered that visitor numbers and tourism should be represented as separate nodes, to show that visitor numbers could be high, but unless tourists are spending money, they are not contributing to the local economy, while increasing the strain on local services.

Action – BBN modelling was used to construct an integrated BBN model from the results of the cycle 3 interviews, supplemented with data and models from the previous cycles. A combined BBN model was produced for review in the final cycles (see Figure 5.3).

Reflection – Cycle 4 interviews comprised one-to-one interviews with three established KIs to discuss a draft combined model (Figure 5.3) before the final cycle. We asked each KI to reflect on the integration of the data into the combined model so it could be amended and refined prior to the final cycle of engagement. However, the KIs struggled with understanding the complexity of the combined BBN.

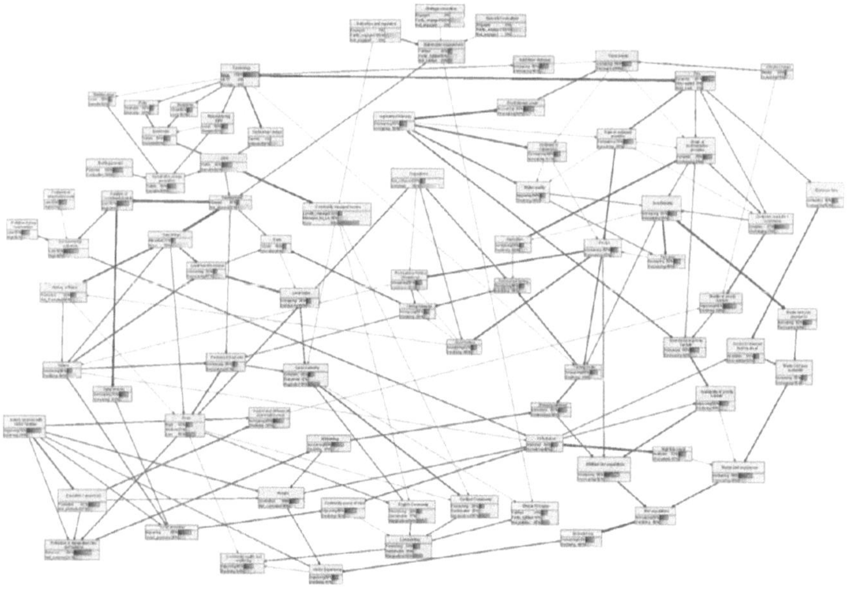

FIGURE 5.3 Visual representation of combined BBN model, integrating data from KIs interviewed in cycles 1–3.

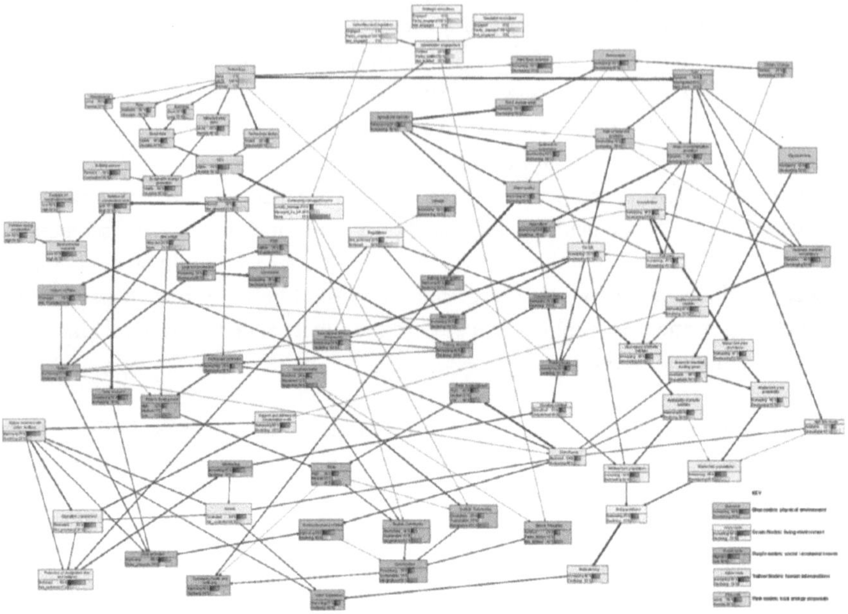

FIGURE 5.4 Integrated BBN model showing nodes, states, and arcs that represent the combined views raised by KIs during previous cycles of engagement. This version of the model shows the likelihood of events if SEG is installed.

Action – BBN modelling was performed to use the data collected in the cycle 4 interviews to amend the combined model before the last cycle of engagement. Possibly reflecting the complexity of the model and consultation fatigue, the KIs made relatively few amendments to the nodes, states, and arcs in the BBN itself. Instead, the KIs tended to focus on considering presentational issues to aid interpretation. At their suggestion, a colour key and title were added to the visual representations to help understanding in the final cycle (see Figure 5.4).

Reflection – Cycle 5 data collection was carried out to assess the KIs' understanding of the combined model. We posted out hard copies of visual representations of BBNs illustrating two scenarios – SEG installed (Figure 5.4) and not installed; and the subsequent likelihood of events. In an accompanying letter, we asked the KIs to review and comment on the visual representations of the combined BBNs models. The letter (copy provided in Appendix III) included explanatory notes and pointers, suggesting the KIs review the nodes, arcs, overall presentation, and clarity in their own time. The KIs responded in a variety of ways to the letter, as summarised in Table 5.4.

KI-A, KI-C, and KI-J sent an email response summarising their thoughts in text, without any annotated visual representations of the BBNs, indicating a personal preference for more formal methods of communication. KI-E emailed their

TABLE 5.4 Summary of responses received in cycle 5

KI	*Date of response*	*Method of response*	*Type of response*
KI-K	9 Aug 2018	Post. Handwritten annotations on visual representations and instructions sheets	Provided comments on presentation and classification of the nodes in categories. Suggested additional nodes and arcs.
KI-A	10 Aug 2018	Email, typed response	Commented on the presentation and the numerical values, focusing on the changes in the likelihood of events calculated from the combined probabilities in the different scenarios.
KI-E	21 Aug 2018	Email typed notes in a table and posted handwritten annotations on the visual representations	Made notes on specific nodes in a table, commenting on the numerical values and impacts. Suggested additional arcs and nodes. Annotations accompanied typed notes, highlighting the position of suggested additional nodes and arcs, and indicating nodes to combine.
KI-C	4 Sept 2018	Emailed typed comments	Commented on presentation issues.
KI-B	19 Sept 2018	One-to-one meeting. Researcher annotated visual representations	Commented on presentation issues and suggested additional aids to interpretation.
KI-F	Oct 2018	Post. Handwritten annotations on instructions sheets	Commented on the presentation and instruction sheets.
KI-J	14 Aug 2019	Email typed notes	Commented on the presentation, suggested additional nodes, and sought clarification on the distinctions between some nodes.

response, but also posted back annotated plans. KI-K fully annotated the visual representations of the BBNs (Figure 5.5) and the explanation notes (Figure 5.6). KI-F annotated the explanation notes in pencil.

In their written contributions, many of the KIs struggled with the complexity of the visual representations. KI-K stated the BBNs were 'a bit confusing', focusing on the presentation issues, specifically the colour coding. KI-A and KI-F focused

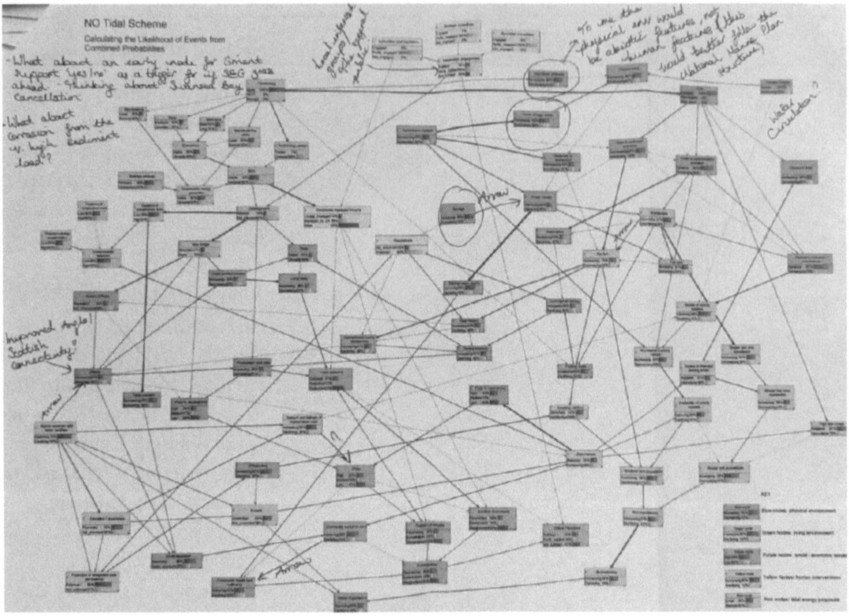

FIGURE 5.5 Annotated BBN returned by KI-K.

on the numerical values which they found 'confusing' (KI-A) and 'seems far too complicated' (KI-F). They both indicated concern that their responses would be unsatisfactory. KI-A apologised, stating 'I'm sorry if my responses are not as in depth as you might have hoped' and KI-F stated that they felt 'really sorry', linking this to anxiety about their own abilities: 'I just can't overcome my frustration at my inability to understand how this [the BBN] works'. Despite my attempts to aid interpretation, KI-F stated that the 'guidance instructions are insufficient'. Agreeing with KI-F, KI-B repeated their suggestions from cycle 4 that there needed to be more interpretation, suggesting that 'layers of priorities' could be shown 'projected in 3D' to 'bring out the areas of interest'.

Discussion

The capacity of BBNs to incorporate qualitative data, be updated, and produce visual representations can enable stakeholders to review and test their prior beliefs to facilitate learning (Stewart et al., 2014). However, stakeholders' views are usually incorporated into BBNs using single, extractive methods with no opportunity for reflection. To seize the opportunity to improve stakeholder engagement using BBNs, we used PAR inspired cycles of action and reflection as the basis for systematic engagement, where BBNs were built iteratively with stakeholders over five cycles of engagement. Each cycle of engagement provided the KIs with the opportunities to review and develop, first their own and then combined BBN models.

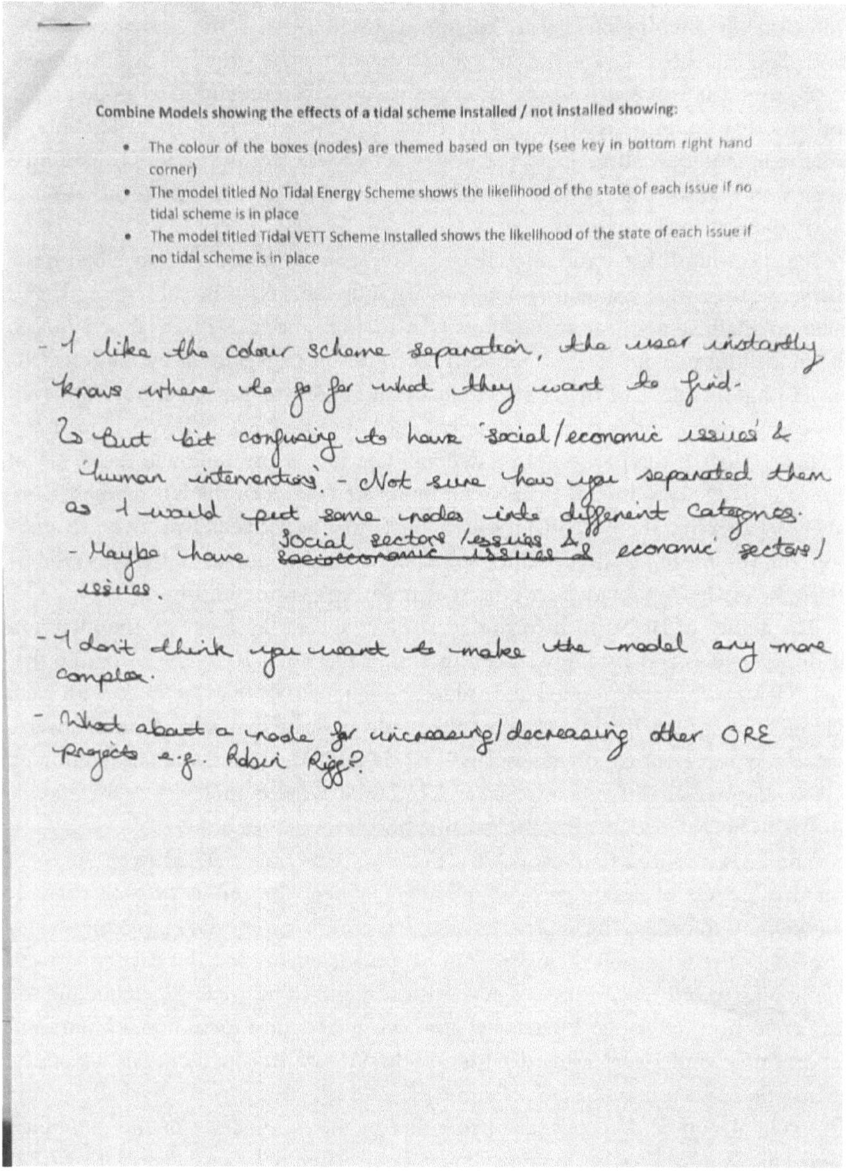
Combine Models showing the effects of a tidal scheme installed / not installed showing:

- The colour of the boxes (nodes) are themed based on type (see key in bottom right hand corner)
- The model titled No Tidal Energy Scheme shows the likelihood of the state of each issue if no tidal scheme is in place
- The model titled Tidal VETT Scheme Installed shows the likelihood of the state of each issue if no tidal scheme is in place

- I like the colour scheme separation, the user instantly knows where to go for what they want to find.
↳ But bit confusing to have 'social/economic issues' & 'human interventions'. - Not sure how you separated them as I would put some nodes into different categories.
- Maybe have ~~socioeconomic issues &~~ social sectors/issues & economic sectors/issues.

- I don't think you want to make the model any more complex.

- What about a node for increasing/decreasing other ORE projects e.g. Robin Rigg?

FIGURE 5.6 Handwritten response from KI-K.

Our results show that reviewing the visual representations of the updated BBNs over cycles of engagement provided our KIs with opportunities to learn, reflect and broaden their thoughts. We found that the KIs used the BBNs and the process of their development to systematically work through their thoughts, deepening their understanding of, and broadening their own knowledge. However, the experience of engaging with models varied across the KIs. Inevitably, as the BBNs were developed with each cycle of engagement with the KIs, the complexity of the modelling increased. Cycles 1 and 2 concentrated on the structure of the models, considering the issues and connections between them represented by nodes, states and arcs that could be relatively easily explained. After cycle 2, event likelihoods were represented as numbers in the nodes. As a result, some of the KIs expressed anxiety and doubted their own capabilities and knowledge, as well as the value of their contributions.

We also found that some knowledge was lost in the process of developing the BBNs. As they were gradually amended with individual KIs, the subsequent BBNs failed to retain some meaning and detail. In addition, because it is a laborious task, the modelling was undertaken between the cycles of engagement by the researchers and independent of the KIs to reduce the burden of participation. However, this meant that the KIs missed experiencing how knowledge was developed and deepened during the process of modelling. The loss of meaning and detail meant that knowledge was not adequately communicated between the KIs by the BBNs, adversely affecting the potential for learning from alternative perspectives. In addition, missing meaning and detail led some KIs to raise concerns about the credibility of the evidence provided by others, aggravating existing hostilities.

The ability of BBNs to incorporate qualitative data did facilitate the inclusion of diverse views and lay knowledge. To ensure that each KI could recognise that their views were incorporated and valued, individual BBNs were built with each KI before being combined. However, the results indicate that incorporating diverse views and representing them visually, first as individual, and then as combined BBNs, emphasised differences between KIs as well as the complexity of the modelling. This heightened anxiety and existing hostilities between the KIs.

The PAR-inspired framework of action and reflection enabled us to apply an iterative process of design, review and adapt, where the results of each cycle of engagement informed the approach taken in the next. After each engagement with our KIs, we reflected on their reactions to specific events and the data collection. In response to our reflections, we made amendments to future engagement and the data collection timetable. Effectively, little cycles of action (design and adapt) and reflection (review) were created within each cycle of engagement. Most notably, after cycle 3, we reflected a lack of enthusiasm for the group events we had planned for cycles 4 and 5. We considered that due to the complexity of the combined model and the lack of enthusiasm expressed, holding a group event with all KIs would run the risk of unwarranted consultation fatigue and unduly increase the burden of participation. This reflection led to action; we amended the data collection timetable by replacing the planned group events with more one-to-one interviews in cycle 4 and with posting hard copies of the integrated models to previously engaged KIs in cycle 5. Such adjustments are characteristic of PAR and co-operative inquiry (Reason, 2002). This enabled us to check both the research design throughout the study and provided opportunities for the adaptations to be made in response to feedback and primary data analysis.

Conclusion

Tidal energy schemes represent opportunities and risks for sustainable ocean management. Considering the complexity of the marine environment and potential impacts, expert and non-expert knowledge are both needed in the consenting process for proposed tidal energy schemes. However, current participatory processes leave stakeholders dissatisfied and often generate conflict. Aiming to improve participation in the consenting process of tidal energy schemes, this study tested the use of advanced statistical models, BBNs, in PAR-inspired phases of action and reflection. Currently, participant data incorporated into BBNs is extractive, typically comprising a single survey with closed questions and little opportunity for feedback. In contrast, PAR prescribes participant-led, reflective practice. To capture the potential benefits for stakeholder engagement we tested the use of BBNs in accordance with the features of PAR. Specifically, we used PAR-inspired cycles of action and reflection to provide a systematic framework for knowledge co-production using BBNs, based on iterative cycles of engagement and modelling.

The results of this study indicate that the features of BBNs used in cycles of engagement could improve participation. The capacity of BBNs to incorporate qualitative data and represent it visually enabled the KIs to review their prior beliefs to learn and broaden their thoughts. The ability to update the models facilitated further development of the modelling. Revisiting the KIs in cycles enabled further learning and reflection on the updated visual representations. Fundamentally, the features of BBNs applied in PAR-inspired cycles encouraged KIs to learn and reflect on their own views. The features could therefore offer some improvement in participatory processes.

It is, however, noted that the increasing complexity of the visual representations of the BBNs as they were developed over the cycles of engagement hampers the potential benefits to participatory co-production of knowledge. KIs became increasingly anxious when trying to understand and interpret the models. The results show that the KIs' anxiety increased their (emotional) burden of participation. This is likely to reduce engagement in practice.

References

Aggidis, G. A., & Feather, O. (2012). Tidal range turbines and generation on the Solway Firth. *Renewable Energy*, 43, 9–17.

Aguilera, P. A., Fernández, A., Fernández, R., Rumí, R., & Salmerón, A. (2011). Bayesian networks in environmental modelling. *Environmental Modelling and Software*, 26(12), 1376–1388. https://doi.org/10.1016/j.envsoft.2011.06.004

Bailey, I., West, J., & Whitehead, I. (2011). Out of sight but not out of mind? Public perceptions of wave energy. *Journal of Environmental Policy and Planning*, 13(2), 139–157. https://doi.org/10.1080/1523908X.2011.573632

Barton, D. N., Kuikka, S., Varis, O., Uusitalo, L., Henriksen, H. J., Borsuk, M., Hera, A. D. la, Farmani, R., Johnson, S., & Linnell, J. D. C. (2012). Bayesian networks in environmental and resource management. *Integrated Environmental Assessment and Management*, 8(3), 418–429. https://doi.org/10.1002/ieam.1327

Becker, A., Plater, A., & Wolf, J. (2017). *The Energy River: Realising Energy Potential From the River Mersey*. [Online] Liverpool: University of Liverpool. Available at: www.liverpool.ac.uk/media/livacuk/instituteofsustainablecoastsandoceans/bluegreenenergy/Energy-River-Report-Final.pdf (Accessed: 9 August 2019).

BEIS (Department of Business, Energy and Industrial Strategy). (2021). *BEIS Public Attitudes Tracker: Energy Infrastructure and Energy Sources*. [Online] UK Government. Available at: www.gov.uk/government/statistics/beis-public-attitudes-tracker-autumn-2021 (Accessed 1 March 2022)

Bell, D., Gray, T., & Haggett, C. (2005). The "social gap" in wind farm siting decisions: Explanations and policy responses. *Environmental Politics*, 14(4), 460–477. https://doi.org/10.1080/09644010500175833

Breu, K., & Peppard, J. (2003). Useful knowledge for information systems practice: The contribution of the participatory paradigm. *Journal of Information Technology*, 18(3), 177–193. https://doi.org/10.1080/0268396032000122141

Burrows, R., Walkington, I., Yates, N., Hedges, T., Chen, D., Li, M., Zhou, J., Wolf, J., Proctor, R., Holt, J., & Prandle, D. (2008). Tapping the tidal power potential of the Eastern Irish Sea. Joule Centre Final Report JIRP10/03.

Burrows, R., Walkington, I. A., Yates, N. C., Hedges, T. S., Wolf, J., & Holt, J. (2009). The tidal range energy potential of the West Coast of the United Kingdom. *Applied Ocean Research*, 31(4), 229–238. https://doi.org/10.1016/j.apor.2009.10.002

Campbell, D., Chilton, S., Clark, S., Fitzsimmons, C., Gazzola, P., Hutchinson, G., Metcalf, H., Roe, M., Rushton, S., Schuchert, P., Slater, M., Speak, S., Stead, S. M., & Sweeting, C. (2012). Bayesian belief networks as an interdisciplinary marine governance and policy tool. *Valuing Nature Network Report*, 1–27.

Carson, L. (2009). Deliberative public participation and hexachlorobenzene stockpiles. *Journal of Environmental Management*, 90(4), 1636–1643. https://doi.org/10.1016/j.jenvman.2008.05.019

Cash, D. W., Clark, W. C., Alcock, F., Dickson, N. M., Eckley, N., Guston, D. H., Jäger, J., & Mitchell, R. B. (2003). Knowledge systems for sustainable development. *PNAS*, 100(14), 8086–8091. https://doi.org/10.1073/pnas.1231332100

Chen, S. H., & Pollino, C. A. (2012). Good practice in Bayesian network modelling. *Environmental Modelling and Software*, 37, 134–145. https://doi.org/10.1016/j.envsoft.2012.03.012

Chwalisz, C. (2017). *The People's Verdict: Adding Informed Citizen Voices to Public Decision-Making*. London: Rowmand & Littlefield International.

Elsawah, S., Guillaume, J. H. A., Filatova, T., Rook, J., & Jakeman, A. J. (2015). A methodology for eliciting, representing, and analysing stakeholder knowledge for decision making on complex socio-ecological systems: From cognitive maps to agent-based models. *Journal of Environmental Management*, 151, 500–516. https://doi.org/10.1016/j.jenvman.2014.11.028

Fiorino, D. J. (1990). Citizen participation and environmental risk: A survey of institutional mechanisms. *Science, Technology, & Human Values*, 15(2), 226–243. www.jstor.org/stable/689860

González, J. S., & Lacal-Arántegui, R. (2016). A review of regulatory framework for wind energy in European Union countries: Current state and expected developments. *Renewable and Sustainable Energy Reviews*, 56, 588–602. https://doi.org/10.1016/j.rser.2015.11.091

Gray, T., Haggett, C., & Bell, D. (2005). Offshore wind farms and commercial fisheries in the UK: A study in stakeholder consultation. *Ethics, Place and Environment*, 8(2), 127–140. https://doi.org/10.1080/13668790500237013

Halcrow Group Ltd, Mott MacDonald and RSK Group plc (2009). *Solway Energy Gateway – Feasibility Study*. Envirolink Northwest.

Hooper, T., & Austen, M. (2013). Tidal barrages in the UK: Ecological and social impacts, potential mitigation, and tools to support barrage planning. *Renewable and Sustainable Energy Reviews*, 23, 289–298. https://doi.org/10.1016/j.rser.2013.03.001

Johnson, S., Logan, M., Fox, D., Kirkwood, J., Pinto, U., & Mengersen, K. (2016). Environmental decision-making using Bayesian networks: creating an environmental report card. *Applied Stochastic Models in Business and Industry*, 33(4), 335–347. https://doi.org/10.1002/asmb.2190

Johnson, S. R., Tomlinson, G. A., Hawker, G. A., Granton, J. T., Grosbein, H. A., & Feldman, B. M. (2010). A valid and reliable belief elicitation method for Bayesian priors. *Journal of Clinical Epidemiology*, 63(4), 370–383. https://doi.org/10.1016/j.jclinepi.2009.08.005

Kindon, S., Pain, R., & Kesby, M. (eds.) (2007). *Participatory Action Research Approaches and Methodologies: Connecting people, Participation and Place*. Abingdon, Oxon: Routledge.

Liao, Y., Xu, B., Wang, J., & Liu, X. (2017). A new method for assessing the risk of infectious disease outbreak. *Scientific Reports*, 7(March 2015), 1–12. https://doi.org/10.1038/srep40084

Low Choy, S., James, A., & Mengersen, K. (2009). Expert elicitation and its interface with technology: A review with a view to designing Elicitator. *18th World IMACS Congress and MODSIM09 International Congress on Modelling and Simulation: Interfacing Modelling and Simulation with Mathematical and Computational Sciences*, Proceedings, July, 4269–4275.

Mackinnon, K., Smith, H. C. M., Moore, F., van der Weijde, A. H., & Lazakis, I. (2018). Environmental interactions of tidal lagoons: A comparison of industry perspectives. *Renewable Energy*, 119, 309–319. https://doi.org/10.1016/j.renene.2017.11.066

Ocean Energy Forum (2016). Ocean Energy Strategic Roadmap: Building Ocean Energy for Europe. [Online] Ocean Energy Forum. Available at: https://webgate.ec.europa.eu/maritimeforum/en/frontpage/1036 (Accessed: 18 June 2020).

Ottinger, G. (2013). Changing knowledge, local knowledge, and knowledge gaps: STS insights into procedural justice. *Science Technology & Human Values*, 38(2), 250–270. https://doi.org/10.1177/0162243912469669.

Pain, R., Whitman, G., Milledge, D., & Lune Rivers Trust. (2012). *Participatory Action Research Toolkit: An Introduction to Using PAR as an Approach to Learning, Research and Action*. Durham: Durham University/RELU/Lune Rivers Trust Durham University.

Pieraccini, M. (2015). Rethinking participation in environmental decision-making: Epistemologies of marine conservation in South-East England. *Journal of Environmental Law*, 27(1), 45–67. https://doi.org/10.1093/jel/equ035

Quero García, P., García Sanabria, J., & Chica Ruiz, J. A. (2019). The role of maritime spatial planning on the advance of blue energy in the European Union. *Marine Policy*, 99(October 2018), 123–131. https://doi.org/10.1016/j.marpol.2018.10.015

Reason, P. (1998). A participatory world. *Resurgence*, 168, 42–44.

Reason, P. (2002). The practice of co-operative inquiry. *Systemic Practice and Action Research*, 15(3), 169–176.

Reason, P., & Bradbury, H. (2001). *Handbook of Action Research: Participative Inquiry and Practice*. SAGE Publications.

Reed, M. S. (2008). Stakeholder participation for environmental management: A literature review. *Biological Conservation*, 141(10), 2417–2431. https://doi.org/10.1016/j.biocon.2008.07.014

Reid, R. S., Nkedianye, D., Said, M. Y., Kaelo, D., Neselle, M., Makui, O., Onetu, L., Kiruswa, S., Ole Kamuaro, N., Kristjanson, P., Ogutu, J., BurnSilver, S. B., Goldman, M.

J., Boone, R. B., Galvin, K. A., Dickson, N. M., & Clark, W. C. (2016). Evolution of models to support community and policy action with science: Balancing pastoral livelihoods and wildlife conservation in savannas of East Africa. *PNAS*, 113(17), 4579–4584. https://doi.org/10.1073/pnas.0900313106

Renooij, S. (2001). Probability elicitation for belief networks: Issues to consider. *Knowledge Engineering Review*, 16(3), 255–269. https://doi.org/10.1017/S0269888901000145

Rydin, Y., Lee, M., & Lock, S. J. (2015). Public engagement in decision-making on major wind energy projects. *Journal of Environmental Law*, 27(1), 139–150. https://doi.org/10.1093/jel/eqv001

Simas, T., O'Hagan, A. M., O'Callaghan, J., Hamawi, S., Magagna, D., Bailey, I., Greaves, D., Saulnier, J. B., Marina, D., Bald, J., Huertas, C., & Sundberg, J. (2015). Review of consenting processes for ocean energy in selected European Union Member States. *International Journal of Marine Energy*, 9, 41–59. https://doi.org/10.1016/j.ijome.2014.12.001

Solway Energy Gateway Ltd (2011). Solway Energy Gateway. [Online] Solway Energy Gateway. Available at: www.solwayenergygateway.co.uk/ (Accessed: 11 April 2017).

Sterling, E. J., Betley, E., Sigouin, A., Gomez, A., Toomey, A., Cullman, G., Malone, C., Pekor, A., Arengo, F., Blair, M., Filardi, C., Landrigan, K., & Porzecanski, A. L. (2017). Assessing the evidence for stakeholder engagement in biodiversity conservation. *Biological Conservation*, 209, 159–171. https://doi.org/10.1016/j.biocon.2017.02.008

Stewart, G. B., Mengersen, K., & Meader, N. (2014). Potential uses of Bayesian networks as tools for synthesis of systematic reviews of complex interventions. *Research Synthesis Methods*, 5(1), 1–12. https://doi.org/10.1002/jrsm.1087

Toomey, A. H., Knight, A. T., & Barlow, J. (2017). Navigating the space between research and implementation in conservation. *Conservation Letters*, 10(5), 619–625. https://doi.org/10.1111/conl.12315

UK Government (2013). Wave and tidal energy: Part of the UK's energy mix. [Online] UK Government. Available at: www.gov.uk/guidance/wave-and-tidal-energy-part-of-the-uks-energy-mix (Accessed 7 December 2019).

UK Government (2021a). Energy trends: UK renewables. [Online] UK Government. Available at: www.gov.uk/government/statistics/energy-trends-section-6-renewables (Accessed 2 January 2022)

UK Government (2021b). UK's path to net zero set out in landmark strategy. [Online] UK Government. Available at: www.gov.uk/government/news/uks-path-to-net-zero-set-out-in-landmark-strategy. (Accessed 3 January 2022)

Uusitalo, L. (2007). Advantages and challenges of Bayesian networks in environmental modelling. *Ecological Modelling*, 203(3–4), 312–318. https://doi.org/10.1016/j.ecolmodel.2006.11.033

6

ASSESSING THE *PROFESSIONALISATION* OF MARINE CITIZEN SCIENCE

Benedict McAteer and Wesley Flannery

Introduction

Marine citizen science has gained momentum as a participatory approach to knowledge production that uses amateurs to contribute to the evidence base that underpins marine policy (Garcia-Soto et al., 2021). The rise of citizen science has been framed as a response to the need for data on environmental changes (Bennett, 2016, Chase and Levine, 2016), the limited scope of government monitoring programmes (Sharpe and Conrad, 2006), and the desire for community participation in environmental management (Pandya, 2012; Thompson, 2016). By increasing monitoring efforts and empowering members of the public to take political action to protect the oceans, marine citizen science has helped marine management to address issues related to climate change (Kelly et al., 2020; Sandahl and Tøttrup, 2020). Citizen science also offers the potential of both broadening the engagement of citizens with governance processes (Turrini et al., 2018) and instilling scientific and environmental learning advancements amongst participants (Haywood, 2016). Due to its capacity to produce multiple policy-relevant outputs, citizen science is advanced by academics and practitioners as a potential solution to governance concerns regarding knowledge rationalisation and the marginalisation of local marine communities (Schewe et al., 2020; McAteer and Flannery, 2022). Although traditionally initiated and operationalised by non-governmental organisations (NGOs) and environmental charities, the growing recognition of marine citizen science has led to a greater involvement of government bodies in the design and implementation of projects (Turbé et al., 2019).

The impact of the shifting positionality of government, from *users* of citizen science data to *co-producers* of initiatives, is not yet meaningfully explored in the literature. Greater government involvement is reflective of a significant change to how citizen science is being operationalised. This shift raises questions about the

DOI: 10.4324/9781003311171-8

capacity of citizen science actors to organise projects independent of government influence. Government investment has had a direct impact on the organisational elements of citizen science projects (Turbé et al., 2019). Conceptualisations of citizen science as an intrinsically collaborative process obfuscate the politics and power dynamics that operate between citizen science actors, these being government, practitioners and volunteers. Learning from co-production literature, it is clear that participatory processes are embedded within unequal arrangements of power (Turnhout et al., 2020). Indeed, scholars have suggested that co-production processes, as opposed to being truly collaborative, are governance mechanisms that define who does and does not participate, what voices matter, and how decision-making may be changed as a result (Lemos et al., 2018). If we are to develop a better understanding of how the full potential of marine citizen science can be realised, it is vital that research critically examines the impact of government investment in projects and how this shapes their capacity to produce knowledge and to instigate action that challenges the *status quo*.

Drawing on primary research, this chapter illustrates how government involvement in citizen science is hindering the agency and transformative potential of projects. In contrast with citizen sciences' participatory and empowerment ethos, our findings reveal that government support for citizen science projects can be understood as a process of enhancing the control that government bodies have over co-production endeavours. Specifically, government investment can significantly limit the radical potential of citizen science knowledge production through a process of professionalisation. Government funding professionalises marine citizen science through the use of strict knowledge production standards, standardised participation practices, short-term funding timeframes, and recruitment models that target specific "types" of volunteers. This chapter illustrates how professionalising projects reduces the capacity of citizen science to empower local knowledge and to transform marine management into more democratic, inclusive, and adaptable processes. A professionalised form of citizen science is also seen to limit the array of participatory pathways that projects deploy and tends to result in the exclusion of volunteers who are motivated by activist desires to challenge prevailing marine management processes. However, the movement toward a professionalised model of citizen science is not solely driven by the government. There is evidence of some practitioners openly pushing to create more professional frameworks of participation, which can exclude certain groups of volunteers and limit their involvement. Moreover, a growing selection of volunteers are seen to be driven by motivations to enhance their professional research traits and skillsets as a means of enhancing career opportunities. Although the promise of citizen science has been built upon the active engagement that it supports, this chapter illustrates how projects are being increasingly moulded around professional scientific structures and principles that limit volunteers to contributory roles. Whilst this can be seen to enhance the policy-relevance of some citizen science and to regulate the way knowledge is generated, it is can also be considered as a form of governmentality that limits the transformative potential of participatory knowledge production.

The following section presents reviews the evolution of citizen science. This includes a discussion on the origins of citizen science, its defining characteristics, and its use within marine governance. Specific gaps in research are noted, particularly concerning the limited critical examination of government involvement with citizen science and the need to assess the emerging arrangements of power that this has initiated. The research findings are divided into two thematic sections. The first assesses the organisational dynamics of citizen science, revealing the impact of government involvement on the design and implementation of initiatives. This is followed by a section that exemplifies how a movement toward professionalisation is narrowing the scope of who can participate in citizen science. There is evidence of an increasingly limited pathway to participation in projects, with many volunteers becoming excluded from initiatives. Projects also appear to be unresponsive to the motivations and desired outcomes of some volunteers. The discussion section interprets what professionalisation mean for the field of citizen science, and co-production more broadly. Recommendations are proposed in relation to how marine citizen science can become more effective in how it co-produces knowledge, how it seeks to instigate transformative change to governance processes, and how it can recruit and retain a wide range of participants.

The evolution of citizen science

Broadly defined, citizen science is a means of participatory research, wherein members of the public voluntarily work with professional scientists, to produce scientific knowledge in a contributory, often collective, manner (Bonney et al., 2009). Citizen science has quickly grown in popularity within the field of marine conservation (Kelly et al., 2020), following a "participatory trend" in scientific knowledge production (Chilvers and Kearnes, 2020). The increased support for participatory research and co-production has created new relationships between civil society, science and government, wherein individuals can have a greater influence on decision-making processes (Albert et al., 2021). The origins of this participatory trend, and in particular the rise of marine citizen science, have been understood as a response to two key factors. First is the urgent need for data on environmental challenges (Bennett, 2016; Chase and Levine, 2016). Second is the growing desire for community participation in marine management (Pandya, 2012; Thompson, 2016). The limited scope of government monitoring programmes to efficiently deal with complex ecological challenges (Sharpe and Conrad, 2006), due to resource, time, and cognitive constraints (Conrad and Daoust, 2008; Vercammen and Burgman, 2019), has led researchers to frame citizen science as a cost-effective means of broadening the scale of data collection processes (Jarvis et al., 2015) and improving the knowledge base that informs conservation management (Jambeck and Johnsen, 2015; Steven et al., 2019).

The participatory nature of citizen science has encouraged scholars and practitioners to promote it as a means of instilling social and learning outcomes within volunteers. These can include enhanced environmental citizenship (McKinley et al., 2017), behavioural change and empowerment (Toomey and Domroese, 2013; Ruiz-Mallén et al., 2016), environmental stewardship (Merenlender et al., 2016), and increased environmental and scientific literacy (Haywood, 2016; Kelly et al., 2022; McAteer et al., 2021). The outcomes that volunteers obtain are dependent upon how they engage with citizen science. Engagement processes within citizen science projects can be divided into three categories – contributory, collaborative and co-produced – that involve various forms and degrees of participation. Contributory projects refer to activities where participants contribute to the collection of data, to enhance datasets (Dickinson et al., 2012). An example of this is the UK's Royal Society for the Protection of Bird's "Beached Bird Survey", where volunteers monitor the frequency of dead and sick birds along beaches and shorelines (McAteer et al., 2021). Collaborative and co-produced projects imply a deeper engagement of participants, whereby participation may also involve problem definition, data analysis and interpretation, and the dissemination of findings (Shirk et al., 2012). The Hudson River Estuary Eel Project is an example of a collaborative citizen science project. The project involves a catch and release monitoring programme that enables volunteers to take on data collection and analysis roles, whilst also facilitating participants with opportunities to co-design research objectives (Ballard et al., 2018). The Reclam the Bay scheme, where local volunteers worked to restore shellfish and maintain clean water in New Jersey's Barnegat Bay, is an example of a co-produced initiative (Bonney et al., 2009). Following one year of coordination by scientists, volunteers took full ownership of the scheme and have helped to initiate similar projects in other parts of North America.

It is the active and collaborative characteristics of citizen science that distinguish it from other forms of participatory research. It has overtaken many of its counterparts – including civic participation, deliberative governance, empowered participatory governance and collaborative policy dialogues – in terms of interactive knowledge co-production (Shirk et al., 2012). Instead of merely participating as subjects within research studies, citizen scientists play integral roles in the development and success of projects (Wiggins and Crowston, 2011). Citizen science supports a conceptualisation of science that can be responsive to the concerns of citizens and can legitimately engage them with scientific knowledge production. Due to this, citizen science is framed by academics and practitioners as a participatory approach that can, through the production of new knowledge, transform conservation management into more transparent, socially relevant, and democratic processes (Couvet and Prevot, 2015; Grossberndt et al., 2021; Loos et al., 2015; Peters and Besley, 2019). In marine governance, where decision-makers are often guided by hegemonic agendas (Tafon, 2018) and informed by the knowledge of dominant stakeholders (Said and Trouillet, 2020), citizen science has been suggested as a potentially transformational solution to unjust and undemocratic processes (McAteer and Flannery, 2022). Although transformation remains poorly defined in citizen science literature, this chapter interprets it as a fundamental form of change that is greater than progressive or incremental shifts. Transformation, in this sense, is a significant reordering, one that challenges existing structures to produce something fundamentally different (Blythe et al., 2018; Geels et al., 2017). Citizen science can provide many of the required conditions for marine governance transformation to be instigated, specifically because of the active participation that it supports and the diverse range of knowledge that it can produce (Bela et al., 2016).

More so than contributory projects, collaborative and co-produced categories are interpreted as being capable of using knowledge to instigate change within environmental management processes (Eitzel et al., 2017). However, as we have argued elsewhere, the three categories of citizen science should not be interpreted as mutually exclusive (McAteer and Flannery, 2022). Contributory, collaborative and co-produced projects can, and should, be interpreted as holding differing degrees of transformational potential. All citizen science projects operate within a unique arrangement of actors, each with differing degrees of power and influence over the evolvement and potential output of projects, each with a range of different opportunities and barriers to instigating transformation (Bela et al., 2016). Thus, if we are to improve our understanding of the wider capacity of citizen science to instigate positive change in marine governance, the power dynamics between participating actors must be examined. This involves decoding the relationship between volunteers, practitioners, and government actors, and must be complemented by assessments of how these arrangements are supporting or inhibiting the development of projects. Bringing power analysis to citizen science could provide practical and conceptual guidance on how the unique challenges that restrict the transformative output of all projects can be overcome.

Politics and power dynamics of citizen science

Although citizen science is underpinned by suggestions of mutuality and equality amongst actors (Chilvers and Kearnes, 2020), the co-production literature has demonstrated how elite actors can use their resources and knowledge to shape participatory efforts to serve their interests and needs (Parkinson, 2012). This is reflected in the work of Akaateba et al. (2018), who, in their study on land reform processes in Ghana, found that co-production can be a conduit for private wealth accumulation within the broader context of weak institutional capacities and poor governance. In such a context, those with power and resources were able to take advantage of co-production processes and undermine efforts to promote more equitable governance. Similar challenges are found in the context of a narwhal co-management programme in Nunavut, Canada. Collaborative attempts sought to revise policy assumptions that did not consider indigenous knowledge, yet compartmentalised views of knowledge from managements actors were seen to constrain the influence of co-management efforts (Dale and Armitage, 2011). Management systems continued to privilege professionally collected, scientific knowledge, and proved to be resistant to change when certain types of knowledge were undermined. Evidence of bias towards elite actors within co-production processes is worrying, as they illustrate how co-production processes can reproduce existing inequalities

(Parkinson, 2012). This form of critical analysis of power dynamics is often missing from the citizen science literature, making it difficult to accurately assess the barriers that prevent projects from realising their potential. Aligning with the argument of Turnhout et al. (2020), we assert that researchers have tended to assume that there is trust and symmetrical power relations within citizen science projects. Rather than following this naïve interpretation of the relationship between competing citizen science actors, we acknowledge the critical need to assess the role of power in the shaping and evolvement of initiatives.

Of the literature that does consider power in citizen science, it is framed as an attribute that some actors have, and others lack (Leach and Fairhead, 2002; Eitzel et al., 2017). This acknowledgement of uneven power relations between citizen science actors presents a simplistic notion of power, whereby elite actors can act unchallenged as they predetermine the problem framing or participatory scope of an initiative (Parkinson, 2012). This framing of power inequalities within citizen science is further compounded by the strong authority that is attributed to scientific expertise vis-a-vis other knowledge systems (Armitage, 2008). In the field of co-production, literature has positioned power as a constraining force that can limit the agency of less powerful actors, namely volunteers (Ottinger, 2010). Power, in this sense, is a force that is exerted in a repressive and constraining manner. This includes the implementation of standardised practices within co-production projects, whereby powerful actors set standards to dismiss knowledge that does not align with management assumptions or is interpreted as being irrelevant to the central project (Ottinger, 2010).

Despite some recognition of power within the citizen science literature, little attention has been paid to empirically assessing the workings of power. We argue that the prevailing perspectives of power within this literature are ontologically and epistemologically grounded in the natural sciences, wherein mere participation is viewed as a means of remedying power inequities. This has contributed to citizen science's inability to identify the deeply embedded structural and societal barriers that can inhibit transformative change. When operating in this depoliticised and conservative manner, the space for citizen science to challenge prevailing governance logics is restricted, and participatory initiatives, often unconsciously, do little more than reinforce the *status quo* of governance frameworks (Gaventa and Cornwall, 2001). For instance, Hampshire et al. (2005) illustrate how shifts and re-negotiations of power relations are possible during participatory research projects but are commonly limited due to the desire of elite actors to define and maintain dominant agendas. In their study, power is revealed as being manifested in various forms at different stages of participatory research projects, with the overall effect of changing the priorities of volunteers to align with those of elite actors (Hampshire et al., 2005). We argue that more explicit considerations of power, one that considers the opportunities and challenges to shifting imbalances, can inform a more accurate understanding of citizen science and the barriers that it must navigate to effectively contribute toward more sustainable and inclusive marine governance.

We contend that exploring the power and political dimensions of marine citizen science is vital to understanding the opportunities and barriers that impact a project's potential to contribute toward more sustainable and inclusive marine governance. This perception led to the creation of a range of research questions that were posed to volunteers, practitioners, and government actors, who are engaged with marine citizen science in the UK and Ireland. These questions investigated how projects are organised, the relationship between participating actors, the degree to which projects are becoming professionalised, the impacts of greater professionalism within citizen science, the way volunteers participate in projects, and the pathways to participation that are facilitated by initiatives. The responses of citizen science actors to these research questions are presented in the following sections. The findings are presented in a thematic manner. Following the presentation of this chapter's research findings, a discussion section is presented that critically interprets the meaning and importance of the responses.

Research findings

The findings presented in this chapter stem from two methods of data collection. First, an online survey, viewed as the most effective means of examining a large group of individuals, was circulated through eight citizen science projects in the UK and Ireland, the majority of which are in partnership with government bodies. The groups examined include Seasearch's Diving Group, the Irish Whale and Dolphin Group's Ferry Survey, Ulster Wildlife's Sea Deep project, Keep Northern Ireland Beautiful's Marine Litter Survey, the Royal Society for the Protection of Birds' Beached Bird Survey, Cloughey and District Community Association's Beach Care Group, Coastwatch Ireland's Coastal Survey, and the British Trust for Ornithology's Wetland Bird Survey. In total, the survey was sent to 737 citizen science volunteers, with 308 volunteers responding. This gives a response rate of 41.7%. The survey was designed to critically assesses volunteers' roles, motivations, outcomes, and experiences of participation. The findings, which were descriptively and thematically analysed, reveal how volunteers' perceptions of the purpose and potential of citizen science are dependent upon their degree of participation. They also highlight how there are different pathways to participation in citizen science, with evidence emerging from the findings suggesting that initiatives commonly support pathways that enable the engagement of volunteers who closely align with the objectives of the respective project. The findings of the survey are discussed throughout this chapter to illustrate how increasingly professionalised structures of participation can result in the active exclusion of specific types of volunteers. They also hint at how professionalisation is limiting the ability of projects to actively respond to the needs and concerns of volunteers.

This chapter is also informed by an analysis of semi-structured interviews (N=29) conducted with key citizen science actors. These actors involve practitioners (project coordinators and managers, NGO officers, and professional scientists) and government actors (government scientists, funding commissioners, planning officers). Collectively, these actors have insight into the internal functions of citizen science projects. Such insight can be the result of direct experience of managing and running initiatives, by financially commissioning them or by being an end-user of citizen science knowledge. Interviewees were drawn up through a process of mapping, ensuring that a variety of actors holding different roles in the realm of marine citizen science engaged with the study and provided insight. Interviewees were prompted to discuss issues relating to their perceptions of citizen science, their relationship with practitioners/government actors, their experience of how projects evolve, and how knowledge is used, produced, and has an impact. An inductive form of thematic analysis was used to analyse the collected interviews. This involved identifying common themes, topics, and patterns of meaning that came up repeatedly. The most common themes regarding the organisational dynamics and professionalisation of citizen science were analysed to reveal commonalities amongst interviewees. As an inductive approach was taken to the analysis of the interviews, the data determined the themes that emerged. This enabled theoretical aspects to develop from the data and to remain grounded in empirical observation.

The findings presented in this chapter reveal the perceptions of key actors in relation to the frameworks of power and politics that function throughout the realm of citizen science, demonstrating how these relations, in a practical manner, shape the operationalisation and potential capacity of projects. Analysis of interview transcripts and survey responses reveal how specific arrangements of power shape the development of citizen science projects, in both constraining and constructive manners. This section is broken down into two sub-sections, each covering a theme that was extracted inductively during the analysis process. To begin, critical discussions on the organisational dynamics of citizen science are presented. This section illuminates the flows and arrangements of power that underpin projects and demonstrates how the relationship between practitioners and government actors can become unbalanced. This section reveals important insight of how government actors can sway the design and scope of projects toward their own ends, by way of objective setting, the implementation of standardised practices and short-term timeframes. This is followed by a sub-section that reviews how professionalisation is contributing to the exclusion of certain types of volunteers. The findings reveal evidence of how volunteer capacity building commonly focuses on the mobilisation of participants who are motivated to develop their professional data collection capabilities. This appears to be to the detriment of other volunteers, specifically those who are driven by activist motivations and interpret their participation as an opportunity to instigate change to marine governance.

Organisational dynamics

Establishing a strong working relationship between practitioners and government actors is a crucial element of any citizen science project. As one departmental interviewee asserts,

> For a project to be successful and to achieve the goals that it sets out to achieve, it's really important that they work with us and vice-versa . . . good communication is key. That goes for when they are applying for funding, when we agree upon the ins and outs of the project, and when it comes to producing the proposed data output.
>
> *(Government manager, #2)*

This suggests that the potential success of a project is dependent upon a practitioner-government actor relationship that is open and responsive to change. It is also suggested that good relationships are crucial for the development of future initiatives. Practitioners state how "there isn't an endless array of funding streams out there, we work within a pretty limited pool. So, it's important to build good networks when possible and to then highlight to those bodies the unique value that citizen science can offer" (NGO officer, #7). This means that both practitioners and government actors must find acceptable balances between competing perceptions of what the objectives of projects should be, how data is to be produced, analysed and communicated, and what the role of volunteers should be. The potential for government actors to overpower project coordinators when moulding the scope of citizen science initiatives is noted by some practitioners. As one practitioner discussed, "Funding can bring certain stipulations, which I totally understand, but they don't always provide the kind of freedom necessary to construct an engaging project for the volunteers" (NGO officer, #6). Those involved with the management of citizen science projects reveal how funders often under-resource volunteer engagement processes and "fail to realise the importance of creating active participation structures that ensure volunteers feel significantly rewarded for their contribution" (Project coordinator, #11). The ability of government actors to sway the development of projects is, therefore, an evident concern for those involved with the coordination of citizen science initiatives. Practitioners allude to the need to make trade-offs when communicating with government actors, suggesting that it can be difficult to simultaneously facilitate scientific rigour, policy influence, and deep citizen engagement in initiatives.

To fully understand the challenges regarding the practitioner-government actor relationship, it is important to critically examine the practical ways through which government actors are seen to be manipulating the development of citizen science initiatives. When analysing the interviews findings, the setting of objectives emerges as one example. Objectives decide both the goals that projects seek to achieve and the research design that is to be constructed and operationalised. Practitioners discuss how the broader objective and remit of a project – for example, to monitor specific marine species, to measure ecosystem decline, or to assess the impact of plastic pollution upon marine ecology – is decided upon prior to the establishment of funding networks. However, the methodology and step-by-step research objectives of a project – for example, how knowledge is to be produced, what dataset or gaps such knowledge will contribute to, and what the practical tasks of volunteers will involve – are, predominantly, decided in collaboration with government actors.

In interviews, practitioners discuss how it is at this stage that government actors can begin to sway the development and operationalisation of initiatives. This is achieved by pushing for the creation of objectives that align with the underlying interests of government actors. These are seen to include "filling specific gaps of knowledge that can support a policy decision . . . or to add scientific evidence to either support or prevent a potential coastal development" (Project coordinator, #11). Due to the legal stipulations embedded within funding agreements, practitioners accept that "there is relatively little that can be done to prevent objectives being taken out of our hands. Attempting to refuse the proposed objectives of funders or to propose alternatives can risk funding falling through, as well as placing long-term support in a delicate position" (NGO officer, #7). Therefore, some practitioners interpret the objective setting stage as a key point in the development of projects. It is at this stage in which initiatives can be taken out of the control of coordinating organisations and become overwhelmingly guided by government.

This general theme of government actors coercing citizen science projects into setting policy-defined objectives, which reflect little emphasis upon the unique potential of projects as participatory initiatives, is noted by several practitioners. As one discusses, "It is a frustrating issue that often limits the scope of citizen science to really get engaged with volunteers" (NGO officer, #8). Indeed, the setting of objectives in this manner has been seen to harm the both the recruitment and retainment of volunteers. Practitioners speak of how

> we have lost a number of really good volunteers because of the creation of purely scientific objectives, and it is frustrating to see that. It can put off potential participants as well, because they won't join an initiative that others feel let down by or excluded from.
>
> *(Project coordinator, #11)*

Interviewees also reveal that the implementation of professional research standards and standardised practices by government actors can play a key role in shaping the potential output of projects. These include data standards, which practitioners appreciate the value of, that ensure scientific quality and to help to increase the trust placed in citizen science datasets. However, practitioners speak less positively about standardised frameworks that mould participation practices and regulate how, and which, volunteers carry out tasks. Standards, as one government actor suggests,

> help to even the terrain on which citizen scientists meet policy . . . Standardised practices help us to coordinate the work of citizen science projects and provide the means for distinguishing relevant and reliable data from the irrelevant and unreliable.
>
> *(Government manager, #1)*

This leads many practitioners to interpret standards as having a dual-role in shaping the scope of projects. A project coordinator discussed how, on one hand, standards serve a "bridging function that can enable our data to have a higher measure of legitimacy among policy and decision-makers" (Project coordinator, #11). At the same time, standards can serve a "policing function that allows those same policy-makers to dismiss some of our knowledge as irrelevant, without taking into consideration the intricacies of what local knowledge represents and how it can vary in composition and structure" (Project coordinator, #11). Such assertions highlight how professional standards can contribute to the establishment of expert authority, thus "policing" the boundaries of projects, and can push initiatives toward the production of knowledge that can seamlessly and efficiently link up with policy thinking.

A further theme revealed in interviews that has implications upon the organisational structure of citizen science projects relates to differing perspectives regarding the duration of initiatives. For practitioners, the relatively short-term nature of government funding was discussed as an impactful issue upon the creative scope of citizen science projects. Often, funding for a project is granted for between one and three years with no guarantee of follow-up financial support. As one practitioner discussed, this can mean that "the initial good work carried out by a project slows down or stops entirely. It can be a real waste because it takes time to get things running as we'd like and to get volunteers on board. Then, before you know it, we have to start looking for new funding streams or be forced to look at how we can transfer volunteers to other projects" (Project coordinator, #11). Expanding on this point, practitioners discuss how project objectives, such as, "empowering volunteers and establishing research protocols that can get the most out of them [volunteers] requires time and ongoing commitment from funders" (NGO officer, #7). However, securing long-term funding networks that evolve and respond to the dynamic needs of a citizen science project is revealed as being increasingly difficult. Government actors discuss the challenge of time-bound funding in similar depth, asserting that they themselves can be restricted by financial boundaries. This can force them to make decisions on which projects to support, and for how long, by considering the wider policy field that they are responsible for. As one government manager asserted:

> We cannot satisfy the funding requirements of all citizen science projects here. So, in our case, we have to think about what type of data do we need at the moment, be it fishing stock information or endangered species recordings or recordings on marine litter. Whatever the most pressing gap in policy is. Then we have a look at what citizen science projects are applying for funding, or who would like to be, and start discussing with them how we can work together.
>
> *(Government manager, #1)*

This suggests that government actors often seek out citizen science initiatives that align with their most pressing challenges. Funding agreements are then shaped around these challenges, meaning that projects are supported to fill specific gaps of knowledge within given time frames.

Participation pathways

Participation in citizen science can be defined on different levels. These levels are associated with volunteer tasks. The responses to survey questions regarding volunteers' degree of participation demonstrates how the roles carried out by citizen science participants are relatively restricted to data collection tasks. Indeed, the vast majority of volunteers (92.9%) stated that participation with their current project involved the collection of data, with a significantly lower number engaging with roles regarding the analysis (10.1%) or dissemination (9.4%) of data. Even fewer respondents (4.5%) took part in tasks linked to co-designing the structure of their current project. Other roles carried out by respondents included delivering training, organising logistical issues, and liaising with local stakeholders. Added to this, 4.9% of respondents noted that their role involved taking part in project meetings arranged with local councillors and government departments. This suggests that lobbying against decision-makers reflects an important aspect of citizen science participation, albeit for a small number of volunteers. In total, the survey findings suggest that a high majority of citizen science volunteers are participating in contributory manners. Volunteers assist with the creation of data on behalf of pre-established and pre-designed programmes. Thus, participation is, predominantly, concerned with filling gaps of knowledge and answering specific, predetermined research questions.

Added to personal and environmental motivations for participation, many respondents stated that community-based drivers played a key role in attracting them to engage with citizen science. Primarily, these drivers are linked to "putting something back into the community . . . to improve our own environment" (Female, 65+, retired). In turn, these desires supported motivations to enhance the role of local knowledge in the realm of decision-making and to improve the manner in which communities can engage with political actors. The findings of the survey suggest that community concerns are of critical importance for a number of volunteers. For example, 59.7% of respondents were in strong agreement that not enough attention is currently paid to local environmental issues, with only 15.6% of respondents in disagreement or feeling indifferent to the notion. Added to this, a total of 89.9% of participants agreed that the production of local knowledge is an important way of engaging with local and national government. Interpreting this, there is an evident belief amongst many respondents that citizen science can collectively play a role in bringing local environmental issues to the attention of decision-makers. In this light, volunteers perceive citizen science as a platform that can reconfigure the degree of importance placed upon local environmental concerns and lead to the development of more responsive relationships between communities and government. However, the degree to which these volunteers are supported by projects to realise these desires, as well as how projects are focusing on community concerns, is not evident. Only one half of all respondents (51.3%) felt that their initial motivations for participating were realised, meaning that many volunteers felt unfulfilled in relation to what they had hoped to achieve. As one respondent summed up,

> I feel empowered by participating in active research like this, but I don't feel that the knowledge that I put forward was truly empowered. I don't think that it has influenced the council's thoughts about the rate of erosion here, certainly not to the degree I had hoped.
>
> *(Male, 35–44, teacher)*

A key outcome of participation, as revealed in the survey findings, regards the manner in which volunteers feel that their co-produced knowledge can most effectively create change to systems of environmental management. This presents an intriguing insight of the mentality of respondents and, in particular, their understanding of the full potential of citizen science. Over half of the respondents (51.3%) feel that lobbying government is the most effective use of knowledge produced through citizen science. This suggests that a large quantity of participants interpret knowledge as a powerful means of challenging the conceptualisations held by decision-makers. In turn, this can lead to policy or management changes that more accurately represent the concerns of citizens. Knowledge, in this sense, is a form of capital that carries power and can influence the actions of decision-makers. On the contrary, slightly under one third of volunteers (30.8%) argued that the knowledge that citizen science produces, if it is to transform the approach of management systems, should act as a means of raising the public awareness of environmental issues. This interpretation suggests that citizen science knowledge, in and of itself, is not enough to change governance regimes. Rather, there is a need to engage with wider society and enhance the environmental education of members of the public. Further still, a small group of respondents (6.5%) felt that the production of citizen science knowledge is most effective when it educates project volunteers. In total, there are clear patterns amongst large proportions of citizen volunteers. Their desire to instigate transformative change through their participation and co-production of knowledge is clear. However, the manner in which increasingly professionalised projects are capable of factoring these perceptions into their objectives and structure of implementation is less clear. Volunteers speak of how "there needs to be a better mechanism to ensure our feedback shapes future programmes" (Female, 65+, retired) and "I don't see how we can change policy when we produce findings in such a one-dimensional way . . . I think a more holistic approach is needed" (Female, 35–44, academic). These responses hint that some volunteers feel unsupported in their participation and that projects are failing to actively respond to their requirements. Such a scenario risks damaging the desire and capacity of volunteers to maximise their potential contribution.

Discussion

This chapter illustrates a process of professionalisation within citizen science initiatives, which covertly empowers government agencies to shape projects and limit their transformative potential. Professionalism is a process whereby citizen science projects, by being increasingly pushed by government actors to follow strict scientific standards and principles, are operating less like truly participatory efforts and more like an extension of government scientific endeavours that merely draws on volunteer labour. This process of professionalisation is interpreted as part of a wider process of narrowing the scope of what citizen science is and what it can achieve. Interview findings reveal how there remains critical differences between the interpretations of practitioners and government actors regarding the role and value of citizen science. Whilst practitioners acknowledge a wide scope of aspects linked to citizen science, from producing knowledge, instilling learning, and social outcomes within volunteers, as well as the potential to encourage increases in the stewardship and civic participation of volunteers, government actors carry a more refined perception of the practice. In the eyes of some practitioners, citizen science is almost exclusively valued by government as a cost-effect means of contributing environmental knowledge, with learning and social outcomes viewed as additional or unintended consequences. This difference of interpretation is crucial, particularly when considering that government actors can have a much greater capacity than practitioners to push their agenda and objectives onto projects. Perhaps, most crucially, a narrowed version of citizen science, where the focus is on supporting projects to produce specific types of knowledge in a professional manner, will drastically cut the transformative potential of the practice. With projects implementing rigid participation processes in line with professional standards of data collection and analysis, the scope for social learning and transformation, as well as the production of alternative forms of knowing, is significantly lessened.

This process of narrowing the scope of citizen science also reveals a broader example of how unequal balances of power can shape the operations of citizen science. Although the granting of government support for citizen science projects may hint at the development of partnerships between coordinating organisations and their volunteers, in reality, it signals the deepening of unequal balances between such actors and the government. As interview findings exemplify, this is achieved through the implementation of strict project objectives, standardised practices and, more generally, a professionalisation of initiatives. It is important to be aware of this reality and to assess how it can be challenged. The professionalisation of citizen science means that projects can, paradoxically, end up reproducing, rather than mitigating, existing unequal arrangements of power that they may originally intend to change. At the basis of this process lies a strong tendency within both literature and practice to depoliticise citizen science. As has been argued within the literature review of this chapter, by failing to address the political and power dimensions of citizen science, the practice is at continued risk of reinforcing and strengthening traditional and professional modes of knowledge co-production and participation. Indeed, the professionalisation of projects can be interpreted as a process of removing the "citizen" dynamic from citizen science.

The findings presented in this chapter support the assertions made by scholars within the wider field of volunteerism, who note the evidence of powerful actors attempting to professionalise volunteer initiatives so that policy and management objectives can be better served. Groninger (2011) reveals how a movement to professionalise volunteers through training, management and oversight, has concerned many volunteer-run organisations in recent years. This movement has been seen to challenge volunteer culture and ideology, with the requirement of fulfilling formal expectations, some that are similar to the tasks that are asked of paid employees, pushing many volunteers away from maintaining their engagement (Hyndman and McDonnell, 2009). Whilst volunteers are beginning to be facilitated with the opportunity to reach new heights regarding their knowledge contribution and output, stemming from the opportunity to match the best practice of government funders with voluntary sector values, we provide further evidence of a growing tension between voluntary tradition and the increasing expectancy and scrutiny to which volunteers are subject to. This chapter reflects on the work of Groninger (2011) whilst exemplifying how volunteers in co-production initiatives can be discouraged, and potentially excluded, as a result of the utilisation of professional practice within projects. While the professionalisation of co-production endeavours can help to enhance the validity of citizen science data, it can also limit the agency of volunteers. We also reveal insight of how the increased use of citizen labour in participatory research is regularly resulting in the creation of mundane methodologies. As Hemment et al. (2011) have discussed, such scenarios are resulting in volunteers carrying out the role of "data drones". The array of professional expectations that can be embedded within co-production projects, particularly when government bodies are involved with the co-design of initiatives, appears to be a significant factor that influences how projects develop, how volunteers are utilised, and the types of knowledge that projects can be produced.

The impact of an increasingly professionalised citizen science should not be understood as a purely top-down, government inflicted issue. In interviews with practitioners, specific pathways to participation appear to be supported, yet not all volunteers align with these pathways. This means that some volunteers can be, effectively, excluded or discouraged from participating in citizen science, should their interests and desired outcomes of participation fall beyond the intentions of practitioners. It is crucial that this trend is challenged and that all types of volunteers are facilitated with a pathway to fully engage and contribute to citizen science. Should projects reach their potential and successfully contribute to the sustainable management of marine environments, it is critical that volunteers are encouraged to maximise their potential contribution and are provided with the necessary support to do so. By paying greater attention to the needs and requirements of volunteers, as well as the aspects of participation and types of participation that they are most interested in, we suggest that citizen science can develop more efficient and inclusive recruitment mechanisms.

The issue of how professionalisation is limiting the scope of who can participate in citizen science is a particularly important finding, especially as doubts regarding the ability of initiatives to avoid participant bias (Gonsamo and D'Odorico, 2014) and to fully engage with marginalised individuals (Walajahi, 2019) remain significant challenges. Such limitations have already been suggested as factors that make it difficult for citizen science to challenge social inequalities (Bela et al., 2016), and we call on practitioners, as well as government actors, to consider how projects can become more inclusive and support a more diverse range of routes to participation. Volunteers who are driven by ideals of activisms, for instance, could be given greater support by practitioners to play their part in improving the capacity of projects to challenge the status quo of marine governance and, through the co-production of alternative knowledge and participation in knowledge dissemination and lobbying tasks, outline the case for change. With over half of the survey respondents suggesting that lobbying government is the most effective use of knowledge produced through citizen science, there is a clear appetite amongst volunteers to mobilise their knowledge to challenge cases of environmental injustice.

Conclusion

Due to the active forms of participation and knowledge co-production that it facilitates, citizen science has a strong potential to benefit the management of the marine environment. As a transdisciplinary method of knowledge creation, it offers the potential to collect valuable data, over large scales, that is simply unattainable through traditional methods of research. Beyond the production of knowledge, however, it also presents the unique capacity to actively engage amateurs in scientific studies. This can enhance the scientific and environmental literacy of volunteers, whilst also enabling citizens, who may otherwise feel marginalised or excluded from governance processes, the opportunity to influence management agendas and voice their opinions about how the marine environment should be sustainably conserved. Despite this potential, the capacity of citizen science to help volunteers to realise these opportunities is becoming increasingly constrained by the professionalisation of projects. The findings presented in this chapter illustrate how professionalisation is pushing projects to focus on contributory research objectives, the generation of knowledge that aligns with pre-determined policy issues, and the engagement of specific types of volunteers. This process, although not unique to citizen science, reveals how participatory endeavours, when funded by government bodies, can be taken out of the control of volunteers and practitioners.

The shift toward professionalisation is harming the very thing that scholars believe provides citizen science its uniqueness: active and open participation. We argue that it is possible to challenge this movement toward professionalisation and, by attempting to reframe the arrangements of power between citizen science actors, practitioners and volunteers can instigate change to how projects operate within marine governance processes. By becoming power-aware and conscious of the barriers to transformation that exist, citizen science projects can develop new approaches and grow to be more efficient and effective in their attempt to actively tackle governance problems. In a practical sense, establishing networks of projects, sourcing collective funding streams, and creating evaluative feedback loops that can allow practitioners and government actors to learn from volunteers are some potential recommendations that can help the continued evolution of citizen science. Furthermore, there is an evident need to better comprehend the mechanisms that have realised successful transformation in citizen science, and similar participatory approaches, so that those processes can be connected to, or embedded within, broader processes of societal transformations. Knowledge co-production should not simply be about doing management better, but about revising assumptions and challenging historic injustice. If we endorse a transformative stance on sustainability, marine citizen science cannot just be about sending around human sensors to collect data. It should serve the purpose of creating the cognitive and emotional preconditions to bring about radical change.

References

Akaateba, M. A., Huang, H., & Adumpo, E. A. (2018). Between co-production and institutional hybridity in land delivery: Insights from local planning practice in peri-urban Tamale, Ghana. *Land Use Policy*, 72, 215–226.

Albert, A., Balázs, B., Butkevičienė, E., Mayer, K., & Perelló, J. (2021). Citizen social science: New and established approaches to participation in social research. Chapter 7. In: Vohland, K. et al. (Eds.), *The Science of Citizen Science*. Cham: Springer Nature, pp. 119–138.

Armitage, D. (2008). Governance and the commons in a multi-level world. *International Journal of the Commons*, 2(1), 7–32.

Ballard, H., Phillips, T., & Robinson, L. (2018). Conservation outcomes of citizen science. In: Hecker, S. et al. (Eds.), *Citizen Science – Innovation in Open Science, Society and Policy*. London: UCL Press, pp. 254–268.

Bela, G., Peltola, T., Young, J. C., Balázs, B., Arpin, I., Pataki, G., . . . & Bonn, A. (2016). Learning and the transformative potential of citizen science. *Conservation Biology*, 30(5), 990–999.

Bennett, N.J. (2016). Using perceptions as evidence to improve conservation and environmental management. *Conservation Biology*, 30(3), 582–592.

Blythe, J., Silver, J., Evans, L., Armitage, D., Bennett, N. J., Moore, M. L., . . . & Brown, K. (2018). The dark side of transformation: Latent risks in contemporary sustainability discourse. *Antipode*, 50(5), 1206–1223.

Bonney, R., Cooper, C. B., Dickinson, J., Kelling, S., Phillips, T., Rosenberg, K. V., & Shirk, J. (2009). Citizen science: A developing tool for expanding science knowledge and scientific literacy. *BioScience*, 59(11), 977–984.

Chase, S. K., & Levine, A. (2016). A framework for evaluating and designing citizen science programs for natural resources monitoring. *Conservation Biology*, 30(3), 456–466.

Chilvers, J., & Kearnes, M. (2020). Remaking participation in science and democracy. *Science, Technology, & Human Values*, 45(3), 347–380.

Conrad, C. T., & Daoust, T. (2008). Community-based monitoring frameworks: Increasing the effectiveness of environmental stewardship. *Environmental Management*, 41(3), 358–366.

Couvet, D., & Prevot, A. C. (2015). Citizen-science programs: Towards transformative biodiversity governance. *Environmental Development*, 13, 39–45.

Dale, A., & Armitage, D. (2011). Marine mammal co-management in Canada's Arctic: Knowledge co-production for learning and adaptive capacity. *Marine Policy*, 35(4), 440–449.

Dickinson, J. L., Shirk, J., Bonter, D., Bonney, R., Crain, R. L., Martin, J., . . . & Purcell, K. (2012). The current state of citizen science as a tool for ecological research and public engagement. *Frontiers in Ecology and the Environment*, 10(6), 291–297.

Eitzel, M. V., Cappadonna, J. L., Santos-Lang, C., Duerr, R. E., Virapongse, A., West, S. E., . . . & Jiang, Q. (2017). Citizen science terminology matters: Exploring key terms. *Citizen Science: Theory and Practice*, 2(1), 1–20.

Garcia-Soto, C., Seys, J. J., Zielinski, O., Busch, J. A., Luna, S. I., Baez, J. C., . . . & Gorsky, G. (2021). Marine Citizen Science: Current state in Europe and new technological developments. *Frontiers in Marine Science*, 8, 621472.

Gaventa, J. & Cornwall, A. (2001). Power and knowledge. In: Reason, P. et al. (Eds), *Handbook of Action Research: Participative Inquiry and Practice*. London: Sage Publications, pp. 70–80.

Geels, F. W., Sovacool, B. K., Schwanen, T., & Sorrell, S. (2017). Sociotechnical transitions for deep decarbonization. *Science*, 357(6357), 1242–1244.

Gonsamo, A., & D'Odorico, P. (2014). Citizen science: Best practices to remove observer bias in trend analysis. *International Journal of Biometeorology*, 58(10), 2159–2163.

Groninger, K. (2011). Volunteering and professionalisation in UK museums. *North Street Review: Arts and Visual Culture*, 15, 23–29.

Grossberndt, S., Passani, A., Di Lisio, G., Janssen, A., & Castell, N. (2021). Transformative potential and learning outcomes of air quality citizen science projects in high schools using low-cost sensors. *Atmosphere*, 12(6), 736.

Hampshire, K., Hills, E., & Iqbal, N. (2005). Power relations in participatory research and community development: A case study from northern England. *Human Organization*, 64(4), 340–349.

Haywood, B. K. (2016). Beyond data points and research contributions: The personal meaning and value associated with public participation in scientific research. *International Journal of Science Education, Part B*, 6(3), 239–262.

Hemment, D., Ellis, R., & Wynne, B. (2011). Participatory mass observation and citizen science. *Leonardo*, 44(1), 62–63.

Hyndman, N., & McDonnell, P. (2009). Governance and charities: An exploration of key themes and the development of a research agenda. *Financial Accountability & Management*, 25(1), 5–31.

Jambeck, J. R., & Johnsen, K. (2015). Citizen-based litter and marine debris data collection and mapping. *Computing in Science & Engineering*, 17(4), 20–26.

Jarvis, R. M., Breen, B. B., Krägeloh, C. U., & Billington, D. R. (2015). Citizen science and the power of public participation in marine spatial planning. *Marine Policy*, 57, 21–26.

Kelly, R., Evans, K., Alexander, K., Bettiol, S., Corney, S., Cullen-Knox, C., . . . & Pecl, G. T. (2022). Connecting to the oceans: Supporting ocean literacy and public engagement. Reviews in Fish Biology and Fisheries, 32(1), 123–143.

Kelly, R., Fleming, A., Pecl, G. T., von Gönner, J., & Bonn, A. (2020). Citizen science and marine conservation: A global review. *Philosophical Transactions of the Royal Society B*, 375(1814), 20190461.

Leach, M., & Fairhead, J. (2002). Manners of contestation: "citizen science" and "indigenous knowledge" in West Africa and the Caribbean. *International Social Science Journal*, 54(173), 299–311.

Lemos, M. C., Arnott, J. C., Ardoin, N. M., Baja, K., Bednarek, A. T., Dewulf, A., . . . & Wyborn, C. (2018). To co-produce or not to co-produce. *Nature Sustainability*, 1(12), 722–724.

Loos, J., Horcea-Milcu, A. I., Kirkland, P., Hartel, T., Osváth-Ferencz, M., & Fischer, J. (2015). Challenges for biodiversity monitoring using citizen science in transitioning social – ecological systems. *Journal for Nature Conservation*, 26, 45–48.

McAteer, B., & Flannery, W. (2022). Power, knowledge and the transformative potential of marine community science. *Ocean & Coastal Management*, 218, 106036.

McAteer, B., Flannery, W., & Murtagh, B. (2021). Linking the motivations and outcomes of volunteers to understand participation in marine community science. *Marine Policy*, 124, 104375.

McKinley, D. C., Miller-Rushing, A. J., Ballard, H. L., Bonney, R., Brown, H., Cook-Patton, S. C., . . . & Soukup, M. A. (2017). Citizen science can improve conservation science, natural resource management, and environmental protection. *Biological Conservation*, 208, 15–28.

Merenlender, A. M., Crall, A. W., Drill, S., Prysby, M., & Ballard, H. (2016). Evaluating environmental education, citizen science, and stewardship through naturalist programs. *Conservation Biology*, 30(6), 1255–1265.

Ottinger, G. (2010). Buckets of resistance: Standards and the effectiveness of citizen science. *Science, Technology, & Human Values*, 35(2), 244–270.

Pandya, R. E. (2012). A framework for engaging diverse communities in citizen science in the US. *Frontiers in Ecology and the Environment*, 10(6), 314–317.

Parkinson, J. (2012). Democratizing deliberative systems. In: Parkinson, J. et al. (Eds.), *Deliberative Systems: Deliberative Democracy at the Large Scale*. Cambridge: Cambridge University Press, pp. 151–172.

Peters, M. A., & Besley, T. (2019). Citizen science and post-normal science in a post-truth era: Democratising knowledge; socialising responsibility. *Educational Philosophy and Theory*, 51(13), 1293–1303.

Ruiz-Mallén, I., Riboli-Sasco, L., Ribrault, C., Heras, M., Laguna, D., & Perié, L. (2016). Citizen science: Toward transformative learning. *Science Communication*, 38(4), 523–534.

Said, A., & Trouillet, B. (2020). Bringing 'Deep knowledge'of fisheries into marine spatial planning. *Maritime Studies*, 19(3), 347–357.

Sandahl, A., & Tøttrup, A. P. (2020). Marine citizen science: Recent developments and future recommendations. *Citizen Science: Theory and Practice*, 5(1), 1–11.

Schewe, R. L., Hoffman, D., Witt, J., Shoup, B., & Freeman, M. (2020). Citizen-science and participatory research as a means to improve stakeholder engagement in resource management: A case study of Vietnamese American Fishers on the US Gulf Coast. Environmental Management, 65(1), 74–87.

Sharpe, A., & Conrad, C. (2006). Community based ecological monitoring in Nova Scotia: Challenges and opportunities. *Environmental Monitoring and Assessment*, 113(1), 395–409.

Shirk, J. L., Ballard, H. L., Wilderman, C. C., Phillips, T., Wiggins, A., Jordan, R., . . . & Bonney, R. (2012). Public participation in scientific research: A framework for deliberate design. *Ecology and Society*, 17(2), 29–48.

Steven, R., Barnes, M., Garnett, S. T., Garrard, G., O'Connor, J., Oliver, J. L., . . . & Fuller, R. A. (2019). Aligning citizen science with best practice: Threatened species conservation in Australia. *Conservation Science and Practice*, 1(10), e100.

Tafon, R. V. (2018). Taking power to sea: Towards a post-structuralist discourse theoretical critique of marine spatial planning. *Environment and Planning C: Politics and Space*, 36(2), 258–273.

Thompson, M. M. (2016). Upside-down GIS: The future of citizen science and community participation. *The Cartographic Journal*, 53(4), 326–334.

Toomey, A. H., & Domroese, M. C. (2013). Can citizen science lead to positive conservation attitudes and behaviors? *Human Ecology Review*, 20(1), 50–62.

Turbé, A., Barba, J., Pelacho, M., Mugdal, S., Robinson, L. D., Serrano-Sanz, F., . . . & Schade, S. (2019). Understanding the citizen science landscape for European environmental policy: An assessment and recommendations. *Citizen Science: Theory and Practice*, 4(1), 1–16.

Turnhout, E., Metze, T., Wyborn, C., Klenk, N., & Louder, E. (2020). The politics of co-production: Participation, power, and transformation. *Current Opinion in Environmental Sustainability*, 42, 15–21.

Turrini, T., Dörler, D., Richter, A., Heigl, F., & Bonn, A. (2018). The threefold potential of environmental citizen science-generating knowledge, creating learning opportunities and enabling civic participation. *Biological Conservation*, 225, 176–186.

Vercammen, A., & Burgman, M. (2019). Untapped potential of collective intelligence in conservation and environmental decision making. *Conservation Biology*, 33(6), 1247–1255.

Walajahi, H. (2019). Engaging the "Citizen" in citizen science: Who's actually included?. The *American Journal of Bioethics*, 19(8), 31–33.

Wiggins, A., & Crowston, K. (2011). From conservation to crowdsourcing: A typology of citizen science. In *2011 44th Hawaii international conference on system sciences* (pp. 1–10). IEEE.

7

THE POWER AND PRECARITY OF KNOWLEDGE CO-PRODUCTION

A case study of SakKijânginnaniattut Nunatsiavut Sivunitsangit (the Sustainable Nunatsiavut Futures Project)

Michael A. Petriello, Melanie Zurba, Jörn O. Schmidt, Katrina Anthony, Nathan Jacque, Caroline Nochasak, Jacqueline Winters, John Winters, Megan Bailey, Eric C. J. Oliver, Paul McCarney, Breanna Bishop, Hekia Bodwitch, Rachael Cadman, and Megan McLaren

Introduction

Climate change and biodiversity loss are the most urgent and dire ecological threats to coastal and marine ecosystems and peoples worldwide (IPCC, 2022). The urgency of these threats is amplified by the reality that nearly three-quarters of the global population lives within 50 km of a coastline (Small and Nicholls, 2003) that is likely experiencing high human pressures and low levels of protection (Williams et al., 2021). Moreover, the magnitude of these threats and consequent societal responses are directly mediated by the social and cultural contexts in which they occur. For example, the risks posed by the climate and biodiversity crises are significantly pronounced in the Arctic, where temperatures are warming at the fastest rates globally due to Arctic amplification (Bekryaev et al., 2010). This, in turn, creates rapid and unpredictable shifts in wildlife populations and changes to sea- and lake-ice conditions that directly alter the ability of Indigenous Peoples and local communities to sustain their coastal cultures and livelihoods (Ford et al., 2012; Hauser et al., 2021; Hinzman et al., 2005; Yletyinen, 2019). Moreover, legacies and ongoing expressions of colonialism tend to privilege Western Scientific Knowledge (WSK) in decision making and policy formation, often overshadowing the rich and empowering history of Inuit Knowledge (IK). Therefore, there is a need for co-developing solutions to climate change with Indigenous communities in the Arctic, such as Labrador Inuit of Nunatsiavut, including mobilizing IK into decision making alongside WSK (Ford et al., 2012; Hirsch et al., 2016; Inuit Tapiriit Kanatami (ITK), 2018a). Climatic changes, long-term cultural shifts and

DOI: 10.4324/9781003311171-9

replacement, and histories of oppression and exclusion in Inuit Nunangat (the Inuit homeland within Canada, which means "the place where Inuit live" in Inuttitut) demand new research methods and frameworks that match the pace of climate change, recognize and redress colonial pasts, and bring together IK and WSK to navigate the varied cultural and institutional landscapes that mediate coastal and marine conservation in Inuit Nunangat (Alexander et al., 2019; Zurba et al., 2022).

The mutual interdependence of marine ecosystems and human societies indicates that innovative and *collaborative* approaches that bridge institutions, social spheres, and knowledge systems are required to achieve sustainable seas in the face of climate change. *Collaboration* among diverse actors at local to global scales is required to support innovative approaches to marine and coastal conservation and management (Hidalgo et al., 2022; Mazor et al., 2013). In its most fundamental form, collaboration refers to people and organizations coming together to solve a common problem. Given the high concentration of the human population that depends on oceans that cover over 70% of the Earth's surface, participants in collaborative research and management in the marine sciences must span multiple scales, institutions, and worldviews such as nation states, industries, academics, and local and Indigenous communities and their knowledges (Hind et al., 2015; Thornton and Scheer, 2012). Specifically, collaborative engagement with Indigenous peoples and local communities helps contextualize environmental concerns, enhance self-determination in research and governance, strengthen the effectiveness of environmental monitoring and management, and empower communities through equitable conservation (Dawson et al., 2021; *Ellam Yua* et al., 2022; *M'sit No'kmaq* et al., 2021). Moreover, collaboration facilitates the cross-fertilization of worldviews, values, and knowledges across academic disciplines, which allows researchers to leverage their disciplinary strengths towards addressing large-scale and complex global sustainability problems through integrative techniques, including those that account for the emotional and relational factors undergirding research (Pohl et al., 2021).

Yet collaboration embodies a concept that is neither simple nor uniform. The scope and success of collaborative approaches vary based on contextual factors such as levels of participation, governance arrangements, power dynamics, and historical legacies of resource use and cultural recognition (Adger et al., 2005; Alvarado et al., 2020; Armitage et al., 2011; Hill et al., 2012). Moreover, collaboration alone may not guarantee the effective integration, exchange, and recognition of different forms of knowledge required for successful conservation and management, adaptive governance, and policy innovation (Cvitanovic et al., 2015). For these reasons, collaboration is often one component of broader approaches to knowledge exchange and generation. Scholars and practitioners are therefore increasingly recommending and ground-truthing knowledge-related concepts and processes through reviews, frameworks, and case studies to fortify collaborative approaches in support of transformative solutions to social and environmental problems (Apetrei et al., 2021; Armitage et al., 2011; Cvitanovic et al., 2015; *Ellam Yua* et al., 2022). Of the many knowledge-related concepts guiding collaborations, knowledge co-production (KCP) is one of the most invoked and explored frames for engaging and bridging different forms of knowledge in marine science, climate research, and sustainability research (Apetrei et al., 2021; Wyborn et al., 2019).

The purpose of this chapter is to critically reflect on KCP through a qualitative case study drawing from nearly two years of our shared experiences as collaborators in the Knowledge Co-Production and Transdisciplinary Approaches for Sustainable Nunatsiavut Futures project (shortened to *SakKijânginnaniattut Nunatsiavut Sivunitsangit* in Inuttitut, *Sustainable Nunatsiavut Futures* Project in English), a six-year multi-partner research programme focused on KCP for marine spatial planning with and for Nunatsiavut communities, Newfoundland and Labrador, Canada. We specifically turn our collective gaze towards the power and precarity of KCP, given both the potential benefits from its breadth of interpretations, concepts, and approaches and the uncertainties they may introduce to collaborations (see "Primer on KCP" below; *Ellam Yua* et al., 2022; Zurba et al., 2022). First, we will present the project background with a positionality statement about our relationships to the research. Second, we will briefly overview the many interpretations of KCP as an approach to transdisciplinary research and practice on the transformative road towards sustainable coexistence between the sea and society. Third, we will summarize and interrogate the origins, approaches, and interpretations of KCP in the project. In particular, we highlight how iterative reflection, critical analysis, relationship building, and trust *during* our KCP process, rather than solely before or after the project, allows partners to question and reassert the very foundations of their work, including the meaning and value of KCP itself. This chapter problematizes KCP to inform what collaboration means in a KCP context and how an emphasis on relationships and emotions can help build a foundation for future work in the Sustainable Nunatsiavut Futures Project and other co-production efforts.

The SakKijânginnaniattut Nunatsiavut Sivunitsangit project background

The region of Nunatsiavut ("Our beautiful land" in English) is 72,250 km^2 (27,896 mi^2) of land and freshwater, with an additional 48,690 km^2 (18,800 mi^2) of tidal waters, that make up the Labrador Inuit Settlement Area (LISA) (Labrador Inuit Land Claims Agreement (LILCA), 2005). The area is one of four Inuit regions in Inuit Nunangat, a term that encompasses the land, water, and ice at the foundation of Inuit culture in Canada (ITK, 2018b; LILCA, 2005). Within the LISA, Labrador Inuit have control over health, natural resources, education, culture, and development in 15,799 km^2 (6,100 mi^2) of land and water, known as Labrador Inuit Lands (LIL). Nunatsiavut is home to an estimated 2,560 residents (Statistics Canada, 2018) who live in five coastal communities (Nain, Hopedale, Postville, Makkovik, and Rigolet) dotting subarctic and post-glacial landscapes with complex and rocky coastlines, including bays, headlands, deep fjords, inland seas, offshore islands, and islets. English is the dominant language; 14% of the population speaks Inuttitut (Statistics Canada, 2018). A more comprehensive background is covered in a recent review (Zurba et al., 2022).

The seeds of the *SakKijânginnaniattut Nunatsiavut Sivunitsangit* project were germinated during a two-day workshop in summer 2019. The purpose of the workshop, which was co-organized by the Nunatsiavut Government (NG), Dalhousie University, and Memorial University of Newfoundland and Labrador, and attended by academics, government representatives, and non-government actors, was to co-develop a proposal for collaborative, community-engaged research that co-produced knowledge about Nunatsiavut's changing coastal ecosystem dynamics. The resulting SakKijânginnaniattut Nunatsiavut Sivunitsangit project focuses on three research themes: understanding environmental changes to the Nunatsiavut coasts, co-developing management planning efforts to confront and cope with these changes, and identifying and assessing KCP processes and outcomes within the project.

The three themes are pursued by four work packages (WPs) that are tasked with different yet interrelated responsibilities in the work system. Work Package 1 (WP 1) focuses on the development and evaluation of KCP processes and outcomes; WP 2 explores community-engaged ocean monitoring for the Nunatsiavut coast; WP 3 spatially maps Nunatsiavut coasts; and WP 4 analyses shifts in species distributions across the coasts. However, the project is pursuing changes to move beyond the current work package structure as of May 2022. Currently, the project brings together over 50 collaborators representing more than 18 partner organizations, including 4 Inuit Research Coordinators (IRCs). Please see our review (Zurba et al., 2022) for a more detailed project background statement.

Positionality

A note on names

Throughout this chapter, we refer to the project by its shortened formal title in Inuttitut, SakKijânginnaniattut Nunatsiavut Sivunitsangit, rather than the English translation "Sustainable Nunatsiavut Futures" or an initialism of the Inuttitut ("SNS") or English versions of the name ("SNF"). This purposeful choice stems from internal conversations about the overuse of abbreviations, initialisms and acronyms in science, which often aim to distil complex topics, ideas, and titles down to several letters for ease of use across the larger scientific community. More importantly, we aspire to recognize and respect the geographic, social, and historical context of our collective experience and attend to the inherently political nature of doing research in and with Indigenous communities. Critically, we are in the process of evaluating a new name for the project in either language. However, at the moment the use of "SNS" or "SNF" in our work context risks diminishing the purpose, place, and intent of our collective endeavour. The initialism reduces our focus on SakKijânginnaniattut (Sustainability) down to an "S", boils its place-based importance in "Nunatsiavut" down to an "N", and confines the temporal focus on Sivunitsangit (Futures) to a one-dimensional "S" or "F". Further, Inuit project partners have explicitly highlighted the importance of all project partners

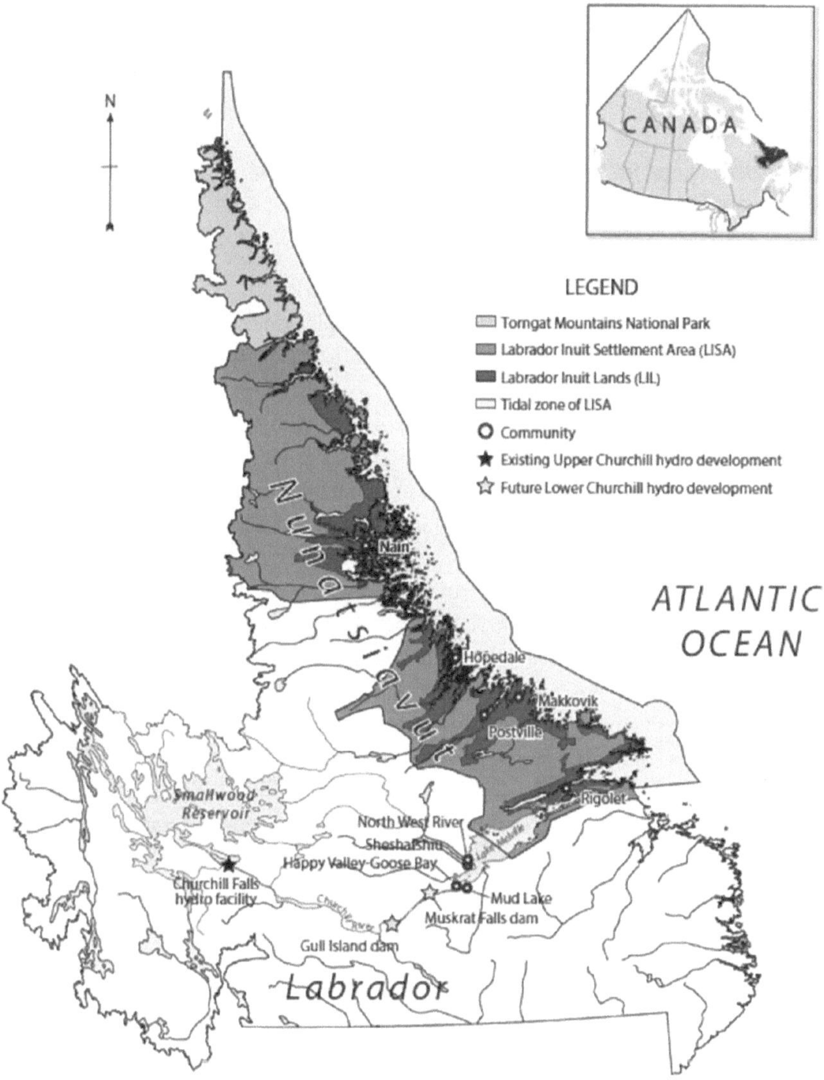

FIGURE 7.1 Map of Nunatsiavut, one of four regions of Inuit Nunangat.

Source: (Courtesy of the Nunatsiavut Government)

meaningfully engaging with correct and appropriate pronunciations of Inuttitut terms, including "Nunatsiavut". It is important that our project respect this request and that we contribute to increasing the normative goal of external researchers correctly referencing the cultural and geographic locations in which their work occurs. Using and always saying the place and name of the project in Inuttitut is also an actively anti-colonial act that recognizes how Indigenous communities have always had names for their regions in Indigenous languages and that these names have been changed throughout history to words in colonizing or official state languages. Therefore, all references to our project in this chapter will use the Inuttitut project title.

SakKijânginnaniattut Nunatsiavut Sivunitsangit and researcher positionality

Research and authorship are relational processes that connect collaborators to one another and the context in which their work is embedded. Given that iterative reflection and context-dependency are two key principles underlying KCP, scholars and practitioners are frequently drawing on their positionality to iteratively refine project methods and goals, detail their individual and collective relationships to their work, cultivate trust among project partners, and enhance awareness of the impact and longevity of project outcomes (Carter et al., 2019; Maclean et al., 2022).

This narrative reflection extends from recurring conversations and activities attended by multiple project partners and hosted by WP 1. As such, our group of authors represents the geographic, professional, and cultural diversity of the overall project, such as IRCs in Nunatsiavut (Anthony, Nochasak, Winters, and Winters); university academics based in the United States, Canada, and Germany (Bailey, Bodwitch, McCarney, Oliver, Petriello, Schmidt, Zurba); doctoral students based in Canada (Bishop, Cadman); and project management based in Canada (McLaren). These roles put us in collaboration with other work packages. However, as members of WP 1, we are committed to understanding and exploring good practices and pathways for KCP in the contexts of SakKijânginnaniattut Nunatsiavut Sivunitsangit, Nunatsiavut, and Inuit Nunangat. Our commitment is underscored by the goal of equitable and respectful acknowledgement and engagement with the diverse backgrounds and knowledge systems represented in the project. Following Carter et al. (2019), the terms "we" and "our" are used to represent our shared experiences unless otherwise attributed to individual authors, which will be marked in the chapter. We additionally position this chapter as a reflexive counterpoint to our recently published inductive and deductive review of KCP studies and context-specific case studies in Nunatsiavut (Zurba et al., 2022).

Transdisciplinarity

As the long-form project title indicates, the project sits squarely within a transdisciplinary research model of KCP. For SakKijânginnaniattut Nunatsiavut Sivunitsangit, transdisciplinary research refers to

> collaborations between different disciplines (e.g. natural, health and social sciences) and academic institutions with communities and other non-academic institutions. [It] engages with multiple groups of people who hold a plurality of perspectives and addresses real-world problems relevant to society.

> [It] aims to produce results that have value to communities, partners and broader audiences rather than a product that addresses problems from only one perspective.
>
> *(Sustainable Nunatsiavut Futures, 2020, p. 1)*

In other words,

> Transdisciplinarity occurs by the interaction of different disciplines, including many forms of collaboration among various sectors, groups and institutions. Transdisciplinary research can also be driven by community interests . . . provid[ing] an opportunity for communities and researchers to learn from each other.
>
> *(Sustainable Nunatsiavut Futures, 2020, p. v)*

Philosophically, SakKijânginnaniattut Nunatsiavut Sivunitsangit is committed to a different paradigm of scientific research from the standard scientific model. We note that doing ethical science requires good research governance, and thus are building towards a governance model that adheres to shared "living" rules, norms, values, and principles that are collectively discussed and refined with all project partners in the context of Nunatsiavut marine spatial planning (Figure 7.2). The "living" values and principles (see KCP activities further in the chapter) filter into project objectives that are co-designed with input from different stakeholders and knowledge systems. The KCP principles, values, and objectives are pursued through the governance vision described in the project background and Figure 7.2.

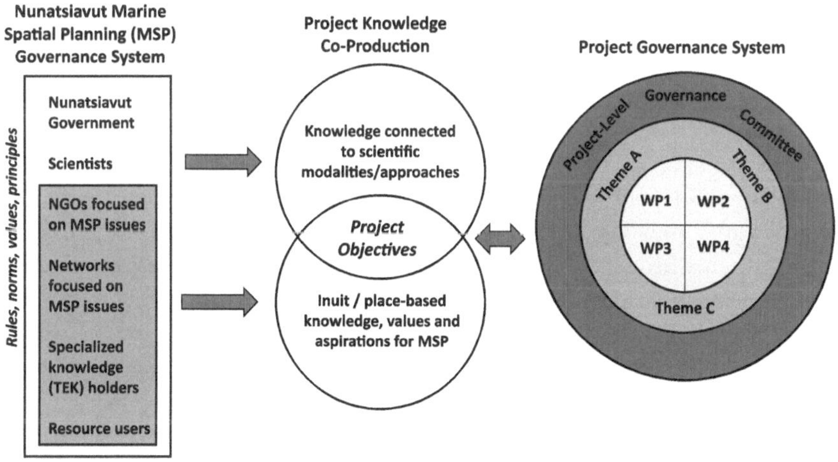

FIGURE 7.2 Conceptual working diagram of the governance framework guiding knowledge co-production (KCP) within the SakKijânginnaniattut Nunatsiavut Sivunitsangit project.

(SakKijânginnaniattut Nunatsiavut Sivunitsangit science meeting, June 8, 2021)

Primer on KCP

Definitions

The global push for community-engaged research and practice in sustainability-related fields has generated a kaleidoscope of KCP definitions, principles, frameworks, and nested concepts (Zurba et al., 2022). Although sorting through the many KCP domains is outside of the scope of this chapter, we first aim to dispel the conceptual fog of KCP definitions using three disciplinarily and contextually distinct definitions. First, in the context of marine science and co-management, Armitage et al. (2011) seminally defined KCP as "the collaborative process of bringing a plurality of knowledge sources and types together to address a defined problem and build an integrated or systems-oriented understanding of that problem" (p. 996). Second, others have drawn from interdisciplinary fields to define KCP as "processes that iteratively unite ways of knowing and acting – including ideas, norms, practices and discourses – leading to mutual reinforcement and reciprocal transformation of societal outcomes" (Wyborn et al., 2019, p. 320). Most recently, *Ellam Yua* et al. (2022) drew from the social-ecological context of Arctic research to define KCP as "a process that brings together Indigenous Peoples' knowledge systems and science to generate new knowledge and understandings of the world that would likely not be achieved through the application of one knowledge system". While distinct in origin and scope, the three definitions demonstrate that KCP definitions, across disciplines and contexts, often depict KCP as *collaborative* and *iterative processes* of uniting *multiple knowledge systems* in pursuit of *solutions-oriented outcomes* (e.g. knowledge and transformations) to *social and ecological problems* across *multiple scales*.

Approaches and tools

The aforementioned conceptual depiction of KCP provides a descriptive template for understanding, framing, and meeting sustainability goals. Many approaches and tools are adopted within the collaborative and iterative processes with the expressed goals of doing and generating science differently (Wyborn et al., 2019). The approaches frequently straddle knowledge divides and strive for cross-cultural and intergenerational knowledge exchange and integration through both remote and land-based activities, traditional knowledge interviews, focus groups, workshops, youth and Elder engagement, and participatory research methodologies (Carter et al., 2019; Zurba et al., 2022). By extension, academic knowledge co-producers are paying growing attention to boundary work in their endeavours because it "includes methodologies to support knowledge sharing and co-creation between research partners as well as work that can translate research outcomes into on-ground action" (Zurba et al., 2019, p. 1024). For example, research in the Indigenous Arctic, including Inuit Nunangat, shows that co-produced tools such as marine resource management indicators and maps can act as boundary objects (i.e. items that facilitate discussion and other types of exchange between groups that are distinct, or even disparate) for bridging Inuit Knowledge and WSK (Bishop et al., 2022; Kourantidou et al., 2020). Importantly, during the formation and early execution of SakKijânginnaniattut Nunatsiavut Sivunitsangit, the use of the term "boundary object" became its own boundary object, whereby academics and practitioners were able to explore different conceptions of what boundaries are and why or why not a term like a boundary object makes sense to different groups. While no single approach is a panacea, they represent a larger toolkit of transformative approaches to ocean sustainability and climate change.

Higher-order domains and principles

KCP definitions, approaches, and tools are the starting and middle points of these efforts. Recent noteworthy syntheses have also aimed to move beyond the dizzying array of prescriptive definitions and approaches and toward higher-order guiding concepts, principles, and modes of KCP. Climate change and sustainability scholars have identified multiple lenses and models of co-production research, including an iterative lens, social learning lens, empowerment lens, social critique model and instrumental model for conceptualizing KCP (Apetrei et al., 2021; Bremer and Meisch, 2017). These higher-order academic frames are complemented by six modes of co-production in practice, of which *researching* solutions is just one (Chambers et al., 2021). Other modes deploy KCP to leverage power from dominant actors (brokering power), elevate marginalized groups through KCP (empowering voices and reframing power), and manage differences in relationships, empowerment, and agency among co-producers (navigating differences, reframing agency). Moreover, paths for carrying out KCP research and practice have been found to be guided by overarching principles such as context-specificity, pluralism, goal-orientation, and interactions among participants (Norström et al., 2020). Yet tensions still arise when adhering to higher-order principles, calling for novel conceptual contributions such as "co-productive agility", or the "willingness and ability of diverse actors to iteratively engage in reflexive dialogue to grow shared ideas and actions that would not have been possible from the outset" (Chambers et al., 2022, p. 2), to manoeuvre KCP landscapes in diverse contexts.

In the context of this case study, SakKijânginnaniattut Nunatsiavut Sivunitsangit collaborators identified four principles for KCP work in Nunatsiavut (Zurba et al., 2022):(1) context dependency, (2) frequent, early, and sustained engagement with IK holders; (3) shared understanding and commitment to KCP and project goals; and (4) empowerment. These findings are buoyed by the recently proposed emphasis on equity in Indigenous Arctic KCP research and practice (*Ellam Yua* et al., 2022). In particular, some have called for scholars, practitioners, and policy makers to pay more attention to the emotional, relational, and trust-related aspects of KCP and knowledge exchange (Cvitanovic et al., 2021; Pohl et al., 2021). This call to action mirrors a recent increasing broader exploration of concepts and theories that frame the social, relational, and interpersonal dimensions of KCP in climate and sustainability research, such as empowerment, equity, trust, and representation (Bremer and Meisch, 2017; Chambers et al., 2022; Chapman and Schott, 2020; Cvitanovic et al., 2021; Maclean et al., 2022, Zurba et al., 2022).

Despite the many forms of KCP referenced earlier, we are not defining KCP in this reflection for three reasons: (1) iterative conversations among our group have shown that there are concerns about the term KCP and its perceived benefits and limitations; (2) our project partners have made a shared commitment to leaving KCP undefined because of its multidimensional nature and multidisciplinary implications; and (3) the absence of a concrete definition of KCP facilitates flexible thinking, practical creativity, and values-based decision-making rather than rigid adherence to "one way" of pursuing KCP. Notably, KCP is not a means to an end; it is a process, and thus the principles that underpin that process instead of a singular definition may be more important in the outcomes KCP projects seek to achieve.

KCP in SakKijânginnaniattut Nunatsiavut Sivunitsangit

The origins and "power" of KCP

The first focused attempts to articulate a shared understanding of KCP in SakKijânginnaniattut Nunatsiavut Sivunitsangit are rooted in the project's 2020 kick-off workshop (Sustainable Nunatsiavut Futures, 2020). Along with the description and definition of transdisciplinarity presented earlier, the project at that time (2020) chose to describe and define KCP to foster and facilitate discussions within this broad conceptual arena. While the project intentionally leaves KCP undefined at this point in time (2022), it began from the viewpoint of KCP as

> a collaborative and social learning experience [that] may involve communities, governments and scientists. As an interactive process, it demands constant awareness of the plurality of perspectives held by actors (those involved in the knowledge co-production). Knowledge co-production can be readily applied to topics that are broad and can embody a range of world views and disciplines (e.g. local knowledge and academic disciplines).
>
> *(Sustainable Nunatsiavut Futures, 2020, p. v).*

This description was accompanied with the concise definition of KCP as "the collaborative and social learning process which involves communities, governments, scientists and institutional learning; this process embraces complexity and it does not classify knowledge into a hierarchy system" (Sustainable Nunatsiavut Futures, 2020, p. 1).

The nuanced yet contained description and definition of KCP provided a springboard for the project's 50 partners to begin interrogating, negotiating, and planning for the inherent complexities of transdisciplinary KCP in the context of rapid climate change in the Arctic and continual recognition of Inuit sovereignty. These processes involved overviews and panel discussions about the Nunatsiavut

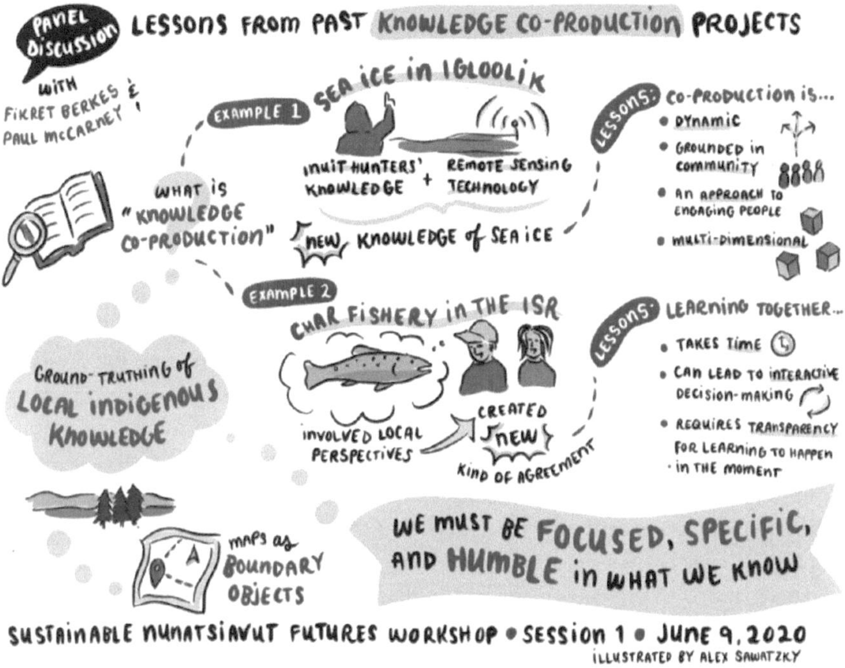

FIGURE 7.3 Graphic depiction of the lessons from past knowledge co-production projects from the SakKijânginnaniattut Nunatsiavut Sivunitsangit Kickoff Workshop panel discussion (June 9, 2020) (Artist: Alex Sawatzky).

context of the project, Indigenous research ethics, past KCP projects, the diversity of KCP tools and methods the project had and needed, and indicators of success (e.g. open communication) and failure (e.g. excessive objectives) in KCP endeavours. Critically, the fundamental question "What is knowledge co-production?" underscored each process (see Figure 7.3) and set the stage for iterative reflection on the meaning and purpose of KCP in SakKijânginnaniattut Nunatsiavut Sivunitsangit thereafter.

The importance of critical reflection on KCP was evident from the beginning of the project. Project partners started from different backgrounds with different levels and bodies of knowledge, such as diverse disciplines across the natural and social sciences (e.g. oceanography and geography), different scales of knowledge (e.g. individual and community knowledge), different institutional settings (e.g. non-governmental organizations, governments, and universities), different career stages (e.g. graduate students and tenured faculty), and different cultural contexts (e.g. Inuk, settler, and international). These diverse "starting points" highlighted the need for a process of mutual learning in which common understandings of KCP, its objectives, and the system in which it is taking place are developed. This process is ongoing and iterative, presenting partners with opportunities to constructively and transparently grapple with KCP as a conceptually diffuse and ever-changing concept rather than as a rigid definition and set of ideas to which the project must strictly adhere. By casting aside static interpretations of KCP, partners are encouraged to question KCP as the foundation of this project. In this spirit, we interrogate the power and precarity of KCP in our project two years in.

Since summer 2020, the project has embarked on numerous formal and informal pathways to co-develop processes and goals around a shared understanding of KCP. The pathways have taken many forms, including values-elicitation exercises, literature reviews (Zurba et al., 2022) and synthesis (this chapter), KCP workshops, co-developed project guidelines, informal coffee chats and virtual get-togethers, and a working group for Early Career Researchers called *IlinniaKatigenniik* ("Learning together" in English). The activities allow project partners to account for the diversity of ways KCP is understood and challenged and how this diversity informs research by individuals in the project. In doing so, they have revealed the importance of relationships, reflection, ethics, respect, authenticity, and humanity to the project.

Through values-elicitation exercises led by Dr Max Liboiron, project members prioritized relational and reflective values such as *placing relationships first*, *open communication*, *being human*, *questioning assumptions*, *having fun*, and *ethical place-based engagement* as fundamental drivers of decision-making and collaboration in the project (Figure 7.4). Other activities, including a project-level KCP workshop led by Dr Matthew Wildcat, Renée Beausoleil, and Mandee McDonald, centred on crafting a vision of KCP for the project. This workshop produced seven KCP themes built on the meanings and aspirations that individual project members' ascribed to KCP (Figure 7.5). The

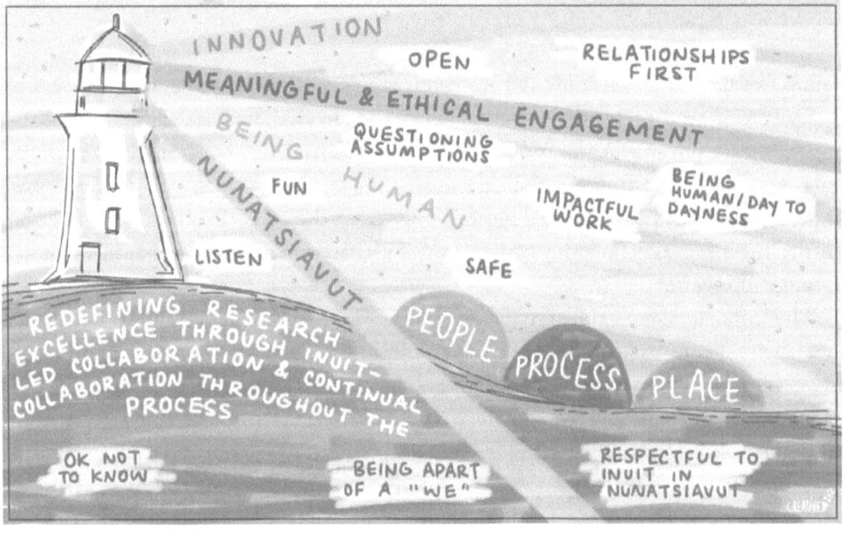

FIGURE 7.4 Graphic representation of the values guiding work package 1 in the Sak-Kijânginnaniattut Nunatsiavut Sivunitsangit (Sustainable Nunatsiavut Futures) Project as of January 19, 2021 (Artist: Ashton Rodenhiser, Minds Eye Creative).

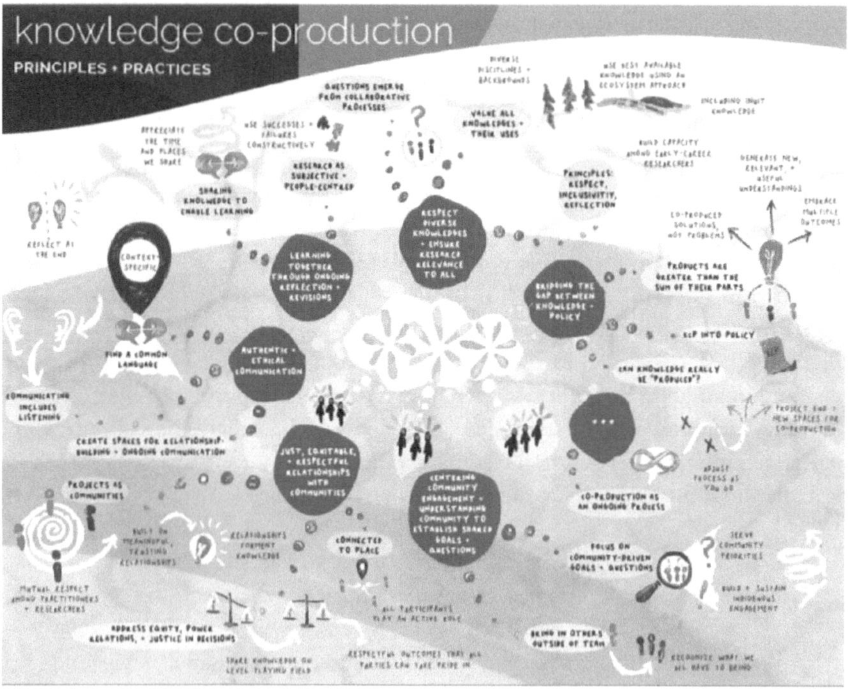

FIGURE 7.5 Graphic representation of the individual and collective vision for KCP in the SakKijânginnaniattut Nunatsiavut Sivunitsangit (Sustainable Nunatsiavut Future) Project as of May 4, 2021 (Artist: Alex Sawatzky).

workshop findings presented a multi-layered image of KCP that aligned with values identified by WP 1, shedding light on the critical role of relationships, learning, and transparency as core values and visions for KCP. Taken together, the values and visions expand the original conception of KCP in the project while reaffirming the philosophical commitment to questioning the concept in our work.

One path through which values and KCP themes are enacted and reaffirmed is through project engagement and support for Early Career Researchers. IlinniaKatigenniik was formed in 2020 flowing directly from these early conversations about the importance of trust and humanity for the project. The group is composed of Early Career Researchers such as students, postdoctoral fellows, and IRCs, including 9 of the 15 chapter co-authors, who focus on building relationships, seeking connections between research interests, and learning new skills together. The group was formed with two ideas in mind. The first was that forming relationships across the project, particularly in the era of COVID, would require a sustained and intentional effort. The second was an acknowledgement of existing and potential power dynamics within the project and a desire to identify clear actions to challenge and productively change power dynamics embedded in all research projects. For example, the group is also intended to frame a pathway for Early Career Researchers to provide direct and collective input into the governance and design of the project in ways that meaningfully challenge traditional hierarchies of project governance that typically privilege the voices of senior faculty. Therefore, IlinniaKatigenniik is a space without hierarchy, where group members can share personal updates, explore their passions, and ask questions. As an active space of mutual learning that draws on the collective skills and interests of its members, IlinniaKatigenniik allows participants to anticipate, navigate, and proactively address the inherent challenges of KCP for Early Career Researchers in marine conservation (e.g. see Rölfer et al. [2022]).

The group's success is due in part to the fact that it has grown organically, with little attention to specific outcomes or "products" beyond creating a community for Early Career Researchers. The group has instead emphasized the process of relationship building by sharing vulnerabilities, personal experiences, humility, and humour. Group members experience the benefits of this process in real time. IlinniaKatigenniik member Rachael Cadman, a PhD candidate and chapter co-author, describes how the group has contributed to her work:

> *Every meeting enriches my work in a variety of ways. Hearing IRCs talk about hunting and traveling trips, or weekends spent with their family, recenters my mind in Nunatsiavut and I remember the real impact that our research has on people's lives. Hearing from group members out collecting data or formulating new ideas, I gain context for my own work. The group has sparked new ideas and given me new collaborators.*

Another group member, a PhD candidate and chapter co-author, Breanna Bishop describes arriving in Rigolet, Nunatsiavut, and meeting fellow IlinniaKatigenniik members Katrina Anthony and John Winters in person for the first time:

> *When I first saw them, I jumped up to hug them as if we had known each other for years. It took me a while to remember that we had never met in person before.*

The work accomplished in this group has enriched research and cultivated a sense of accountability and trust among group members. The group emphasizes the importance in seeing any process of knowledge production as a relational process that must both understand underlying relationships and actively work to establish new relationships suited to the particular context in which research occurs (e.g. defined by the actors, knowledge holders, histories, and places research takes place). While these activities may not traditionally be considered the "co-production of knowledge", our experience has been that they are the bedrock of working and *learning* together.

Reflections on "precarity" in KCP

The activities carried out so far capture momentary snapshots of "living" values, visions, and principles of KCP in SakKijânginnaniattut Nunatsiavut Sivunitsangit.

In other words, they may change and should be treated as embodying a specific moment in time in the project history.

Yet the values and visions also pinpoint key areas of critical analyses and uncertainty about KCP supported by collaborative relationships founded on trust, honesty, shared humanity, and communication.

First, within our efforts to generate, or co-produce, new knowledge, we have seen that these processes are informed by the different contexts and concepts we as individuals bring to the table. In turn, we have come to recognize the indispensable value of relationship building as a precursor to bridging, sharing and integrating individual experiences and knowledge in KCP spaces. For example, interpersonal dynamics and different cultural norms of communication may act as unseen barriers to "KCP". Caroline Nochasak, IRC and chapter co-author, noted that a "timid attitude" could be a weakness of KCP in that

> *We can work more on not wavering if you have a question or if you have a point to bring up. We all want a supportive and growing environment. Asking/answering questions people have can help the greater group towards the main goal.*

Observations about different forms of communication appear to stem from cultural and institutional differences between Inuit and non-Inuit partners and academic and non-academic perspectives. In traditional academic environments, scholars are often trained to assert their points through analytical dialogue and "expertise". This argumentative approach can be at odds with Inuit norms that may lean towards gentler or more understated forms of delivery and disagreement when asserting different views, particularly with respect to sensitive topics such as resource access that are rooted in colonial legacies. In turn, these realities risk inadvertently drowning out Inuit voices in the project when juxtaposed with academic jargon, such as the phrase "knowledge co-production".

Second, open and non-judgemental conversations about these cultural differences in workshops and informal group settings (e.g. coffee chats) have fostered significant levels of trust. This trust has led many partners to independently suggest the project may benefit from moving beyond KCP or substituting it with a new term to represent our form of collaboration. In particular, discussions about the name and meaning of KCP and analogous phrases have generated concerns and uncertainties with diverse project partners. These concerns have taken two forms: The phrases "knowledge co-production" and "co-production of knowledge" are inaccessible to many, whether academic or non-academic; and the phrases implicitly focus on an end product, rather than a commitment to a process of working together.

As phrases, "KCP" and its variations are often perceived as academic jargon. The term originates from the social sciences, meaning that a subset of academics in the project is comfortable discussing the KCP literature and discourse. On the one hand, the frequent references to, calls for, and in-depth conversations about KCP can be alienating for many project team members, such as natural scientists who are eager to contribute to KCP but intimidated by its conceptual weight. On the other hand, social scientists may be assumed to have a full understanding of this transdisciplinary domain of social inquiry, placing its failure or success in the hands of few individuals who are equal members in the mutual learning processes framing the project. This tension is productive in that it catalyses ongoing and iterative reflection that is philosophically aligned with KCP as a normative stance and research process. Yet the tension similarly reveals how KCP can establish a precarious collaborative foundation when it is purposefully and iteratively invoked as a cornerstone of how and why we work together.

The collaborative barriers posed by "KCP" are also maintained by its focus on "production". As one of the keywords in this initialism, "production" elicits thoughts of a factory production line; it is very mechanical and exploitative. It also implicitly focuses on an end "product", rather than a commitment to a process of working together. Partners at multiple levels of the project have raised the question, "To what degree is new knowledge 'produced' through this process?", largely because many perceive our work as processes of weaving different knowledge systems. Dr Eric C. J. Oliver, a Labrador Inuk, professor of oceanography, and chapter co-author, noted:

> *The "pointy end" of all this transdisciplinary/KCP work is the* process, *the* hard work *that we are all putting in and getting value from, is the process, not the* product. *If we get the* process *right, then the* product/*outcome should flow naturally from it.*

Other Inuit project team members have stated that IK is not "produced" but an extension of worldviews and experiences built on the land, water, and ice they call home. For these reasons, some project members, such as Mary Denniston, NG Environmental Protection Analyst, project member, and Labrador Inuk, have suggested new terms such as "knowledge unification" may more appropriately depict the unique nature of our collaborative experience and goals.

Third, diverse communication styles and ambiguities around the term "KCP" and its focus on "products" are amplified by real-world time constraints, events, and different paces in which partners and sub-projects operate. The catalyst for SakKijânginnaniattut Nunatsiavut Sivunitsangit was to find new and transformative paths for science in Nunatsiavut. Yet the ambitious goals of sustainability transformations, transdisciplinary research, and broad structural change require substantial time investments that will likely surpass academic and government funding cycles and allotments. For example, this project is operating on a six-year timescale framed by the realities of the COVID-19 pandemic. As noted by Megan McLaren, project manager and co-author:

> *A foundational aspect of KCP is having and developing solid relationships. Figuring out how to build new relationships, in the timescale of a research project, without having time face-to-face, is a huge challenge [due to travel restrictions, other logistic*

> *constraints]. Even more so when many project team members are living in an area with poor internet connectivity, so video calls can be a challenge. This is a challenge both in relation to the core project team, but also in establishing community representation and voices in the project.*

In other words, limited in-person communication and collaboration challenge the ability of team members to foster their aspired values for this KCP system, such as building trust and connection. A suite of geographic, seasonal, technological, and public health barriers (e.g. international team members and COVID-19) contribute to the inadvertent misalignment or purposeful reprioritization of work flows for concurrent projects, such as cancelled field visits to Nunatsiavut from COVID-19 (Petriello, Zurba) and prioritizing accelerated field schedules over other seasonally independent work to avoid unsafe ice conditions from climate change (Anthony, Nochasak, Winters, Winters). Taken together, these examples show how pluralism, "production", and real-world problems are both key components of KCP and challenges to enacting its potential.

Conclusion

Our case study does not prescribe a specific way of "doing" KCP. However, it demonstrates why KCP is a powerful yet precarious term and concept in collaborative spheres. We have shown that the SakKijânginnaniattut Nunatsiavut Sivunitsangit project has embarked on a KCP path paved with efforts to establish, maintain, and affirm relationships with project partners across social sectors and knowledge systems. This approach to KCP is similar to the recently proposed mode 5 of co-production (*Navigating differences*) in that this project has currently "placed a stronger emphasis on managing processes of relating together, learning and empowerment over producing and transferring scientific knowledge about human-ecosystem interactions" (Chambers et al., 2021). Furthermore, the processes of questioning the KCP assumptions to maximize the project contributions to community suggests the project is adopting an *empowerment lens* to frame its philosophy and objectives, which Bremer and Meisch (2017) describe as a lens that "looks at the ways co-production recognizes and empowers traditional environmental knowledge (TEK) systems" (p. 10). Lastly, the values, visions, approaches, and reflexivity outlined in our case study suggest that the multiple actors in this project support and reaffirm the importance of *co-productive agility* (Chambers et al., 2022) in KCP for sustainable seas.

Conversely, the case study also reveals that KCP may not be the most appropriate frame for all projects, even those that are founded on notions of co-producing knowledge, transdisciplinarity, and transformative solutions. At this stage in the project, we have found that "precarity" in KCP emerges through efforts to formulate shared understandings of its meanings, the social science lens and assumptions through which KCP is defined and operationalized, and the realities that place-based collaborative research does not occur in a vacuum sealed off from the world around it. Our observations and experiences echo recent considerations for carrying out KCP through global networks, such as how team members are supported and convened through KCP (Schneider et al., 2021). Early Career Researchers, for example, may be more likely than other partners to confront distinct barriers to productive participation with knowledge co-production at different project scales, from individual to institutional levels (Rölfer et al., 2022). We recommend that research groups that are drawing from, or developing, KCP models leave room for deconstructing and/or reconstructing the term and its contributions to their collaborative efforts *during* their work together rather than as an afterthought. In this way, we aspire for our case study to be used to foster growth, reflexivity, and humanity in other collaborative projects.

Acknowledgements

SakKijânginnaniattut Nunatsiavut Sivunitsangit (the Sustainable Nunatsiavut Futures Project) is supported in part by the Canada First Research Excellence Fund ("CFREF") through the Ocean Frontier Institute ("OFI"). We thank project leadership for their openness to reflexive discussions. We also kindly acknowledge several partners for their thought-provoking commentary and insights into KCP and its placed-based components over the course of this project, including Amanda Bates, Anna Metaxas, Rodd Laing, Mary Denniston, Liz Pijogge, and all participants in the Work Package 1 values-elicitation and project-level KCP workshops. Special thanks go to Max Liboiron, Matthew Wildcat, Renée Beausoleil, and Mandee McDonald for steering the project towards shared values and visions, and Alex Sawatzky and Ashton Rodenhiser for their dynamic illustrations. Lastly, we are deeply indebted to the Nunatsiavut Government for their committed support and providing the map in Figure 7.1.

References

Adger, W.N., Brown, K., & Tompkins, E.L. (2005). The political economy of cross-scale networks in resource co-management. *Ecology and Society*, 10(2), 9. Retrieved from www.ecologyandsociety.org/vol10/iss2/art9.

Alexander, S.M., Provenchar, J.F., Henri, D.A., Taylor, J.J., Lloren, J.I., Nanayakkara, L., Johnson, J.T., & Cooke, S.J. (2019). Bridging Indigenous and science-based knowledge in coastal and marine research, monitoring, and management in Canada. *Environmental Evidence*, 8, 36. Retrieved from https://doi.org/10.1186/s13750-019-0181-3.

Alvarado, C.M.M., Rendon, A.Z., & del Socorro Vázquez Pérez, A. (2020). Integrating public participation in knowledge generation processes: Evidence from citizen science initiatives in Mexico. *Environmental Science & Policy*, 114, 230–241. Retrieved from https://doi.org/10.1016/j.envsci.2020.08.007.

Apetrei, C.I., Caniglia, G., von Wehrden, H., & Lang, D.J. (2021). Just another buzzword? A systematic literature review of knowledge-related concepts in sustainability science. *Global Environmental Change*, 68, 102222. Retrieved from https://doi.org/10.1016/j.gloenvcha.2021.102222.

Armitage, D., Berkes, F., Dale, A., Kocho-Schellenberg, E., & Patton, E. (2011). Co-management and the co-production of knowledge: Learning to adapt in Canada's arctic. *Global Environmental Change*, 21, 995–1004. Retrieved from https://doi.org/10.1016/j.gloenvcha.2011.04.006.

Bekryaev, R.V., Polyakov, I.V., & Alexeev, V.A. (2010) Role of polar amplification in long-term surface air temperature variations and modern Arctic warming. *Journal of Climate*, 23(14), 3888–3906. Retrieved from https://doi.org/10.1175/2010JCLI3297.1.

Bishop, B., Oliver, E.C.J., & Aporta, C. (2022). Co-producing maps as boundary objects: Bridging Labrador Inuit knowledge and oceanographic research. *Journal of Cultural Geography*, 39(1), 55–89. Retrieved from https://doi.org/10.1080/08873631.2021.1998992.

Bremer, S., & Meisch, S. (2017). Co-production in climate change research: Reviewing different perspectives. *WIREs Climate Change*, 8(6), e482. Retrieved from https://doi.org/10.1002/wcc.482.

Carter, N.A., Dawson, J., Simonee, N., Tagalik, S., & Ljubicic, G. (2019). Lessons learned through research partnerships and capacity enhancement in Inuit Nunangat. *Arctic*, 72(4), 381–403. Retrieved from https://doi.org/10.14430/arctic69507.

Chambers, J.M., Wyborn, C., Klenk, N.L., Ryan, M., Reid, R.S., Riechers, M., Serban, A., Bennett, N.J., Cvitanovic, C., Fernández-Giménez, M.E., Galvin, K.A., Goldstein, B.E., Klenk, N.L., Tengö, M., Brennan, R., Cockburn, J.J., Hill, R., Munera, C., Nel, J.L., Österblom, H., Bednarek, A.T., Bennett, E.M., Brandeis, A., Charli-Joseph, L., Chatterton, P., Kurran, K., Dumrongrojwatthana, P., Paz Durán, A., Fada, SJ., Gerber, J-D., Green, J.M.H., Guerrero, A.M., Haller, T., Horcea-Milcu, A-I., Leimona, B., Montana, J., Rondeau, R., Spierenburg, M., Steyaert, P., Zaehringer, J.G., Gruby, R., Hutton, J., & Pickering, T. (2021). Six modes of co-production for sustainability. *Nature Sustainability*, 4(11), 983–996. Retrieved from https://doi.org/10.1038/s41893-021-00755-x.

Chambers, J.M., Wyborn, C., Klenk, N.L., Ryan, M., Serban, A., Bennett, N.J., Brennan, R., Charli-Joseph, L., Fernández-Giménez, M.E., Galvin, K.A., Goldstein, B.E., Haller, T., Hill, R., Munera, C., Nel, J.L., Österblom, H., Reid, R.S., Riechers, M., Spierenburg, M., Tengö, M., Bennett, E., Brandeis, A., Chatterton, P., Cockburn, J.J., Cvitanovic, C., Dumrongrojwatthana, P., Paz Durán, A., Gerber, J-D., Green, J.H.M., Gruby, R., Guerrero, A.M., Horcea-Milcu, A-I., Montana, J., Steyaert, P., Zaehringer, J.G., Bednarek, A.T., Curran, K., Fada, S.J., Hutton, J., Leimona, B., Pickering, T., & Rondeau, R. (2022). Co-productive agility and four collaborative pathways to sustainability transformations. *Global Environmental Change*, 72, 102422. Retrieved from https://doi.org/10.1016/j.gloenvcha.2021.102422.

Chapman, J., & Schott, S. (2020). Knowledge coevolution: Generating new understanding through bridging and strengthening distinct knowledge systems and empowering local knowledge holders. *Sustainability Science*, 15(3), 931–943. Retrieved from https://doi.org/10.1007/s11625-020-00781-2.

Cvitanovic, C., Hobday, A.J., van Kerkhoff, L., Wilson, S.K., Dobbs, K., & Marshall, N.A. (2015). Improving knowledge exchange among scientists and decision-makers to facilitate the adaptive governance of marine resources: A review of knowledge and research needs. *Ocean & Coastal Management*, 112, 25–35. Retrieved from http://dx.doi.org/10.1016/j.ocecoaman.2015.05.002.

Cvitanovic, C., Shellock, R.J., Mackay, M., van Putten, E.I., Karcher, D.B., Dickey-Collas, M., & Ballesteros, M. (2021). Strategies for building and managing 'trust' to enable knowledge exchange at the interface of environmental science and policy. *Environmental Science & Policy*, 123, 179–189. Retrieved from https://doi.org/10.1016/j.envsci.2021.05.020.

Dawson, N.M., Coolsaet, B., Sterling, E.J., Loveridge, R., Gross-Camp, N.D., Wongbusarakum, S., Sangha, K.K., Scherl, L.M., Phan, H.P., Zafra-Calvo, N., Lavey, W.G., Byakagaba, P., Idrobo, C.J., Chenet, A., Bennett, N.J., Mansourian, S., & Rosado-May, F.J. (2021). The role of Indigenous peoples and local communities in effective and equitable conservation. *Ecology and Society*, 26(3), 19. Retrieved from https://doi.org/10.5751/ES-12625-260319.

Ellam Yua, Raymond-Yakoubian, J., Daniel, R.A., & Behe, C. (2022). A framework for co-production of knowledge in the context of Arctic research. Negeqlikacaarni kangingnaulriani ayuqenrilnguut piyaraitgun kangingnauryararkat. *Ecology and Society*, 27(1), 34. Retrieved from https://doi.org/10.5751/ES-12960-270134.

Ford, J.D., Bolton, K.C., Shirley, J., Pearce, T., Tremblay, M., & Westlake, M. (2012). Research on the human dimensions of climate change in Nunavut, Nunavik, and Nunatsiavut: A literature review and gap analysis. *Arctic*, 65(3), 289–304. Retrieved from www.jstor.org/stable/41758936.

Hauser, D.D.W., Whiting, A.V., Mahoney, A.R., Goodwin, J., Harris, C., Schaeffer, R.J., Laxague, N.J.M, Subramaniam, A., Witte, C.R., Betcher, S., Lindsay, J.M., & Zappa, C.J. (2021). Co-production of knowledge reveals loss of Indigenous hunting opportunities in the face of accelerating Arctic climate change. *Environmental Research Letters*, 16(9), 095003. Retrieved from https://doi.org/10.1088/1748-9326/ac1a36.

Hidalgo, M., Bartolino, V., Coll, M., Hunsicker, M.E., Travers-Trolet, M., & Browman, H.I. (2022). 'Adaptation science' is needed to inform the sustainable management of the world's oceans in the face of climate change. *ICES Journal of Marine Science*, 79(2), 457–462. Retrieved from https://doi.org/10.1093/icesjms/fsac014.

Hill, R., Grant, C., George, M., Robinson, C.J., Jackson, S., & Abel, N. (2012). A typology of Indigenous engagement in Australian environmental management: Implications for knowledge integration and social-ecological system sustainability. *Ecology and Society*, 17(1), 23. Retrieved from http://dx.doi.org/10.5751/ES-04587-170123.

Hind, E.J., Alexander, S.M., Green, S.J., Kritzer, J.P., Sweet, M.J., Johnson, A.E., Amargós, F.P., Smith, N.S., & Peterson, A.M. (2015). Fostering effective international collaboration for marine science in small island states. *Frontiers in Marine Science*, 2, 86. Retrieved from https://doi.org/10.3389/fmars.2015.00086.

Hinzman, L.D., Bettez, N.D., Bolton, W.R., Chapin, F.S., Dyurgerov, M.B., Fastie, C.L., Griffith, B., Hollister, R.D., Hope, A., Huntington, H.P., Jensen, A.M., Jia, G.J., Jorgenson, T., Kane, D.L., Klein, D.R., Kofinas, G., Lynch, A.H., Lloyd, A.H., McGuire, A.D., Nelson, F.E., Oechel, W.C., Osterkamp, T.E., Racine, C.H., Romanovsky, V.E., Stone, R.S., Stow, D.A., Sturm, M., Tweedie, C.E., Vourlitis, G.L., Walker, M.D., Walter, D.A., Webber, P.J., Welker, J.M., Winker, K.S., & Yoshikawa, K. (2005). Evidence and implications of recent climate change in northern Alaska and other Arctic regions. *Climate Change*, 72(3), 251–298. https://doi.org/10.1007/s10584-005-5352-2.

Hirsch, R., Furgal, C., Hackett, C., Sheldon, T., Bell, T., Angnatok, D., Winters, K., & Pamak, C. (2016). Going off, growing strong: A program to enhance individual youth and community resilience in the face of change in Nain, Nunatsiavut. *Études/Inuit/Studies*, 40(1), 63–84. Retrieved from https://doi.org/10.7202/1040145ar.

Intergovernmental Panel on Climate Change (IPCC). (2022). *Climate Change 2022: Impacts, Adaptation, and Vulnerability*. Contribution of Working Group II to the Sixth Assessment Report of the Intergovernmental Panel on Climate Change. Pörtner, H.-O., Roberts, D.C, Tignor, M., Poloczanska, E.S., Mintenbeck, K., Alegría, A., Craig, M., Langsdorf, S., Löschke, S., Möller, V., Okem, A., & Rama, B. (Eds). Cambridge University Press. Retrieved from www.ipcc.ch/report/ar6/wg2/downloads/report/IPCC_AR6_WGII_FinalDraft_FullReport.pdf.

Inuit Tapiriit Kanatami (ITK). (2018a). National Inuit Strategy on Research. Retrieved from www.itk.ca/national-strategy-on-research-launched/.

Inuit Tapiriit Kanatami (ITK). (2018b). About Canadian Inuit. Retrieved from www.itk.ca/about-canadian-inuit/#nunangat.

Kourantidou, M., Hoover, C., & Bailey, M. (2020). Conceptualizing indicators as boundary objects in integrating Inuit knowledge and western science for marine resource management. *Arctic Science*, 6(3), 279–306. Retrieved from http://dx.doi.org/10.1139/as-2019-0013.

Labrador Inuit Land Claims Agreement (LILCA). (2005). Nunatsiavut Government. Retrieved from https://caid.ca/AgrLabInuLCA.pdf.

Maclean, K., Woodward, E., Jarvis, D., Turpin, G., Rowland, D., & Rist, P. (2022). Decolonising knowledge co production: Examining the role of positionality and partnerships to support Indigenous-led bush product enterprises in northern Australia. *Sustainability Science*, 17(2), 333–350. Retrieved from https://doi.org/10.1007/s11625-021-00973-4.

Mazor, T., Possingham, H.P., & Kark, S. (2013). Collaboration among countries in marine conservation can achieve substantial efficiencies. *Diversity and Distributions*, 19(11), 1380–1393. Retrieved from https://doi.org/10.1111/ddi.12095.

M'sit No'kmaq, Marshall, A., Beazley, K.F., Hum, J., joudry, s., Papadopoulos, A., Pictou, S., Rabesca, J., Young, L., & Zurba, M. (2021). "Awakening the sleeping giant": Re-Indigenization principles for transforming biodiversity conservation in Canada and beyond. *FACETS*, 6(1), 839–869. Retrieved from https://doi.org/10.1139/facets-2020-0083.

Norström, A.V., Cvitanovic, C., Löf, M.F., West, S., Wyborn, C., Balvanera, P., Bednarek, A.T., Bennett, E.M., Biggs, R., de Bremond, A., Campbell, B.M., Canadell, J.G., Carpenter, S.R., Folke, C., Fulton, E.A., Gaffney, O., Gelcich, S., Jouffray, J.-B., Leach, M., Le Tissier, M., Martín-López, B., Louder, E., Loutre, M.-F., Meadow, A.M., Nagendra, H., Payne, D., Peterson, G.D., Reyers, B., Scholes, R., Speranza, C.I., Spierenburg, M., Stafford-Smith, M., Tengö, M., van der Hel, S., van Putten, I., & Österblom, H. (2020). Principles for knowledge co-production in sustainability research. *Nature Sustainability*, 3(3), 182–190. Retrieved from https://doi.org/10.1038/s41893-019-0448-2.

Pohl, C., Klein, J.T., Hoffmann, S., Mitchell, C., & Fam, D. (2021). Conceptualising transdisciplinary integration as a multidimensional interactive process. *Environmental Science & Policy*, 118, 18–26. Retrieved from https://doi.org/10.1016/j.envsci.2020.12.005.

Rölfer, L., Ilosvay, X.E.E., Ferse, S.C.A., Jung, J., Karcher, D.B., Kriegl, M., Mafaziya Nijamdeen, T.W.G.F., Riechers, M., & Walker, E.Z. (2022). Disentangling obstacles to knowledge co-production for early-career researchers in the marine sciences. *Frontiers in Marine Science*, 9, 893489. Retrieved from https://doi.org/10.3389/fmars.2022.893489.

Schneider, F., Tribaldos, T., Adler, C., & Biggs, R. (O)., de Bremond, A., Buser, T., Krug, C., Loutre, M-F., Moore, S., Norström, A.V., Paulavets, K., Urbach, D., Spehn, E., Wülser, G., & Zondervan, R. (2021). Co-production of knowledge and sustainability transformations: A strategic compass for global research networks. *Current Opinion in Environmental Sustainability*, 49, 127–142. Retrieved from https://doi.org/10.1016/j.cosust.2021.04.007.

Small, C., & Nicholls, R.J. (2003). A global analysis of human settlement in coastal zones. *Journal of Coastal Research*, 19(3), 584–599. Retrieved from www.jstor.org/stable/4299200.

Statistics Canada. (2018, June 18). Nunatsiavut [Inuit region], Newfoundland and Labrador. Aboriginal Population Profile, 2016 Census. Statistics Canada Catalogue no. 98-510-X2016001. Ottawa. Retrieved from https://www12.statcan.gc.ca/census-recensement/2016/dp-pd/abpopprof/details/page.cfm?Lang=E&Geo1=AB&Code1=2016C1005083&Data=Count&SearchText=Nunatsiavut&SearchType=Begins&B1=All&GeoLevel=PR&GeoCode=2016C1005083&SEX_ID=1&AGE_ID=1&RESGEO_ID=1.

Sustainable Nunatsiavut Futures. (2020). Knowledge co-production and transdisciplinary approaches for Sustainable Nunatsiavut Futures: Kickoff Workshop series. Ocean Frontier Institute. Retrieved from https://docplayer.fi/221866316-Knowledge-co-production-and-transdisciplinary-approaches-for-sustainable-nunatsiavut-futures.html.

Thornton, T.F., & Scheer, A.M. (2012). Collaborative engagement of local and traditional knowledge and science in marine environments: A review. *Ecology and Society*, 17(3), 8. Retrieved from http://dx.doi.org/10.5751/ES-04714-170308.

Williams, B.A., Watson, J.E.M., Beyer, H.L., Klein, C.J., Montgomery, J., Ruting, R.K., Roberson, L.A., Halpern, B.S., Grantham, H.S., Kuempel, C.D., Frazier, M., Venter, O., & Wenger, A. (2021). Global rarity of intact coastal regions. *Conservation Biology*, e12874. Retrieved from https://doi.org/10.1111/cobi.13874.

Wyborn, C., Datta, A., Montana, J., Ryan, M., Leith, P., Chaffin, B., Miller, C., & van Kerkhoff, L. (2019). Co-producing sustainability: Recording the governance of science, policy, and practice. *Annual Review of Environment and Resources*, 44, 319–346. Retrieved from https://doi.org/10.1146/annurev-environ-101718-033103.

Yletyinen, J. (2019). Arctic climate resilience. *Nature Climate Change*, 9(11), 805–806. Retrieved from https://doi.org/10.1038/s41558-019-0616-4.

Zurba, M., Maclean, K., Woodward, E., & Islam, D. (2019). Amplifying Indigenous community participation in place-based research through boundary work. *Progress in Human Geography*, 43(6), 1020–1043. Retrieved from https://doi.org/10.1177/0309132518807758.

Zurba, M., Petriello, M.A., Madge, C., McCarney, P., Bishop, B., McBeth, S., Denniston, M., Bodwitch, H., & Bailey, M. (2022). Learning from knowledge co-production research and practice in the twenty-first century: Global lessons and what they mean for collaborative research in Nunatsiavut. *Sustainability Science*, 17(2), 449–467. Retrieved from https://doi.org/10.1007/s11625-021-00996-x.

8

STAKEHOLDERS' NORMATIVE NOTIONS OF SUSTAINABILITY

A survey for the co-design of a sustainable future of the Western Baltic fishery system

Viola Schaber, Marie-Catherine Riekhof, Michael Stecher, Rudi Voss, and Stefan Baumgärtner

Introduction

Fisheries provide livelihoods for many people, are of enormous economic importance across the globe and are part of many socio-cultural traditions (FAO, 2020). Besides commercial fishing, recreational fishing has gained in importance and sometimes even surpasses commercial fishing in some of these aspects (Cooke and Cowx, 2006; Ihde et al., 2011). Together with increases in world population and consumption, fish and seafood removals have increased four-fold over the past 50 years (Ritchie and Roser, 2021; Crona et al., 2016). Especially when fishery management fails, stocks tend to be overfished (Hilborn et al., 2020), with a share of 35% (and rising) of global fish stocks being exploited above sustainable levels in 2017 (FAO, 2020). In addition to unsustainable fishing levels, climate change and socio-economic developments put additional pressure on marine ecosystems. To counteract this unsustainable trend, the United Nations (UN) formulated 17 Sustainable Development Goals (SDGs), all of which are connected to natural resource use. Returning fisheries to sustainable levels is a difficult challenge for fisheries management as fisheries are embedded in complex marine social-ecological systems (Lade et al., 2015). Therefore, fisheries management increasingly aims for stakeholder participation and co-design, which has become a fundamental component of many states' and local agencies' fisheries legislations worldwide (NOAA, 2015, Commission of the European Communities, 2013). The involvement of stakeholders is thought to secure access to local social-ecological knowledge of fishers to complement scientific data as well as to increase the legitimacy and support for management (Aanesen et al., 2014). The Western Baltic Sea (WBS) is an interesting case study to examine the challenges for returning to a sustainable use of fishery resources, as its social-ecological system is comparatively simple: the number of species harvested is relatively small (HELCOM, 2018c), user groups

DOI: 10.4324/9781003311171-10

are clearly defined, and regulation is straightforward as all bordering countries are members of the European Union employing a common fisheries policy. In addition, recreational fishing plays an important role (e.g. in 2020 for cod: 30% of the total catches originated from recreational catches) and is already included into fisheries management (ICES, 2021b). Existing management measures in the framework of the European Common Fisheries Policy (CFP) with its overarching goal to achieve the Maximum Sustainable Yield (MSY) (EU, 2013) have partly been unsuccessful, and many stocks are in a bad state. In the case of the European Union, advice on catch opportunities is given by the International Council for the Exploration of the Sea (ICES). ICES use a precautionary approach and the concept of safe biological limits to define the state of fish stocks (ICES, 1998). Stocks outside such limits suffer increased risk of low recruitment, causing impaired stock productivity and hence reduced harvesting potential.

While the role of stakeholders' different views on "sustainability" in general and on sustainability assessment in the building sector in particular has been recently taken up in the scientific literature (e.g. Soma et al., 2018; Oen et al., 2010), an explicit discussion of the normative dimension is rare. Withycombe-Keeler et al. (2015) and van der Heel (2018) are notable exceptions. Van der Heel (2018) stressed the need for more explicit engagement with the normative and political dimensions of sustainability research; survey data revealed that sustainability researchers generally acknowledge the value-laden and political nature of their work, yet perspectives on what this means and how to deal with such dimensions vary. To address the problem of freshwater shortages in Phoenix, Arizona, Withycombe-Keeler et al. (2015) suggested a transition to sustainable water governance based on different simulated scenarios including normative values and preferences derived from a stakeholder survey. In a questionnaire survey about direct and indirect impacts on benthic habitats through the capture fishing sector (conducted among others in the Baltic Sea), Soma et al. (2018) found that stakeholder preferences vary across European regions and stakeholder groups.

In this chapter, we explore the notion of sustainability as a normative goal for fisheries management from a societal perspective, using the (German) WBS as a case study. We aim to understand how a "sustainable development of the WBS" is perceived from different stakeholders of the Western Baltic fishery. Sustainable development refers to an ideal conception of how the WBS should be maintained and utilized in the long term – now and in the distant future – from a societal perspective. In particular, sustainability means ensuring opportunities for human use and income as well as achieving good ecological conditions. Political decisions and measures should be directed towards achieving this ideal.

To operationalize the idea of a sustainable development of the WBS, we employ the concept of stochastic viability (Béné et al., 2001; Baumgärtner and Quaas, 2009; Doyen and De Lara, 2010; Béné and Doyen, 2018). This allows us to inquire about the different components of sustainable development in a structured way, taking uncertainty explicitly into account. Using the concept of stochastic viability returns quantitative results which can be directly utilized in fisheries management. In addition, this scientifically guided elicitation process represents views from the different stakeholder groups in a consistent way, providing a base for further discussion, and for co-designing a sustainable future for the WBS fishery. The elicitation process itself can be interpreted as co-producing insights for fishery management: science provided a clear and unified terminology regarding "sustainability" incorporated in a questionnaire, and stakeholders provided their insights.

We make use of the transdisciplinary set-up of the research project marEEshift, as a part of which a meeting with stakeholder groups relevant to the Western Baltic fishery was held. At this meeting, we carried out a questionnaire-based survey among stakeholders to elicit their specific ideas of different aspects of stochastic viability. In such a co-production process of transdisciplinary approaches, challenges emerge from the selection of stakeholder groups and their influence on the outcomes. One fundamental challenge that emerged from the study in that regard is that individual stakeholders may or may not represent their respective group well, since it is very difficult to gauge whether a stakeholder speaks for him- or herself, or on behalf of the whole group. Additionally, the seeming liberty to make choices based on a conceptual "ideal world" appeared to be confounded by the participants values that seemed to be based on their knowledge and experienced based perception of the problem area.

The WBS – a socio-ecological system in transition

The Baltic Sea is an ecologically unique inland sea of the North Atlantic Ocean and comprises one of the world's largest bodies of brackish water (Figure 8.1). It is subject to a multitude of anthropogenic impacts imposed by about 85 million

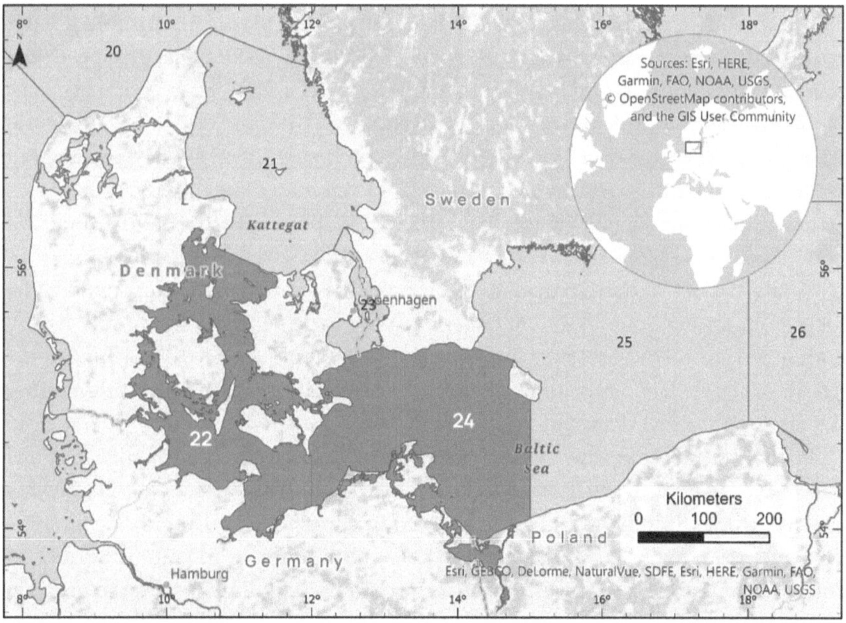

FIGURE 8.1 The Western Baltic Sea (dark grey) with its ICES subdivisions for management.

people living in its 14 bordering countries. Human activities related to the Baltic Sea range from fish and shellfish harvesting, aquaculture, tourism and recreation, renewable energy production and transport infrastructure to shipping (HELCOM, 2018a). The Baltic Sea contributes to human health and wellbeing and is of great socio-cultural, environmental and economic importance (Ahtiainen and Öhman, 2014; Hasler et al., 2016; HELCOM, 2018b). The Baltic Sea can be regarded as an example of negative impacts of human activities and climate change on ecosystem health, which, in turn, has a negative impact on its economic contribution (ability to provide goods and services) and also affects the general wellbeing of citizens (HELCOM, 2018a). Among the provisioning services, provision of food in terms of fisheries is highly important. In total, 230 fish species (including 30–40 freshwater species) have been reported in the Baltic Sea (including the transition areas to the North Sea – Kattegat and Öresund; ICES, 2020), of which only a few are of economic importance.

The WBS is a comparatively small area in the south-west of the Baltic Sea (Figure 8.1) with distinct ecological dynamics and socio-economic characteristics. As for the whole Baltic Sea, tourism plays an important role for the local economy. Food web dynamics in the shallow Western Baltic differ from the larger and deeper Central Baltic, and it is home to regional cod (Gadus morhua) and herring (Clupea harengus) stocks. These two stocks, along with plaice (Pleuronectes platessa), form the backbone of the German fishery in the Western Baltic.

The German fishery in the WBS consists of a small artisanal fishing fleet, mainly composed of 12 m gillnet cutters and a few larger (up to 40 m) boats (BMEL, 2020; Döring et al., 2020; Papaioannou et al., 2012, 2014). In addition to these fleet segments, recreational cod and herring fishing also plays a crucial role (Hyder et al., 2017). Between 2004 and 2006, 113,000 to 147,000 anglers fished in the coastal waters of the Baltic Sea (Mecklenburg-Western Pomerania and Schleswig-Holstein), catching fish for personal consumption (BFAFI, 2007); in 2014/2015, about 161,000 anglers were identified (Weltersbach et al., 2021). They invest about €118 million annually for their angling-related activities. They have to be considered an important part of the fishery, as they take about one-third of the total cod quota in the case of Western Baltic cod (Hyder et al., 2017; ICES, 2021b).

While catches of the German Western Baltic fishery are comparatively low, the socio-cultural-economic value of fisheries to the local coastal communities is very high (e.g. HELCOM, 2018b; Döring et al., 2020). Local employment opportunities are supported, and tourism in the area is boosted in coastal fishing communities (Döring et al., 2020; Papaioannou et al., 2014).

In the past decades, the marine environment has deteriorated significantly for cod and herring reproduction (Mackenzie et al., 2007; Köster et al., 2017; Voss et al., 2019), and the Western Baltic cod and herring stocks are below safe biological limits (ICES, 2021a, b). As a consequence, strict catch limitations have been enforced for the commercial as well as the recreational fishery. Beyond the negative effects on the economic situation of the fisheries sector – the size of the German coastal gillnet fleet has decreased by more than 50% (e.g. Möllmann et al., 2021) over the last three decades – these measures have damaged the livelihoods of coastal fishers and have a negative impact on their cultural identity.

Eliciting sustainability conceptions during engaged stakeholder workshop

Selection of survey participants and implementation

Based on a stakeholder mapping by Schwermer et al. (2021), representatives of relevant stakeholder groups in the Western Baltic fisheries were identified. In a second step, individuals who were interested in collaborating with the *marEEshift* project were identified within each of the stakeholder groups, resulting in a list of 61 possible participants. These included potential representatives from science, various practitioners and cooperatives, including commercial and recreational fishers (i.e. anglers), angling associations, fishing communities/protection associations and angling magazines, administration, politics and various NGOs.[1]

For the "Western Baltic Summit" workshop, two invitations were sent out in advance to these 61 people, the second containing a more detailed description and agenda of the workshop (see Appendices A and B). The "Western Baltic Summit" took place in November 2019 in Hamburg (Germany), which was easy to reach for most participants. The workshop was held in a small venue to establish a relaxed professional atmosphere. In total, 21 people attended the workshop and answered the questionnaire (see Table 8.1). We evaluated the selection of participants by asking for a self-assignment to different groups right at the beginning of the questionnaire (see further).

After a welcome reception, the leading scientists of the *marEEshift* project presented the aims and scope of the project in a short and concise manner. This input was given to communicate the research questions to the stakeholders and lay the foundation for a subsequent open discussion. The discussion was oriented along several guiding questions (see Appendix C), was guided by a professional moderator, and took approximately 60 minutes. Key discussion points were captured on a flipchart.

The survey took place immediately after the discussion. Before the questionnaire was handed out, we introduced and explained the survey (for key points, see Appendix D). In particular, we stated the survey aims, delineated the region under consideration, roughly explained the concept of sustainability and pointed to both the economic and ecological dimensions of the problem. Notably, we asked participants to focus on sustainability as an ideal objective and – to this end – ignore potential restrictions of conflicts and problems of implementation ("Imagine we could achieve everything that we wanted"). Also, participants were asked to take a non-partisan perspective rather than raising partisan claims, and they were also instructed in how to technically fill in and return the questionnaire.

Twenty participants answered the survey directly and individually filling out the questionnaire in a timeframe ranging from 15 to 25 minutes. One participant submitted the questionnaire electronically after the workshop. After the time reserved for answering the survey, lunch was provided, and there were further discussions and feedback in smaller groups.

Sustainability under uncertainty: stochastic viability

We wanted to better understand the stakeholders' normative ideas of how the Western Baltic fishery system *should* look like. To capture the normative idea of strong ecological-economic sustainability under uncertainty about future developments, we built on the concept of stochastic viability (Béné et al., 2001; Baumgärtner and Quaas, 2009; Doyen and De Lara, 2010; Béné and Doyen, 2018). The basic idea of stochastic viability is that the continued existence of certain ecological-economic system components and functions is guaranteed for a determined time period with a sufficient probability. To specify stochastic viability for a given ecological-economic system, one therefore needed to specify the following: (i) which ecological and which economic services flow from ecological and economic stocks, and (ii) which levels of ecological and what economic stocks, should be maintained (iii) over what time horizon (iv) at what level of certainty, that is, at what minimum probability?

We supposed that stakeholders when thinking normatively about sustainability have, perhaps only implicitly, a concept of stochastic viability in mind. The survey questions are designed such as to explicate their specific concept of stochastic viability of the WBS system. While this concept captures the norm of sustainability, that is, how the future development of the WBS system *should* be, the management question of *how to implement* such a development subject to the various actual constraints is a conceptually different and independent question. With the survey questions we aimed at explicating stakeholder's normative concept of sustainability, and not any opinions about potential implementation.

Questionnaire

The questionnaire was formulated in German and consisted of a short introduction, which emphasized its normative nature, followed by seven questions (see Appendix E). Our aim was to keep the questionnaire as short as possible and at the same time collect all the information that, in addition to the stakeholder survey, also contains quantitative parameters that can be used in later quantitative analyses. We chose the questions accordingly and asked for minimum, optimum and maximum levels whenever needed. The first question asked the participants to self-assign themselves to a stakeholder group by offering several options, and the last question asked for further comments. Questions 2–5 aimed at specifying the different components of stochastic viability for the Western Baltic from a stakeholders' perspective, that is, of determining (i)–(iv) defined earlier. In terms of service flows, we asked in Question 2 for the ideal number of commercial fishers, the ideal catches for anglers, as well as the relative distribution of total harvest between commercial fishers and anglers.

Question 3 focused on economic stocks in terms of port infrastructure and distribution channels. Question 4 was related to management principles and thus to the ecological/biological stocks in terms of fish stocks and the entire ecosystem. Finally, Questions 5 and 6 related to the time frame and the level of certainty in fisheries management.

Questionnaire on the sustainable development of the Western Baltic Sea

[Introduction text, not translated, see Appendix E]

1) Which of the following groups do you belong to or represent?

- Commercial fishers, full-time
- Commercial fishers, part-time
- Recreational fishers, non-commercial
- Nature conservation
- Administration & Politics
- Commerce
- Fish processing
- Tourism
- Science
- Other group: ____________________

2) The ideal use of the western Baltic Sea from a societal perspective refers, among other things, to the roles of commercial and recreational fishers.

a) From a societal perspective, how many commercial fishing enterprises (main occupation) from Germany should permanently fish the Western Baltic Sea?

Minimum: ______________ Ideal: ___________ Maximum: ___________

b) From a societal point of view, how many full-time employees should there be in these commercial fishing enterprises in addition to the owner?

(The answer does not have to be in whole numbers, e.g., 0.5 means a half-time position.)

Minimum: ______________ Ideal: ___________ Maximum: ___________

c) From a societal point of view, how many part-time commercial fishers from Germany should permanently fish the Western Baltic Sea, in addition to the number of commercial fishing enterprises mentioned above?

Minimum: ______________ Ideally: ___________ Maximum: ___________

d) For recreational fishers, in addition to the fishing experience and the size of the fish caught, the number of fish caught is important. From a societal point of view, how many fish per day and species should an individual angler be allowed to take from the Western Baltic Sea in the context of sustainable use?

Cod (Gadus morhua): Minimum: _____ Ideally: ______ Maximum: _____

Herring (Clupea harengus): Minimum: ____ Ideally: ______ Maximum: _____

Plaice (Pleuronectes platessa): Minimum: ____ Ideally: ____ Maximum: _____

Sprat (Sprattus sprattus): Minimum: _____ Ideally: ______ Maximum: _____

Whiting (Merlangius merlangus): Minimum: ___ Ideally: ___ Maximum: ___

Sea trout (Salmo trutta): Minimum: _____ Ideally: ______ Maximum: _____

Else. _______ Minimum: _____ Ideally: ______ Maximum: _____

Else. _______ Minimum: _____ Ideally: ______ Maximum: _____

e) Commercial and recreational fishers fish the same fish stocks in the Western Baltic Sea. In what proportion should the two groups ideally use the fish stocks? That is, what relative proportion (in percent) of the total catch per species should be taken by commercial and recreational fishers?

	Share of catch	
	Commercial fishers	*Recreational fishers*
Cod:	%	%
Herring:	%	%
Plaice:	%	%
Sprat:	%	%
Whiting:	%	%
Seatrout:	%	%
Other: ________________	%	%
Other: ________________	%	%

3) Sustainable development of the Western Baltic Sea also includes an idea of how fishing should be organized. This is reflected e.g. in the type and number of fishing ports as well as the marketing possibilities of the fish.

a) From a societal point of view, how many fishing ports should there be permanently on the German Baltic coast that . . .

(i) . . . are geared to the needs of commercial fishing?
Minimum: ________ Ideally: ________ Maximum: ________

ii) . . . are geared to the needs of recreational anglers?
Minimum: ________ Ideally: ________ Maximum: ________

b) A variety of distribution channels are available to commercial fishers to market the fish. From a societal perspective, what percentage of fish should be marketed through which distribution channel?

______% Direct marketing (e.g. restaurants, sales to local residents and vacation guests)
______% Cooperatives
______% Wholesale market
______% Other: ____

4) Sustainable development also involves maintaining fish stocks and the entire ecosystem in a certain state. From a societal perspective, which of the following principles should ideally be applied? (Please select one answer.)

The ecosystem and fish stocks should . . . .

- be permanently maintained at current levels.
- permanently reflect the condition before industrialized fishing began in the mid-20th century.
- be in such a condition that all native species are permanently conserved.
- be in such a condition that the biologically maximum possible amount of fish can be taken per year on a permanent basis.
- be in such a condition that permanently the profit of the commercial fishery is maximized.
- be in such a condition that ideal use by recreational fishers is permanently possible.
- be in such a condition that a good ecological status of the Baltic Sea (i.e. with regard to eutrophication, pollutant load and biodiversity) is achieved.
- *Other:* ____

5) Sustainable use and conservation of ecosystems also refers to future generations. For how many years into the future should we consider the use and conservation of the Western Baltic Sea?

______________ years

6) Future developments are always uncertain. It may not be possible to achieve the minimum levels you mentioned in questions 2 and 3 every year, despite all efforts. The certainty with which the levels will be achieved can be increased by specified, but these are costly.

With what degree of certainty should all minimum levels mentioned in questions 2 and 3 be met each year within a time horizon of 10 years? (100% = absolute certainty, 0% = complete uncertainty).

____________ % certainty

7) Do you have any other comments or additions on what sustainable development of the Western Baltic Sea should look like from an overall societal perspective?

____

Stakeholder views on the sustainable development of the Western Baltic Sea

We present the survey participants' answers according to self-selected groups (hereafter referred to as "stakeholder groups") the corresponding survey participants affiliated themselves with (Table 8.1).

Seven respondents represented non-commercial recreational fishers, three of whom also assigned themselves to the commercial and tourism sectors. Furthermore, two respondents represented the stakeholder of full-time commercial fishers, one person represented science, six participants represented the field of nature conservation and four persons represented the field of administration and politics. One person self-identified in both areas nature conservation and administration and politics. This person only answered Question 5 in numerical terms (the answer was counted to nature conservation) and otherwise put down comments.

Ecosystem services: recreational fishing

The provision of food is an important ecosystem service provided by the Baltic Sea. For individual anglers, in addition to the fishing experience and the size of the fish caught, the number of fishes caught is particularly important.

We asked, how many fish per day per species an individual angler should be allowed to catch in an ideal world regarding a sustainable use of the WBS Sea. Related to the viability concept, we asked for the minimum, optimum and maximum number. We observed the following three results: First, this question was only answered by half of all participants. While the answers from the group of recreational fishers were almost complete, only 1–2 participants from the other

TABLE 8.1 List of survey respondents, by self-allocation into stakeholder groups they represent*

Participant "P"	*Stakeholder Group*
1	Recreational fishers, non-commercial; commercial; tourism
6	Recreational fishers, non-commercial; commercial; tourism
4	Recreational fishers, non-commercial (non is crossed out); tourism
2	Recreational fishers, non-commercial
7	Recreational fishers, non-commercial
12	Recreational fishers, non-commercial
16	Recreational fishers, non-commercial
20	Other groups (working group fishery)
3	Commercial fishers, full-time
19	Commercial fishers, full-time
5	Nature conservation
10	Nature conservation
17	Nature conservation
18	Nature conservation
21	Nature conservation
8	Nature conservation, Administration and Politics
9	Administration & Politics
11	Administration & Politics
13	Administration & Politics
14	Administration & Politics
15	Science**

*Original question: "Which of the following groups do you belong to or represent?"
**Since only the results of the different stakeholder groups are relevant in the context of this work, the results of the participant from the scientific community have not been considered here.

groups answered this question. Second, the responses within the recreational fishers group were quite similar in terms of the optimal values being consistently between the minimum and maximum values. Third, the participants from administration and politics and especially from nature conservation on average consider a lower removal of fish from the sea to be more sensible than the two groups of fishers. One exception is the response for herring of one participant from administration and politics, who did not provide a numerical value, but rather a quantity of 10 kg (this value was converted into numbers by us on the basis of an average weight and corresponds to about 133 individuals). Numerical values for other fish species, such as sprat, whiting and sea trout, which were also included in the questionnaire, were only fully provided by the recreational fishers and the nature conservation stakeholders (see Appendix F).

The fact that only half of the participants answered these questions suggests that many participants found it difficult to provide specific numerical values in relation to catching fish. As far as the almost complete responses to that question from within the group of recreational fishers and the similarity of responses within this

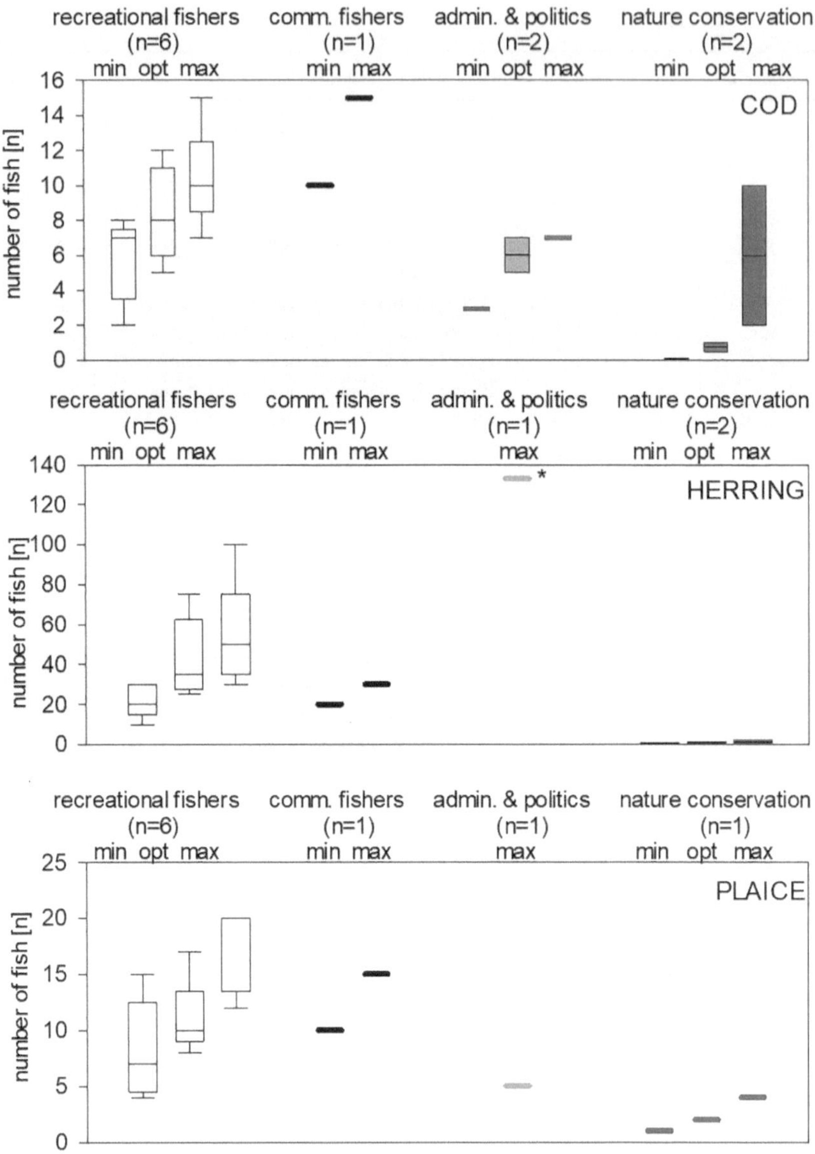

* Original information in the questionaire "10 kg".

FIGURE 8.2 "How many fish should an individual angler be allowed to catch per day from the Western Baltic Sea?" Minimum (min), optimum (opt) and maximum (max) numerical values per target species are given. Answers are grouped by stakeholder groups. The box corresponds to the range in which the middle 50% of the answers are located. The lower line shows the absolute minimum value, the upper line the absolute maximum value, the line inside the box is the median.

group are concerned, we hypothesize that this is due to this group in particular being directly affected by fishing regulations measured in numbers.

Furthermore, we received interesting comments representing the different norms and values of the stakeholder groups. The questionnaire asked for a minimum, an optimum and a maximum value. With the exception of sprat (no answer), Participant P6 did not enter a numerical value for the maximum but commented with "only own consumption", which conveys the value that everyone should have the right to self-sufficiency, that is, to catch fish for their own consumption, but not for sale. Participant P19, representing the group of commercial fishers, conveyed the value of the freedom of each individual and commented that the ideal catches of a species should be determined by the angler himself "as much as he can/wants". Additionally, a third approach became apparent, where several participants (from all stakeholder groups except commercial fishers) referred to scientific fish stock assessments and made the removal of fish dependent on the results of such research.

For example, this was evident in commenting that catches should be managed according to "limit reference points" for spawning stock biomass (P16), as currently used in the assessment of the International Council for the Exploration of the Sea (ICES), or following scientifically advised catch scenarios when the stock status required (P14). In addition, one respondent (P9) argued in more detail that there should be no limit on catches except for temporal restrictions (no catch) during the spawning season of the corresponding fish species. The participant additionally raised the question whether taking fish also meant killing them – referring to sport fishers who sometimes catch fish and put them back into the sea ("catch and release"). This point obviously was not adequately addressed by the introduction and formulation of the questionnaire.

Distribution of ecosystem services among commercial and recreational fishing

Since commercial and recreational fishers largely exploit the same stocks in the WBS, we asked in what proportion the two groups should ideally use the fish stocks and share the services provided by the fish stocks (Figure 8.3).

We observed the following: First, all stakeholder groups allocated shares to both user groups. Second, shares allocated to the groups were positive for all considered species. This means that all groups, including the nature conservation group, agreed on the normative goal of a sustainable use of the sea, as in contrast to a complete protection by, for example, a no-take scenario. Third, all stakeholders ranked both groups in a qualitatively similar way, with an overall larger share allocated to the commercial fishery (40%–80% of the total harvest). In comparison, less than 40% of the total fishery harvest should be caught by recreational fishers. Fourth, commercial and recreational fishers each put a higher relative emphasis on their own group specifically related to cod catches. These results are highly interesting, as they show converging ideas on norms within and between stakeholder groups. The first two observations suggest that

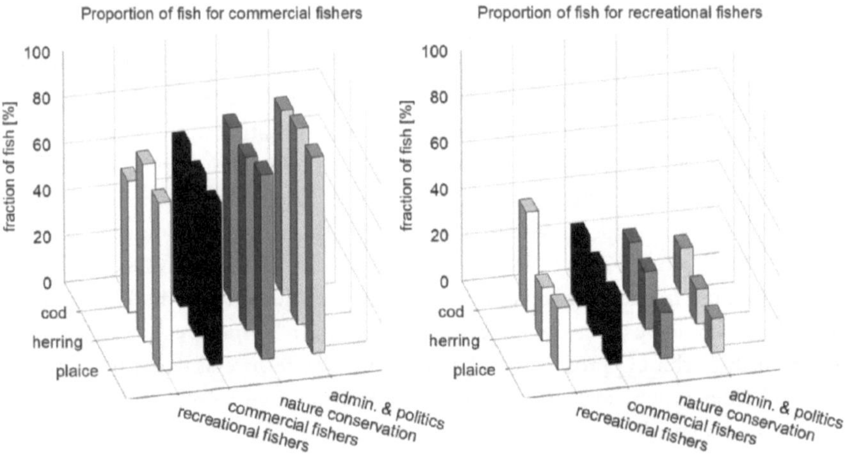

FIGURE 8.3 "What relative proportion (%) of the total harvest per species should be caught by commercial fishers and what by recreational fishers?" Answers grouped by stakeholder group and target species.

all stakeholder groups acknowledge the existence right of fisheries in general and for both user groups in an ideal world.

Economic stocks: marketing infrastructure

Sustainable development of the WBS also includes a vision of how fisheries should be organized, for example, in terms of infrastructure and thus in terms of economic stocks. This is reflected in the different marketing opportunities for fish. We asked what proportion of the fishery yields should be marketed through which distribution channel from a societal perspective. The three most important distribution channels in terms of allocated shares are: (i) direct marketing, such as restaurants or direct sale from the boat to tourists and consumers, (ii) marketing via cooperatives and (iii) the wholesale market. Table 8.2 shows how the different stakeholder groups would distribute the landed fish to the various marketing options in an ideal world.

We highlight two observations. First, all stakeholders were of the opinion that fish should be sold either directly or through cooperatives. The wholesale market or other marketing possibilities (see Appendix F) only played a minor role. Second, there is an interesting and clear discrepancy between the stakeholders "commercial fisheries" and "administration and politics". The commercial fishers opted to market the fish directly (95 %) and only to a small extent make it available to the cooperatives (5 %) as a marketing opportunity. "Administration and politics" chose the more traditional way of marketing the fish via cooperatives (between 40% and 80%, mean 63%).

The first observation shows, again, a general consensus on how an ideal state of the WBS fishery from an economic perspective would look like – the option of direct marketing – although chosen as an ideal marketing channel with different priorities – may be a way forward to support fisheries at a regional level.

TABLE 8.2 "From a societal perspective, what proportion of fish should be marketed through which distribution channel?"*

stakeholder groups	*distribution channel (average %) (Stdev)*		
n=*number of answers*	*direct marketing*	*cooperatives*	*wholesale market*
recreational fishers (n=7)	39 (24)	33 (25)	15 (19)
commercial fishers (n=1)	95	5	0
nature conservation (n=3)	53 (6)	37 (12)	10 (10)
administration and politics (n=3)	30 (17)	63 (21)	7 (6)

*Answers grouped by stakeholder group for the three channels that resulted to be most important for the survey participants and are given in average percentages and Standard deviations (in brackets).

The second observation could be interpreted in terms of different values, namely individual freedom versus a more rule-based approach with support structures, and thus related to more security. It could also be related to different types of transaction costs: For administration and politics, it may be easier to distribute a quota to few cooperatives rather than too many fishers with potentially marketing strategies.

Ecological stocks: management principles

Sustainable development also primarily involves maintaining fish stocks and the entire ecosystem in a good, that is, healthy state. One question asked which overall principle (e.g. Good Environmental Status [GES] of the Baltic Sea, or management according to the Maximum Sustainable Yield [MSY] principle; see Table 8.3) should ideally be applied from a societal point of view. To this question we received 18 valid answers.

Here, seven observations can be clearly identified. First, almost everyone (16 answers) chose the principle of achieving "Good Environmental Status" (defined as "The environmental status of marine waters where these provide ecologically diverse and dynamic oceans and seas which are clean, healthy and productive" by the EU MSFD [Marine Strategy Framework Directive]) in the Baltic Sea, which involves solutions to over-fertilization, pollution and biodiversity conservation. Second, the principle with the second largest number of selections and thus the dominant conception of sustainable fish stocks was Maximum Sustainable Yield. The MSY stock size is defined as the stock level that allows the maximum catch that can be taken from a fish stock forever, thus exploiting its maximum growth potential. Third, the principles of "preserving native species", "MEY" (Maximum Economic Yield), and "renaturation of areas" (specified in "other") were only seldomly selected.

Fourth, in administration and politics, all participants from this group answered and made identical choices. All chose both MSY and GES. Fifth, present level (i.e. status quo) was never selected. Sixth, within the group of recreational fishers, the range of answers was most diverse. Also, half of them chose "ideal use for recreational fishers" as answer. Seventh, the fact that participants gave more than one answer, although they were specifically asked to provide only one answer is an interesting observation and leads to the assumption that there seems to be no single objective, but a need for multiple objectives, and trade-offs.

TABLE 8.3 "Which of the following principles should ideally be applied from a societal perspective?" Answers grouped by stakeholder groups.

number of answers (n) per principles.	*recreational fishers (n=6)*	*commercial fishers (n=1)*	*nature conservation (n=6)*	*admin.& politics (n=4)*	*total (n)*
present level					**0**
pre-industrial condition	1				**1**
preserve native species	2	1	1		**4**
MSY (max. sustainable yield)	3	1	1	4	**9**
MEY (max. economic yield)			1		**1**
ideal use for recreational anglers	3				**3**
Good Environmental Status (GES)	6	1	5	4	**16**
renaturation of areas	1				**1**

The first and second, as well as the fifth, observations suggest a certain agreement between the different groups: MSY and GES are desirable, while this is not the case for the current situation. Both criteria, MSY and GES, are currently relevant in the management plans of these areas (European Commission 2020, ICES, 2021a), which may have triggered the choice of these options, especially for the administration & politics group. In contrast, one could also argue that both are relevant in management plans because the different groups agree on these principles. Regarding point three, somewhat surprisingly MEY was not selected, although it implies higher profits as well as lower catches and should thus be preferable for everyone (Voss et al., 2014; Voss et al., 2022). This implies that the participants were unfamiliar with this concept. In part, this might have to do with the wording, since "highest possible profit" doesn't sound as conservationist as "sustainable yield". The reason why the principles of "preserving native species", and "renaturation of areas" (specified in "other") were only seldomly selected could be due to the fact that "Good Environmental Status" might already imply these aspects.

Time horizon

Sustainable use and conservation of ecosystems also refers to future generations. The stochastic viability concept considers an explicit time horizon of finite length. Another question was therefore related to the time horizon that should be considered when utilizing and conserving the WBS. We asked "For how many years into the future should we consider the use and conservation of the Western Baltic Sea?" (Figure 8.4).

Two interesting results emerge. First, the representatives of all groups chose a positive finite number of 20 years or more. Second, there is a large variation of responses within all stakeholder groups and also between them in terms of maximum numbers, with the variation being highest for recreational fishers. Both nature conservation and administration and politics had an upper bound of 100 years. A reference to this upper level has also been reported in Schwermer et al. (2021) and may be related to a general principle in constitutional states that no use-rights are guaranteed for more than 99 years.

Regarding the second observation, the maximum value of 200 years was chosen by a stakeholder from the group of recreational fishers and thus exceeds the

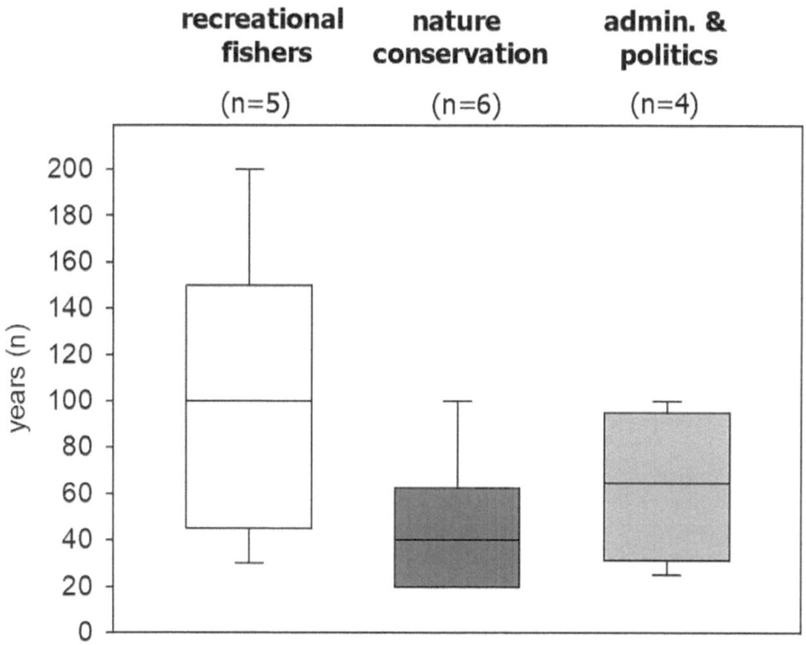

FIGURE 8.4 "For how many years in the future should we consider the use and conservation of the Western Baltic Sea?" Answers grouped by stakeholder group.

maximum values of the other groups by 100 years. The intra-group deviations were also highest here and values varied between 30 and 200 years.

The first observation suggests that respondents agree with the general idea of the stochastic viability concept that uncertainty increases with longer time horizons. The comments support this interpretation. For example, P14 mentioned that all considerations beyond 25 years are too doubtful due to high uncertainties. In a similar direction, P5 commented that for periods longer than 20 years, there is a lack of knowledge of future ecosystem relationships. Interestingly, commercial fishers did not answer this question but argued with a rather daring assessment that "everything beyond a few decades is astrology" (P19).

Certainty levels

Future developments are always uncertain. It may be that, despite all efforts, the minimum levels specified by the participants (see Figures 8.2 and 8.3; Table 8.2 and for more details, see Appendix F) cannot be reached every year. The certainty with which the levels will be achieved can be increased by measures. However, measures are associated with costs. Accordingly, the concept of stochastic viability explicitly considers uncertainty. To obtain an idea about the desired degree of certainty, we asked with what degree of certainty these stated minimum levels should be met each year within a ten-year time horizon (Figure 8.5).

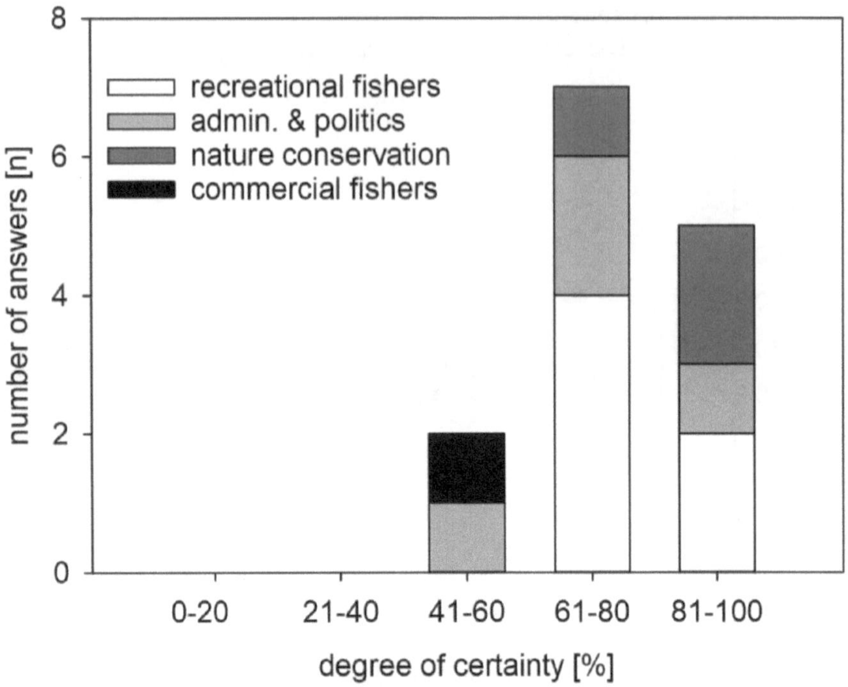

FIGURE 8.5 "With what degree of certainty should the minimum requirements be met each year within the next 10 years?" (100% = absolute certainty, 0% = complete uncertainty). Answers grouped by stakeholder group.

Two main statements can be made here: First, the mean value of all answers (79%) shows that there is an interest in an overall high level of certainty. Second, there is considerable heterogeneity between groups: Commercial fishers showed the lowest values (approx. 50%), indicating that they (as representatives) would accept conditions below the desired minimum on average every second year. The "nature conservation" group revealed highest values, and two participants selected 100% (absolute certainty). While many ideas on principles converge between stakeholder groups, the degree of acceptable uncertainty is quite different between them. This might be a major source of conflict in a situation when the ecological situation is bad and fish stock sizes are low. The conflict arises, as different principles will directly translate into politics via setting annual catch limits. Higher degrees of certainty imply less (or even no) catch, to quickly restore minimum levels, while less demand for certainty enables more flexible recovery pathways.

Insights from and challenges of stakeholder involvement

Based on the results and our interpretation of the answers to the individual questions, we identified five overarching substantive aspects. First, legitimacy of all the different stakeholder groups seems to be a broad consensus: For example, catch opportunities were acknowledged for both recreational and commercial fishers, a "healthy" ecosystem was aimed for (MSY, GES) and future generations of users were considered. This also included consensus on marketing strategies. All this may be a result of already on-going dialogue formats (e.g. Baltic Sea Advisory Council, BSAC) und provide a base for future policy steps.

Second, it is often not clear to which extent the current situation influences or is even mixed-up with the normative ideas of stakeholders. Many respondents chose "MSY" as the principle that should be applied to manage the system. MSY is a goal of the European Common Fishery Policy and in that sense represents the normative status quo. "MEY" was picked only once, although it has scientifically been shown to be more effective in securing viable fisheries and high stock sizes (e.g. Voss et al., 2022). On the one hand, this might indicate that communication of scientific results to the public and wording of scientific concepts ("sustainable" vs. "economic") needs to be improved. On the other hand, the normative goal of stakeholder groups may just be MSY and not MEY, although this would not be rational. Related, one could also argue that MSY is reflected in actual management because so many stakeholders agree on this goal. Another observation that provides evidence for mixing normative goals with the current situation is the relatively low level of certainty specified by commercial fishers. In other studies, fishers have shown a preference for income smoothing (EC, 2007). Thus, the result may suggest that commercial fishers anticipate even lower catch quotas if minimum requirements should be met with a high degree of certainty. That being said, economic preferences for income smoothing are typically elicited on an individual level. Hence, a different interpretation would be that fishers aim for certainty as individuals, while from a societal perspective they allow more uncertainty and also more flexibility.

Third, answers varied not only between stakeholder groups, but also within groups. Thus, it is unclear whether an individual's answers can really be considered as representative for a stakeholder group, or only reflect that particular individual's preferences. This may be related to our choice of respondents and whether they understood themselves as representatives. The division into groups is usually somewhat artificial – groups may be diverse themselves and may not have a predetermined joint position, or no position at all, on the topics touched upon in the questionnaire. In this case, respondents may have had no choice other than resorting to their own personal view. One could argue that the diversity within groups calls for a broader dialogue to understand what a sustainable development of the WBS fishery looks like.

Fourth, answers may convey differences in underlying values: Many answers can be related to "individual freedom" or a "rules-based approach" in terms of how the WBS should be used (see especially result section on recreational fishing and marketing infrastructure). These two opposing approaches could be related to differences in underlying world views, for example, "egalitarianism", "hierarchy" and "individualism" as used by Chuang et al. (2020) to understand British peoples' attitudes to mobility. Further research may help to better understand the interrelations between underlying world views, normative ideas about the WBS and the current state of the system.

Fifth, the observed patterns of response rates to the different questions might reflect a cognitive bias known as the "availability heuristic": Individuals answer not all questions equally, or on the basis of which questions they care about most, but according to the relative ease with which they can answer the question (Tversky and Kahneman, 1973). For instance, mostly recreational fishers answered the questions related to recreational catches and the other respondents rejected putting down catch numbers or were unable to do so. Also, all four respondents from the administration and politics group answered identically with respect to the ideal sustainability principle – they all chose MSY and GES, which is the administration's official position they frequently encounter. This might explain why the agreement on ideal states between the different groups is greater for the ecological realm than the societal realm (e.g. MSY vs. marketing channel), since all stakeholder groups are concerned with the ecological realm, for which clear management goals exist.

Conclusion

Our study revealed a fundamental challenge inherent in transdisciplinary approaches, namely the choice of relevant stakeholders and their influence on outcomes. Individual stakeholders may or may not represent their respective group well – it is very difficult to gauge whether a stakeholder speaks for him- or her-self, or on behalf of the whole group. In fact, they may relate to several groups, and groups may be diverse themselves. Since the inclusion of everyone is usually not possible, some kind of bias is very likely to occur.

We also note that separating an imaginary, normative goal for the system from the current system state seems to be difficult for most stakeholders and may point to a general problem. On the one hand, normative goals cannot be set without taking some current system characteristics into account. On the other hand, already accounting for potential implications of achieving a certain goal (e.g. temporal closure of the fishery) when thinking about the normative ideal induces a bias, that is, the "true" normative goal may not be reported. Our results suggest that a clear separation of the normative goal and the current system state is difficult, but necessary.

Furthermore, the stakeholders participating in the workshop stressed the importance of clear and unified terminology regarding concepts like "sustainability". With our approach of using stochastic viability, we provide a suggestion to make "sustainability" more tangible. In terms of comparing different normative views of stakeholder groups, we have to observe quite some variation in answers within stakeholder groups as well as some doubt whether answers always reflect purely normative views, meaning that our results have to be interpreted with care. Still, the advantage of our study is that we provide concrete values such that outcomes can directly be related to management choices that need to be quantifiable in order to be monitorable (e.g. total allowable catch). Related to this, one can also use our results for the quantification of a model to examine management options and recommendations quantitatively (e.g. see Doyen et al., 2012). Obviously, the limitations discussed must be accounted for.

While we observe considerable distinctions between different stakeholder groups' perspectives on the sustainable development of the WBS, all respondents seem to acknowledge the legitimacy of the stakes held by the different actors in to the region. Thus, there is common ground on how to sustainably use and manage the WBS, and a well-designed transdisciplinary approach with broad exchange between different stakeholders and scientists is an important step towards steering the WBS into a sustainable future.

Acknowledgements

We thank the members of the marEEshift project for their support, and especially Heike Schwermer, Robert Arlinghaus and Moritz Drupp und Björn Bos for their comments on the questionnaire. We thank also KMS (Kiel Marine Science, CAU Kiel) for their support.

Funding

The research leading to these results has received funding from the German Ministry for Education and Research (BMBF) under the projects SpaCeParti (03F0914A), marEEshift (01LC1826C) and balt_ADAPT (03F0863C).

Appendix A–F – supporting information

Supplementary data associated with this article is available at http://ssrn.com/abstract=4135478.

Note

1 NGO (Non-Governmental Organisation): Greenpeace, MSC (Marine Stewardship Council), NABU ("Naturschutzbund Deutschland e.V.", Living and breathing nature conservation), WWF (World Wide Fund For Nature).

References

Aanesen, M., Armstrong, C. W., Bloomfield, H. J. and Röckmann, C. (2014). What does stakeholder involvement mean for fisheries management? *Ecology and Society*, 19(4), 35. http://dx.doi.org/10.5751/ES-06947-190435

Ahtiainen, H., & Öhman, M. C. (2014). *Ecosystem Services in the Baltic Sea: Valuation of Marine and Coastal Ecosystem Services in the Baltic Sea.* TemaNord. https://doi.org/10.6027/TN2014-563

Baumgärtner, S., & Quaas, M. (2009), Ecological-economic viability as a criterion of strong sustainability under uncertainty. *Ecological Economics*, 68(7), 2008–2020.

Béné, C., Doyen, L., & Gabay D. (2001). A viability analysis for a bio-economic model. *Ecological Economics*, 36(3), 385–396. https://EconPapers.repec.org/RePEc:eee:ecolec:v:36:y:2001:i:3:p:385-396

Béné, C., & Doyen, L. (2018). From resistance to transformation: A generic metric of resilience through viability. *Earth's Future*, 6(7), 979–996. https://doi.org/10.1002/2017EF000660

BFAFI. (2007). Report of a pilot study of the National Fisheries Data Collection Program corresponding to Commission Regulation (EC) (Report No. 1581/2004, 7. Appendix XI (Section E):3.

BMEL. (2020). Bericht an die Europäische Kommission nach Artikel 22 der Verordnung (EU) Nr. 1380/2013 über das Gleichgewicht zwischen den Fangkapazitäten und den Fangmöglichkeiten der deutschen Fischereiflotte im Jahr 2020. Flottenbericht. www.portal-fischerei.de/bund/fischereiflotte.

Chuang, F., Manley, & E., Petersen, A. (2020). The role of worldviews in the governance of sustainable mobility. *Proceedings of the National Academy of Sciences*, 117(8), 4034–4042. https://doi.org/10.1073/pnas.1916936117

Commission of the European Communities (EC). (2013). Regulation (EU) No 1380/2013 of the European Parliament and of the Council. Brussels. https://eur-lex.europa.eu/legal-content/EN/TXT/?uri=celex%3A32013R1380

Cooke, S. J., & Cowx, I. G. (2006). Contrasting recreational and commercial fishing: Searching for common issues to promote unified conservation of fisheries resources and aquatic environments. *Biological Conservation*, 128(1), 93–108. doi.org/10.1016/j.biocon.2005.09.019

Crona B. I., Daw, T. M., Swartz, W., Norström, A. V., Nyström, M., Thyresson, M., Folke, C., Hentati-Sundberg, J., Österblom, H., Deutsch, L., & Troell, M. (2016). Masked, diluted and drowned out: How global seafood trade weakens signals from marine ecosystems. *Fish and Fisheries*, 17(4), 1175–1182. https://doi.org/10.1111/faf.12109

Döring, R., Berkenhagen, J., Hentsch, S., & Kraus, G. (2020). Small-Scale Fisheries in Germany: A Disappearing Profession? In: Pascual-Fernández, J., Pita, C., Bavinck, M. (eds) *Small-Scale Fisheries in Europe: Status, Resilience and Governance*. MARE Publication Series, Vol. 23. Springer, Cham. https://doi.org/10.1007/978-3-030-37371-9_23

Doyen, L., & De Lara, M. (2010). Stochastic viability and dynamic programming. *Systems & Control Letters*, 59(10), 629–634.https://doi.org/10.1016/j.sysconle.2010.07.008ff. ffhal-00453499f

Doyen, L., Thébaud, O., Béné, C., Martinet, V., Gourguet, S., Bertignac, M., Fifas, S., & Blanchard, F. (2012). A stochastic viability approach to ecosystem-based fisheries management. *Ecological Economics*, 75, 32–42. https://doi.org/10.1016/j.ecolecon.2012.01.005

EC (European Commission). (2007). Council Regulation (EC) No. 1098/2007 establishing a multi-annual plan for the cod stocks in the Baltic Sea and the fisheries exploiting those stocks, amending Regulation (ECC) No 2847/93 and repealing Regulation (EC) No 779/97. http://data.europa.eu/eli/reg/2007/1098/oj

EU (2013). Regulation (EU) No 1380/2013 of the European Parliament and of the Council of 11 December 2013 on the Common Fisheries Policy, amending Council Regulations (EC) No 1954/2003 and (EC) No 1224/2009 and repealing Council Regulations (EC) No 2371/2002 and (EC) No 639/2004 and Council Decision 2004/585/EC. http://data.europa.eu/eli/reg/2013/1380/oj

European Commission (2020). Report from the Commission to the European Parliament and the Council on the implementation of the Marine Strategy Framework Directive (Directive 2008/56/EC) Brussels, 25.6.2020 COM(2020) 259 final; {SWD(2020) 60 final} – {SWD(2020) 61 final} – {SWD(2020) 62 final}. Brussels: European Commission, 307.

FAO (2020). The State of World Fisheries and Aquaculture 2020. In Brief. Sustainability in Action. Rome. https://doi.org/10.4060/ca9229en

Hasler, B., Ahtiainen, H., Hasselström, L., Heiskanen, A.-S., Soutukorva, Å., & Martinsen, L. (2016). Marine Ecosystem Services: Marine ecosystem services in Nordic marine waters and the Baltic Sea – possibilities for valuation. https://doi.org/10.6027/TN2016-501

HELCOM. (2010). Ecosystem Health of the Baltic Sea 2003–2007: HELCOM Initial Holistic Assessment. *Baltic Sea Environment Proceedings* No. 122. www.helcom.fi/stc/files/Publications/Proceedings/bsep122.pdf

HELCOM. (2018a). Economic and social analyses in the Baltic Sea region – HELCOM Thematic assessment 2011–2016. www.helcom.f/baltic-sea-trends/holistic-assessments/state-of-the-baltic-sea-2018/reports-and-materials.

HELCOM. (2018b). State of the Baltic Sea – Second HELCOM holistic assessment 2011–2016. Baltic Sea Environment Proceedings 155. www.helcom.fi/baltic-sea-trends/holistic-assessments/state-of-the-baltic-sea-2018/reports-and-materials/

HELCOM. (2018c). Status of coastal fish communities in the Baltic Sea during 2011–2016 – the third thematic assessment. Baltic Sea Environment Proceedings No. 161

Hilborn, R., Amoroso, R. O., Anderson, C. M., Baum, J. K., Branch, T. A., Costello, C., de Moor, C. L., Faraj, A., Hively, D., Jensen, O. P., Kurota, H., Little, L. R., Mace, P., McClanahan, T., Melnychuk, M. C., Minto, C., Osio, G. C., Parma, A. M., Pons, M., . . . Ye, Y. (2020). Effective fisheries management instrumental in improving fish stock status. *Proceedings of the National Academy of Sciences of the United States of America*, 117(4), 2218–2224. https://doi.org/10.1073/pnas.1909726116

Hyder, K., Weltersbach, M. S., Armstrong, M., Ferter, K., Townhill, B., Ahvonen, A., Arlinghaus, R., Baikov, A., Bellanger, M., Birzaks, J., Borch, T., Cambie, G., de Graaf, M., Diogo, H. M. C., Dziemian, L., Gordoa, A., Grzebielec, R., Hartill, B., Kagervall, A., . . . Strehlow, H. V. (2017). Recreational sea fishing in Europe in a global context – participation rates, fishing effort, expenditure, and implications for monitoring and assessment. *Fish and Fisheries*, 19(2). https://doi.org/10.1111/faf.12251

ICES. (1998). Report of the Study Group on the Precautionary Approach to Fisheries Management. ICES CM 1998/ACFM:10. ICES.

ICES. (2020). Ecosystem overviews – Baltic Sea ecoregion. https://doi.org/10.17895/ices.advice.7635.

ICES. (2021a). Fisheries overviews Baltic Sea ecoregion. https://doi.org/10.17895/ices.advice.9139.

ICES. (2021b). Cod (Gadus morhua) in subdivisions 22–24, western Baltic stock (western Baltic Sea). ICES Advice on Fishing Opportunities, Catch, and Effort – cod.27.22–24. Baltic Sea Ecoregion, https://doi.org/10.17895/ices.advice.7744

Ihde, T. F., Wilberg, M. J., Loewensteiner, D. A., Secor, D. H., & Miller, T. J. (2011). The increasing importance of marine recreational fishing in the US: Challenges for management. *Fisheries Research*, 108(2–3), 268–276. https://doi.org/10.1016/j.fishres.2010.12.016

Köster, F., Huwer, B., Hinrichsen, H-H., Neumann, V., Makarchouk, A., Eero, M., Dewitz, B. V., Hüssy, K., Tomkiewicz, J., Margonski, P., Temming, A., Hermann, J-P., Oesterwind, D., Dierking, J., Kotterba, P., & Plikshs, M. (2017). Eastern Baltic cod recruitment revisited – dynamics and impacting factors. *ICES Journal of Marine Science*, 74(1), 3–19. https://doi.org/10.1093/icesjms/fsw172

Lade, S. J., Niiranen, S., Hentati-Sundberg, J., Blenckner, T., Boonstra, W. J., Orach, K., Quaas, M. F., Österblom, H., & Schlüter, M. (2015). An empirical model of the Baltic Sea reveals the importance of social dynamics for ecological regime shifts. *Proceedings of the National Academy of Sciences of the United States of America*, 112(35), 11120–11125. https://doi.org/10.1073/PNAS.1504954112

MacKenzie, B. R., Gislason, H., Möllmann, C., & Köster, F. W. (2007). Impact of 21st century climate change on the Baltic Sea fish community and fisheries. *Global Change Biology*, 13(7), 1348–1367. https://doi.org/10.1111/J.1365-2486.2007.01369.X

Möllmann, C., Cormon, X., Funk, S., Otto, S., Schmidt, J.O., Schwermer, H., Sguotti, C., Voss, R., & Quaas, M. (2021). Tipping point realized in cod fishery. *Scientific Reports*, 11 (14259). https://doi.org/10.1038/s41598-021-93843-z

NOAA. (2015). *Introduction to Stakeholder Participation. Office for Coastal Management*. Charleston, SC: NOAA Office for Coastal Management Agency. Available at: https://coast.noaa.gov/data/digi-talcoast/pdf/stakeholder-participation.pdf

Oen, A. M. P., Sparrevik, M., Barton, D. N., Nagothu, U. S., Ellen, G. J., Breedveld, G. D., Skei, J., & Slob, A. (2010). Sediment and society: An approach for assessing management of contaminated sediments and stakeholder involvement in Norway. *Journal of Soils and Sediments*, 10(2), 202–208. https://doi.org/10.1007/S11368-009-0182-X

Papaioannou, E. A., Vafeidis, A. V., Quaas, M. F., Schmidt, J. O., & Strehlow, H. (2014). Using indicators based on primary fisheries data for assessing the development of the German Baltic small-scale fishery and reviewing its adaptation potential to changes in resource abundance and management during 2000–09. *Ocean & Coastal Management*, 98, 38–50. http://dx.doi.org/10.1016/j.ocecoaman.2014.06.005

Papaioannou, E. A., Vafeidis, A. T., Quaas, M. F., & Schmidt, J. O. (2012). The development and use of a spatial database for the determination and characterization of the state of the German Baltic small-scale fishery sector. *ICES Journal of Marine Science*, 69(8), 1480–1490. https://doi.org/10.1093/icesjms/fss096

Ritchie, H., & Roser, M. (2021). "Biodiversity". Published online at OurWorldInData.org. Available at: https://ourworldindata.org/biodiversity (Online Resource)

Schwermer, H., Blöcker, A. M., Möllmann, C., & Döring, M. (2021). The 'Cod-Multiple': Modes of existence of fish, science and people. *Sustainability*, 13(21). https://doi.org/10.3390/su132112229

Soma, K., Nielsen, J. R., Papadopoulou, N., Polet, H., Zengin, M., Smith, C. J., Eigaard, O. R., Sala, A., Bonanomi, S., van den Burg, S. W. K., Piet, G. J., Buisman, E., & Gümüş, A. (2018). Stakeholder perceptions in fisheries management – Sectors with benthic impacts. *Marine Policy*, 92, 73–85. https://doi.org/10.1016/J.MARPOL.2018.02.019

Tversky, A., & Kahneman, D. 1973. Availability: A heuristic for judging frequency and probability. *Cognitive Psychology*, 5(2), 207–232. https://doi.org/10.1016/0010-0285(73)90033-9

Van der Heel, S. (2018). Science for change: A survey on the normative and political dimensions of global sustainability research. *Global Environmental Change*, 52, 248–258. https://doi.org/10.1016/j.gloenvcha.2018.07.005

Voss, R., Quaas, M. F., Stiasny, M. H., Hänsel, M.C., Stecher J. P., Guilherme A., Lehmann, A., Reusch, T. B. H., & Schmidt, J. O. (2019). Ecological-economic sustainability of the Baltic cod fisheries under ocean warming and acidification. *Journal of Environmental Management*, 238, 110–118. https://doi.org/10.1016/j.jenvman.2019.02.105

Voss, V., Quaas, M., & Neuenfeldt, S. (2022). Robust, ecological – economic multispecies management of Central Baltic fishery resources. *ICES Journal of Marine Science*, 78(1), 169–181. https://doi.org/10.1093/icesjms/fsab251

Voss, R., Quaas M. F., Schmidt, J. O., & Hoffmann, J. (2014). Regional trade-offs from multispecies maximum sustainable yield (MMSY) management options. *Marine Ecology Progress Series*, 498, 1–12. https://doi.org/10.3354/meps10639

Weltersbach, M. S., Riepe, C., Lewin, W.C., & Strehlow, H.V. (2021). Ökologische, soziale und ökonomische Dimensionen des Meeresangelns in Deutschland (Thünen Report No. 83, ISBN 978-3-86576-221-4). Johann Heinrich von Thünen-Institut. https://doi.org/10.3220/REP1611578297000

Withycombe-Keeler, L., Wiek, A., White, D. D., & Sampson, D. A. (2015). Linking stakeholder survey, scenario analysis, and simulation modelling to explore the long-term impacts of regional water governance regimes. *Environmental Science & Policy*, 48, 237–249. https://doi.org/10.1016/j.envsci.2015.01.006

PART III
Insights from the case studies

Case studies, often undertaken at the local level, are a classical research setup in transdisciplinary projects. The emphasis on the local context moves into focus the tangible and intangible social constructions of a place and enables us to understand what factors shape marine management and decision-making on the ground. In research settings with an explorative character, qualitative methods of data collection are a valuable means for understanding the relationships between different actors, their perceptions of the situation under study, and the impact these perceptions have on their actions. Based on an ethnographic methodology applied at the Portuguese coast, J. Sá Couto illustrates the significance of local, embodied, and empirical knowledge from local communities for researching situated perceptions of seascapes (Chapter 9). Understanding the manifold challenges that small-scale fishing communities face is an important step towards sustainably managing local marine resources and connected human practices. In this context, an ethnographic approach can deliver a toolbox of methods for investigating local stakeholder knowledges and integrating it into scientific inquiry as well as processes of policy making.

The local environmental knowledge of fishing communities also plays a key role in the case study presented by Ramos and colleagues in Chapter 10. Focusing on the small-scale fishery of a coastal lagoon in Uruguay, the authors used group discussions and qualitative interviews to investigate fishing practices and the socio-economic situation of the coastal residents. Representative for many fishing communities worldwide, this case shows how specific management measures around the lagoon influence the livelihoods of the fishers living off the marine resources in the area. Moreover, the case highlights well how the integration of local environmental knowledge into management and policy processes empowers communities as a whole and legitimises them as actors with a voice. This integration not only adds important technical and ecological information to such processes, but at the

DOI: 10.4324/9781003311171-11

same time puts into value the heritage and experiences of generations of marine practitioners.

The integration of local knowledge into management is an important process at all times; however, it becomes especially crucial in cases of environmental disasters. In Chapter 11, Machado and colleagues reflect on the impacts of an oil spill disaster at the Brazilian coast in 2019 on local fishing communities. Their methodology of a Community-Based Participatory Research (CBPR) approach enabled a 'dialogue of knowledge' with the different actors participating in their research and unveils the vast effects of the disaster on fishers in the region and their families. The value of this research approach, on the one hand, is its suitability for the study of environmental disasters which, as the authors argue, are both complex and transdisciplinary in nature. On the other hand, CBPR constitutes a valuable tool in building and maintaining networks of researchers and non-academic participants who engage in short- or long-term partnerships; a tool that has the potential to aid the investigation of other environmental disasters, their impacts on local residents, and the decision-making on the most effective interventions in the aftermath.

In this section's final chapter, Florido-del-Corral and Abbot-Jiménez analyse a participatory process aiming at the declaration of a Marine Reserve of Fishing Interest (MRFI) around the port of Conil de la Frontera, Spain. Based on a cooperation between social anthropologists, biologists and professional fishers, the authors present the difficulties of reaching a management decision that represents the viewpoints of all actors involved, from academics to environmentalists to the fishing sector. As a key point hindering consensus-finding in Conil emerges the dispute over the area as a social space shared by recreational and professional small-scale fishers. Furthermore, this case study highlights the varying values given by different actors to diverse forms of knowledge. Since the national authorities involved in the discussion attach higher value to scientific, 'objective' knowledge, the local communities perceive their experiential, personal, and 'subjective' knowledge to be underrepresented in the debate about the fishing reserve. Although the value of traditional ecological knowledge has often been proven also in academic literature (and by many chapters in this book), this superimposed hierarchy of knowledge leads to a feeling of being judged and overlooked by the stakeholders involved in the processes – a feeling that constitutes a heavy burden for transdisciplinarity and participatory approaches.

9

SMALL-SCALE FISHERS' KNOWLEDGE FOR OCEAN SUSTAINABILITY

An ethnography in Setúbal, Portugal

Joana Sá Couto

Introduction

Globally, oceans have been the target of growing attention. Year 2021 marked the beginning of the United Nations Decade of Ocean Science for Sustainable Development, and the postponed Ocean Conference is to take place in Portugal. The role of the ocean in the earth system and the ocean-climate nexus has been a key factor in the growing importance of ocean science worldwide. In 2019, the Intergovernmental Panel on Climate Change released a Special Report on the Ocean and Cryosphere in a Changing Climate giving visibility to the dependence of communities worldwide on the ocean and cryosphere and their vulnerability to climate change related issues (IPCC, 2019). Most recently, in the IPCC WGII Sixth Assessment Report released in 2022, it is stated that small-scale fisheries (SSF), livelihoods, and jobs are the most vulnerable to climate-driven changes (IPCC, 2022). In Europe, SSF represent only 8% of the global SSF catch, however, they display enormous cultural and social importance (TBTI, 2016). Adding to environmental challenges, SSF face other social, political, and economic pressures, such as competition with bigger-scale fisheries, lack of funding, unjust top-down policies, and even spatial competition with other economic activities. Since the publication of the EU Blue Growth Agenda in 2012, oceans are framed as crucial for economic growth. This strategy is based on five pillars: renewable ocean energy, aquaculture, coastal and maritime tourism, mineral resource exploration, and blue biotechnology. Small-scale fishing is omitted or downplayed due to its low growth potential, which exacerbates the uncertain future of these communities (Ayilu et al., 2022; Hadjimichael, 2018).

Within this international context, the Portuguese increased its focus on its extensive shoreline and Exclusive Economic Zones (EEZ) towards climate change mitigation and adaptation, among other environmental concerns. The National

DOI: 10.4324/9781003311171-12

Strategy for the Sea 2021–2030 (ENM, 2021) aims to lay the strategic guidelines for future maritime policies and actions giving high priority to the reinforcement of economic, social, and environmentally sustainable fishing activities and sustainable development of coastal communities. Under this scope, it refers to the necessity of stimulating the attractiveness of the fisher profession and to consider the role of fishing communities in a circular economy, through financial aid for young fishers and the adoption of new technologies, exempting older fleets of needed aid (Resolução do Conselho de Ministros n.º 120/2021). These ambitious objectives in the ENM 2021–2030 can only be fulfilled through in-depth knowledge of the reality of fishing communities, and, from there, conceptualising a sustainable ocean management paradigm. However, fisheries have often been disregarded in Portuguese policy. As discussed in this chapter, fishing communities are an intrinsic part of the marine ecosystem and need to be considered in an integrated approach that can lead to better management policies, essential to a more sustainable ocean. SSF are also a great example of possible bridging between social and natural sciences, and needed transdisciplinarity, if approached as a more-than-human-context. Only then it is possible to challenge the commodification of sea resources and labour, "but also the liberal, dualistic epistemologies and knowledge practices that shape hegemonic forms of environmental regulation and control" (Bresnihan, 2016). This case study shines a light on a fishing community in an urban context in Portugal, to reflect on the importance of SSF as a human relationship with the ocean.

Case study

The fishing community of Setúbal

Setúbal is one of the most important Portuguese cities. Its characteristic geography, pronounced by the Sado River Bay, shelters the city, rendering it a prime condition for port activity. This has been the main factor for the area's development, through activities such as freight traffic, naval construction and repair, or salt production, due to its sunny climate (Quintas, 1998).

By the 18th century, it was considered one of the main fishing docks in the country. Throughout the 19th century, this vibrant fishing activity supplied the inordinate development of the canned fish industry (Quintas, 1998). Setúbal benefits from a great richness in terms of biodiversity, and fosters several protected areas, including the Natural Reserve of the Sado Estuary; to the west, the Professor Luíz Saldanha Marine Park, also part of the NATURA 2000 network; and the Arrábida Natural Park (Martins et al., 2005). The earliest fishing-related archaeological vestiges date back to the Iron Age (Martins et al., 2005) confirming an ancient fishing tradition and the dependence of the city's growth and development (albeit mostly unplanned) on fishing, and later, the fish processing industry. Indeed, there was a remarkable growth related to the fishing and fish processing industry in the 19th century. However, in the latter half of the century, the fishing sector entered a crisis from which it never recovered (Garrido, 2018; Lopes and Perreira, 2015).

Nowadays, the Setúbal fishing community is heterogeneous in its identity, due to the differences between elder fishers and new generations, but also due to the multiple types of fishing gear used (Amorim, 2015). It is a small-scale community, given the small dimension of boats, and declining number of fishers. Despite the marked decline of fishing activity, it persists and has a clear social and territorial expression. The fishing dock is located in one of the most important avenues of the city, and it is a display sight for restaurants that promote fresh fish in their windows. The town uses its fishing tradition as a tourist attraction, even though the focus is placed much more on the fish than on the fisherfolk. The fisher as a picturesque symbol is referenced as a manner of highlighting the fish and the fishing tradition of the town, while in practical terms there is still a generalised deprecatory image, already seen in 19th-century press (Ferreira and Santos, 2020). This image is worsened by the successive international critical narratives of fishing activities, which do not consider the difference between industrial fishing and SSF, nor the importance of fishing communities.

Covid-19 placed an even bigger burden on this vulnerable community that has been suffering from enormous challenges related to climate change and its socio-economic status. The lack of protection of these communities, despite their importance for the food provision of thousands, is paradoxical (Bennett et al., 2020; Pita et al., 2020). Additionally, the valuing of other maritime economic activities such as the tourism sector and aquaculture led to the alienation of small-scale fishing communities. The process of gentrification of the city is visible in everyday life. Setúbal remains an important place for transdisciplinary analysis due to the several vulnerabilities it faces: it is a coastal city, urbanized, and undergoing deep changes as a part of the Lisbon Metropolitan Area. It is simultaneously industrialized yet focused on appealing to tourism through its natural landscape. The remaining fishing community exists in an increasingly disputed place of the urban fabric, which creates a unique context and combination of factors of analysis.

Increasing challenges for Portuguese fisheries

Portugal's integration into the European Economic Community (EEC), in 1986, was a turning point in what concerns fisheries management. Access to long-distance fishing was reduced, European subsidies were used to reduce fishing fleets, and maximum capture quotas led to a situation where currently, despite its fishing tradition, Portugal imports 60% of the fish consumed in the country (Álvares, 1986; Garrido, 2018; Madureira and Amorim, 2001; Pittae Cunha, 2011). Portuguese fisheries management obeys norms established by the EU common fisheries policy, based on maintaining and preserving resources with measures, such as a definition of the total allowable catch. There are still regulations on mesh size, seasonal interdictions, as well as mandatory discards of several species as a population control measure (Garrido, 2018). Generally, fisheries management is based on the precautionary principle, using biomass harvest information to measure fishery success (Anderson, 2015; Costa et al., 2022), and through the naturalization of the narrative of Hardin's tragedy of the commons (Bresnihan, 2016; Hardin, 1975).

Furthermore, with the industrialization of fishing, the focus of regulators is on the effects of their measures on profit, management, and regulation costs, considering fishers as profit-seeking rational economic agents (Anderson, 2015). While it is true that the proletarianization of the sector has changed the way in which fishers deal with their activity and with the environment, this does not always imply a lack of ecological sensibility (Bailey, 2018; Howard, 2017; Martins, 1999; Oneto, 2008). In fact, fishing activity is at its core a relationship between humans in the environment, that is, more often than not, economically irrational (Bailey, 2018).

Portugal has one of the largest European EEZs with 1.7 million km^2, which is about 48% of the total marine waters under EU jurisdiction. Despite this considerable area, the Portuguese fishing fleet is composed mostly of small-scale fishing vessels, where 90% of the registered units have a length up to twelve meters and a reduced gross tonnage, with a combined fishing value of only 14% of the national total. With aging and poorly maintained small-scale fishing fleet, the growing political and economic focus on the blue economy may seem paradoxical. However, the Portuguese economic strength in what concerns maritime matters is not fisheries. In the most recent Maritime Economy Monitoring Report, it is stated that between 2012 and 2019, the number of registered fishers decreased by 12%, the number of fishing vessels suffered a loss of 6%, and captures were reduced by 5% (DGPM, 2020).

Environmental challenges such as climate change are impacting fisheries in direct and indirect ways, increasing the ever-present uncertainty in this activity (Daw et al., 2009). In Europe, water temperature changes have strongly impacted maritime fauna distribution over the past three decades (Barange et al., 2018). According to the SIAM II project (Santos and Miranda, 2006), beyond the temperature problem, fishing has been affected by alterations in wind direction and intensity, precipitation, and river flow. In Setúbal, the environmental and economic challenges are manifold: rising sea levels, surface water temperatures, droughts, soil impermeabilization, and resulting floods and heatwaves (Santos and Miranda, 2006). In parallel, there is a relevant cultural dimension since this town is characterized by an important fishing tradition and affinity to the sea. This community is therefore essential to reflect on environmental matters and should be considered when discussing climate change mitigation and adaptation plans. Culture is inherent to the forms of production, consumption, lifestyle, and social organization of a given community, and this understanding is crucial for the implementation of effective adaptation measures. Beyond that, climate change has a deep impact on cultures, at various levels (Adger et al., 2013). In the case of coastal zones, as the seaside changes, communities change with it, building new cultural meanings and adaptation strategies, aiming to increase their resilience, as they are not only human groups in a changing environment, but also because humans and the environment change in synchronicity (Gillis, 2012). Coastal zones have a long history of testing the human potentials and limitations, such as independence, hard work, and the vigilance and capacities necessary to read and respond to warnings with agility and balance (Griffith, 1999).

However, environmental degradation in the town of Setúbal is not limited to climate change impacts: this is a town with heavy industries (cement, paper, automobile, metallurgical) and all the pollution they entail. Supplied by these industries, the Port of Setúbal has considerable strategic importance at the national level and has been increasing the value and volume of its commercial trade. In fact, Setúbal is a town with a set of specific circumstances that make it valuable to tackle the necessity for transdisciplinary science. In face of the growing degradation of environmental conditions and the social fabric of this fishing community, this chapter should also serve as a call for action in valuing SSF communities.

Fieldwork within Setúbal's fishers

The methodology chosen for data collection was ethnography (Bryman, 2012; Heyl, 2001; Shagrir, 2017; Spradley, 1979). Fieldwork was conducted intensively between 2017 and 2018 and has continued intermittently since. It has been conducted with the fishing community, and by attending relevant events in the city related to sustainability, ocean and fishing, and municipality fish-related campaigns. Ethnography has been used to push beyond disciplines and participant observation, integrating methods and theories, but also allowing the ethnographer to remain "fully participatingly human" (Pina-Cabral, 2017: x). When one seeks to approach sustainability in bio-socio-environmental systems, this analysis entails materials, environmental elements, and the body in being with the community experiencing the becoming, as some authors suggest (Bateson, 2000; Crowther, 2019; Descola and Pálsson, 2002; Ingold, 2004, 2011; Viveiros de Castro, 2012). Therefore, the ethnographer's method is being there, learning, allowing themes to emerge in the field that go beyond disciplines (Crowther, 2019; Ingold, 2011), but also allowing space for empathy, the senses, and feelings in the construction of ethnographic knowledge (Davies and Stodulka, 2019; Pussetti, 2016).

During fieldwork, social and economic tensions were at play in the urban space, in the discourses, and economic strategies. There are several ways in which the municipality highlights the fresh fish from local seas, without any real safeguarding of fishers, fostering distrust of institutions. Ethnography allows for a higher level of trust and free dialogue, which, between words and silences, helps to better understand the position of these fishing communities. Speeches about climate change are common in everyday life, as it is undeniable evidence that affects fishing, and the weather seems to be a usual topic of conversation in most social contexts. Statements as to the lack of seasons, drought, the need for strong storms that will "stir up the bottoms and fatten the fish", as well as the differences in the distribution of species and their evolution throughout the year are common between fishers and other members of the community, such as dock workers. Some express concerns related to the lag between the definition of closed seasons and the actual state of the shoals, which often open when a certain species is not yet "ready" for catchment or leads to catching "large and fat fish" when it is closed – which end up being discarded due to regulatory compliance.

This local knowledge is learned through socialization and by doing. This type of embodied learning is extremely important and characteristic of these communities due to the affinity to the environment and to the sea life. These use small boats, with gears that allow them to quickly verify what is caught when pulled into the boat, unlike industrial fishing practices. Each crew member has a specific role, entailing tasks learned through observation and repetition of movements, crucial to the crew, in an organized way, and unison with their vehicle and gear.

In this community, there are older artisanal fishermen (50 to 80 years old, low education attainment, started working very young and usually with family members) who, due to their lifelong dedication to the activity, are able to describe both the undergoing changes and also differences in daily life before current technological advances. Most of these fishermen have a deep ecological awareness and knowledge about their environment. They can perfectly describe the seabed, the best places to fish, and why. Some describe how they used to be guided only by the stars, and always found their way home. The problem of fish stocks is a constant concern since these fishermen do not want to lose their livelihood, but there is a permanent contradiction: on the one hand, it is necessary to protect as much as possible some species in notorious decline, on the other hand, it is necessary to make ends meet.

Some fishermen criticize some of the current gear, which they call murderous. This is mostly related to nylon nets, nowadays dominant in the community. Likewise, it was possible to observe, on a summer afternoon, the display of a video on social media of an industrial fishing boat, which raised comments about the destruction of the seas, but also about the profit that comes from this type of fishing. The fishermen speeches mirror what seems to be the inevitable path for fisheries: the neglect of small-scale communities, in favour of large investments in industrial fishing and on-board preparation of fish, and aquaculture. However, SSF can be much more environmentally sustainable. These fishermen catch relatively small amounts of fish, and, due to the size of boats and gear, they can easily avoid and, if it happens, remove animals caught in bycatch, including cetaceans and birds. In parallel, some use gear with little environmental impact, such as longlines. However, it is important to mention the recurring discourses about the precariousness of the profession: many do not want their children to embark on the activity. This is due to the lack of conditions for making a living out of fishing and the rapid decline of the community as evident by the decrease in the number of fishing boats and the increase of recreational boats on the dock.

The sustainability of SSF is linked to local knowledge about the natural environment, and it is reflected in the fishing gear. In recent years, there has been a growing concern about the problem of plastic pollution in the sea. This added to the heated discourses that consider them agents of ecosystem erosion, mainly due to lost and otherwise discarded fishing gear. During fieldwork, when confronted with the issue of plastic pollution, fishermen often referred to the damaging discharges made by the heavy industries that they felt didn't face as much criticism since they are not as visible – "but we know when it happens". Also, there are techniques for reusing fishing gear that date to times when it was made of cork, cotton, and other materials instead of plastic. These can mend a broken cable, or stitch together a whole broken net, with little to no new materials. These techniques were used to avoid waste and increase the lifetime of a material that is expensive and, currently, made of a more environmentally damaging material. Small-scale artisanal fishermen still know and use these techniques and are eager to teach them. Many of the younger fishermen who know them do not practice them, because they are afraid of working longer hours without the compensation they deserve. Others do not learn them at all. These are techniques that they can only learn from other fishers. The loss of this knowledge, as a consequence of the stratification of the community, results in more discarded fishing gear, which in this case is plastic, with all the problems that this implies.

Valuable information is found in the observation of behaviours and silences, of what is often not verbalized. From the perplexity that some researcher is interested in them, to the accounts of the way of living that is part of their identity. Once trust is established, it is possible to understand the fishing community as a fishing net, as Jentoft (2000a) puts it, in which each node between the meshes is a crucial element of the mesh, and each one interconnects with the others. We can still consider other elements, such as the practical implications of how the fish auction went the day before, how the weather influences the decision to go out to sea, and at which time of the day.

Discussion

The fishing community at a tipping point

The fishing community in Setúbal is at a tipping point. Fish is a valued commodity, but the work of fishers is not recognized, nor their struggles (Howard, 2017). Through ethnography, it was possible to understand their social condition, and, through the discourses of elder and younger fishers, it is possible to also understand the impacts of the fishing crisis in this community. The modernization of the fishing sector has not only turned work considerably lighter (physically, with the help of engines and new materials) but also contributed to an alienation from the activity. Increasingly, fisheries management has led to a transfiguration of fishers, the proletarianization of the sector and alteration of their relationship with the means of production. Even if there is a recognition that fishing is a way of life, and a genuine pleasure in going out to sea, there is also the necessity for profit-making in a growth-centred economy, in what is an economic activity of food production.

However, the proletarianization of the sector carries risks. In an economic sector where the fisher is the least rewarded cog in the wheel, who fishes and is often unable to sell the product, pay with dignity to the crew, as well as covering for the costs of material, fuel, and other expenses, the relationship with labour is transformed. The fishing class currently, due to its precarity and difficulty to make ends meet, cannot be expected to relate to the activity in the same way – what Bernstein (2010) called the commodification of subsistence. It is a transfiguration caused by the capitalist system in a worker's class that has been increasingly forgotten, taken as obsolete, and demanded to be larger-scale, industrialized, and profitable. The alienation from the means of production is a condition of the economic system that alienates not only from the work but also from fellow workers (Engels and Marx, 1972; Howard, 2017) risking the notion of community. And for communities to be sustainable they need policies that nurture this "We" (Jentoft, 2020a). As noted before, in fisheries management, fishing communities have been disregarded. However, they are essential for ocean sustainability. Fisherfolk are more than strategic market agents, they are people who live within communities, raised deeply immersed in cultural and social systems which give meaning to their lives (Jentoft, 2000a; van Ginkel, 2008). The current fisheries management system and the increased proletarianization of the sector contribute to stratify, dismantle, and disintegrate communities, not nurture them, resulting in the kind of actors portrayed in Hardin's Tragedy of the Commons (Hardin, 1975; Jentoft, 2000a; Jentoft, 2020a).

Despite the heterogeneity in the community, there is an important know-how in fishing, through socialization, daily life, and the repetition of tasks and activities. It is through socialization that an individual becomes a fisher, not only for the simple fact of being able to execute tasks but also through the internalization of norms, values, attitudes, interests, knowledge, and capacities necessary to become a member of the group, to be accomplished in the activity and to legitimize their labour (van Ginkel, 2008). Furthermore, this knowledge stems from their abilities, from their incorporated capacities to develop awareness and surpass challenges, built through a history of involvement with the seascape and its elements.

The increase of knowledge is an aspect inherent to the growth of a being within its environment (Ingold, 2004). This knowledge can be defined in several ways. Some authors may use traditional knowledge (Calafati, 2006; Thornton and Scheer, 2012) or local knowledge (Antweiler, 1998; Lam et al., 2020), while some may use local ecological knowledge (LEK) (Aswani et al., 2018; Houde, 2007; Freitas et al., 2018) but all are used concerning the need for valuing fishers' empirical knowledge (Delicado et al., 2012; McClanahan et al., 2009; Murray et al., 2005; Silvano and Valbo-Jørgensen, 2008; Wilson et al., 2013).

This chapter intends to show that knowledge in SSF goes beyond ecological issues, meshing with other matters of social life, and piecing together a community that is intrinsically connected to dynamic knowledge systems crucial for a sustainable seascape. Here, local knowledge is defined as all types of knowledge derived from the experience of an individual within a community. It is important to note that most of this knowledge is embodied, where the body acts as one with the mind, in an act that can hardly be verbalized. The body itself is the knowing subject (Tanaka, 2011). It is, therefore, valuable knowledge from those who deal with the sea daily, with all the challenges it carries. Currently, this knowledge is in danger of being lost with the decline of small-scale communities. The necessary and long overdue recognition of this knowledge and its integration in socio-ecological models, which can be used in ocean management, is essential. It also allows for greater participation of fishers in the decisions that, due to an external logic of economic growth, that have been constraining the way in which they relate to their labour.

Social sciences have a fundamental role here, and ethnography is a method of local knowledge collection by excellence. The knowledge gained from time within the field should fuel the socio-ecological models that are considered when creating policies. However, ethnography, mainly with vulnerable communities, requires time to gain trust. Despite the important role of mediation an ethnographer can have in the conjugation of transdisciplinary knowledge and fostering local participation in the management of policies that affect them, the current way of doing science does not allow sufficient time for this evolution. Scientific research is carried out with limited-time funding (Burgess et al., 2020) – for example with the MAR-Gov project, which attempted a participated management of a protected area – and as such come to an end before the dilution of social tensions even starts to be resolved (Martins, 2013; Stratoudakis et al., 2015; Vasconcelos et al., 2011) It is important to challenge the way of doing ocean science: to think of the ocean not as devoid of people and within a broader time scale, but also in its own analysis, since it deals with global and national problems, but with implications and solutions at the local scale. These communities are part of the seascapes and should be regarded as such.

As aggravating physical problems on the Portuguese coast, among climate change impacts and coastal erosion, there are fragilities at the social and administrative levels accounted for (Schmidt et al., 2012). It is a coastal strip occupied in an unsustainable manner, demonstrating a lack of maritime culture (Peralta, 2006), increasing coastal physical vulnerability, with a demographic pressure that exacerbates social vulnerability. At the administrative level, coastal management has been criticized, for being an "institutional model highly dispersed and incoherent, incapable of defining appropriate orientations for a more sustainable relationship of Portuguese society with its coast" (Schmidt et al., 2012: 36, translation by the author). This is related to a generalized difficulty in generating an efficient dialogue between scientific knowledge and concrete political and management actions (Dessai and Trigo, 2001).

Fisheries management sustainability

Exacerbated by Covid-19, the fishing sector is facing a crisis for in which a top-down, command-and-control regulation system might strangle small-scale fishing into disappearance (Bresnihan, 2016; Howard, 2017; Jentoft, 2000b). This crisis is simultaneously an ecological, socioeconomic, intellectual, and ethical crisis (Pauly, 2019). As mentioned earlier, fisheries management is decided at European and national levels, failing to consider the heterogeneity of European and Portuguese fishing communities, or even heterogeneity within a community itself

(Jentoft, 2000a). However, every fishing community has their own history of crisis. Addressing this can be considered a wicked problem (Jentoft and Chuenpagdee, 2009). Given their complexity, possible solutions for these problems can be achieved through the decentralization of management in the fishing sector, promoting a bottom-up ocean governance, for which fishers' local knowledge is essential, since a solution for a wicked problem is highly context-dependent and impossible to standardize (Jentoft and Chuenpagdee, 2009).

Scientific models that are the basis for many of the fishing management policies do not convincingly integrate the human being, nor its knowledge. However, people that come into contact with the sea everyday know it well, even though this knowledge is local, empiric, and embodied.

Jentoft (2020b) denounced an omission of SSF in the Sustainable Development Goals and urged rethinking Life above water (Jentoft, 2020b). But science alone isn't enough to address the many challenges in ocean governance (Jentoft, 2000a). Fisherfolk need to be involved (Ostrom, 1990), also because managing the commons in a scarcity logic leads to a neoliberal management paradigm that does not take into account social and cultural factors, proving to be counterproductive (Bresnihan, 2016). Furthermore, ocean governance needs to be anchored in transdisciplinary knowledge, going beyond natural or social sciences, integrating knowledge from stakeholders (Chuenpagdee and Jentoft, 2019). Ethnography can be a way to bridge the gap between disciplines since it collects empirical knowledge within communities necessary to understand the landscape in all its elements. By being there and observing natural elements, materials and the humans the researcher interacts with, while maintaining openness for the field to lead the way of inquiry, in a transdisciplinary ethnography, as Crowther (2019) refers, valuable data emerges. This data stemming from this type of research is vast, so it requires various disciplines to be fully understood (Crowther, 2019). Transdisciplinarity defies boundaries in theory and methodology to understand complex issues such as environment-human relations. Only with an in-depth knowledge of the context, it is possible to seek the reparation of social fabric and ensuring social and environmental justice – in this case, in fisheries.

Fisheries management is mainly grounded in resource conservation, yet simultaneously on an efficient capture and capital accumulation under a blue growth logic, while fishing communities become more stratified, and eventually risking disappearance (Bresnihan, 2016; Hadjimichael, 2018; Howard, 2017; Jentoft, 2000a). Despite many references to the necessity of maintenance of these communities for the protection of fishing stocks, many of them, including in Setúbal, keep losing their core identity. After many years of social decline, repairing the social fabric of fishing communities is also about repairing the trust of these people, something that requires time, as ethnography reveals. This is not possible without also addressing the mismanagement of the coastal areas in Portugal (Schmidt et al., 2012), the lack of communication between institutions, conflicts between interest groups, and inconsistencies in sectoral approaches (Haas et al. 2022). Recognizing local knowledge as a pillar of fishing culture and its importance to ocean sustainability and community identity is essential to urgently rethink fisheries management, before this knowledge is lost (Bailey, 2018; Jentoft, 2000a), and before the interests of large economic groups appropriate the ocean of which small-scale fishing communities depend (Barbesgaard, 2018).

Conclusions

The ocean is both a source of concern and an important means of connection in an increasingly modern and global world. The discourse of economic growth and environmental issues such as the reduction of stocks, climate change, and marine pollution have placed the ocean on the international agenda. The ocean is more than a rich body of water. It is a source of oxygen and life. It is also a place from which people live on and by. This shows us the need to integrate human activities in ocean science. Co-production of ocean transdisciplinary knowledge needs local, embodied, and empirical knowledge from local communities in order to better understand the seascape and its importance for the preservation of small-scale fishing communities. The Portuguese fishing community of Setúbal faces social, economic, environmental, and political challenges. Current management strategies do not nurture small-scale communities, and their social fabric is in danger of tearing with no repair. Through ethnography, it was possible to understand how local, embodied, and empirical knowledge remains crucial in everyday life but is at risk due to the decline of the community and the lack of support from institutions. Also, ethnography, which at its core is a methodological approach capable of integrating different methods and theories, allows the researcher to gather different kinds of data in a more-than-human context of analysis. This data can only be understood through various disciplines in a transdisciplinary way. Ethnography can be a methodological laboratory of experimentation of what transdisciplinary ocean science can produce in the transition to sustainability. This ocean sustainability does not seem possible by continuing the current management paradigm. This chapter suggests the need to rethink fisheries management and value small-scale communities and their knowledge, capable of being integrated into scientific models that are considered by policymakers through ethnographic data gathering and transdisciplinary analysis. Both the maritime habitats and small-scale fishing communities are at a tipping point and need to heal together, free from the outdated pressures of a top-down agenda of growth. It is urgent to highlight the importance of involvement transdisciplinary methodologies and understanding the condition of small-scale fishing communities and their role in a sustainable relationship with the ocean. In this time of climate action urgency, it seems clear that the paradigm in which we relate to oceans needs to shift. Small-scale communities have local knowledge crucial for the sustainability of our oceans: this is a call to heal communities in order to heal the ocean.

Acknowledgements

The research in this chapter was partly funded by the Portuguese Foundation for Science and Technology (SFRH/BD/144542/2019). I would like to thank the editors of the book, Vera Köpsel and Sílvia Gómez, for letting me join this journey. A warm thanks to every single person who interacted with me in my fieldwork.

References

Adger, W. N., Barnett, J., Brown, K., Marshall, N., & O'Brien, K. (2013). Cultural dimensions of climate change impacts and adaptation. *Nature Climate Change*, *3*(2), 112–117. https://doi.org/10.1038/nclimate1666

Álvares, P. (1986). *Portugal na CEE: A indústria – A Agricultura – A Pesca – Os trabalhadores – Os Investimentos – Os Fundos – O Presente e o Futuro*. Mem Martins, Portugal: Publicações Europa-América.

Amorim, V. I. (2015). Marés de Incerteza: Etnografia do presente liminar na comunidade piscatória de Setúbal. Master's Dissertation in Anthropology. Lisboa: ISCTE-IUL.

Anderson, L. G. (2015). The application of basic economic principles to real-world fisheries management and regulation. *Marine Resource Economics*, *30*(3), 235–249. https://doi.org/10.1086/681279

Antweiler, C. (1998). Local knowledge and local knowing. An anthropological analysis of contested" cultural products' in the context of development. *Anthropos*, *93*(4/6), 469–494.

Aswani, S., Lemahieu, A., & Sauer, W. H. (2018). Global trends of local ecological knowledge and future implications. *PLoS One*, *13*(4), e0195440. https://doi.org/10.1371/journal.pone.0195440

Ayilu, R. K., Fabinyi, M., & Barclay, K. (2022). Small-scale fisheries in the blue economy: Review of scholarly papers and multilateral documents. *Ocean & Coastal Management*, *216*, 105982. https://doi.org/10.1016/j.ocecoaman.2021.105982

Bailey, K. M. (2018). *Fishing lessons*. University of Chicago Press.

Barange, M., Bahri, T., Beveridge, M. C., Cochrane, K. L., Funge-Smith, S., & Poulain, F. (2018). Impacts of climate change on fisheries and aquaculture: synthesis of currrent knowledge, adaptation and mitigation options. Rome: FAO. www.fao.org/policy-support/tools-and-publications/resources-details/ru/c/1152846/

Barbesgaard, M. (2018). Blue growth: Savior or ocean grabbing? *The Journal of Peasant Studies*, *45*(1), 130–149. https://doi.org/10.1080/03066150.2017.1377186

Bateson, G. (2000). *Steps to an ecology of mind: Collected essays in anthropology, psychiatry, evolution, and epistemology*. University of Chicago Press.

Bennett, N. J., Finkbeiner, E. M., Ban, N. C., Belhabib, D., Jupiter, S. D., Kittinger, J. N., . . . & Christie, P. (2020). The COVID-19 pandemic, small-scale fisheries and coastal fishing communities. *Coastal Management*, *48*(4), 336–347. https://doi.org/10.1080/08920753.2020.1766937

Bernstein, H. (2010). *Class dynamics of Agrarian change*. Halifax: Fernwood.

Bresnihan, P. (2016). *Transforming the fisheries: Neoliberalism, nature and the commons*. University of Nebraska Press.

Bryman, A. (2012). *Social research methods*. Oxford University Press.

Burgess, M. G., Carrella, E., Drexler, M., Axtell, R. L., Bailey, R. M., Watson, J. R., . . . & Wilcox, S. (2020). Opportunities for agent-based modelling in human dimensions of fisheries. *Fish and Fisheries*, *21*(3), 570–587. https://doi.org/10.1111/faf.12447

Calafati, A. G. (2006). "Traditional knowledge" and local development trajectories. *European Planning Studies, 14*(5), 621–639. https://doi.org/10.1080/09654310500500148

Chuenpagdee, R., & Jentoft, S. (2019). *Transdisciplinarity for small-scale fisheries governance. Analysis and practice*. Springer Nature.

Costa, A., Rocha, A., Pereira, B., Maia, C., Chaves, C., Silva, C., . . . & Stratoudakis, Y. (2022). *Estado dos stocks em 2020 e aconselhamento científico para a sua gestão em 2021* (Relatórios Científicos e Técnicos do IPMA nº32). Lisboa: Instituto Português do Mar e da Atmosfera. https://www.ipma.pt/pt/media/noticias/documentos/2021/ESTADO_DOS_RECURSOS_EXPLORADOS_2020-2.pdf

Crowther, R. (2019). *Wellbeing and self-transformation in natural landscapes*. Palgrave Macmillan.

Davies, J., & Stodulka, T. (2019). Foreword: Pathways of affective scholarship. In Stodulka, T., Dinkelaker, S., & Thajib, F. (Eds.) *Affective dimensions of fieldwork and ethnography* (pp. 1–6). Springer.

Daw, T., Adger, W. N., Brown, K., & Badjeck, M. C. (2009). Climate change and capture fisheries: Potential impacts, adaptation and mitigation. In K. Cochrane, C. De Young, D. Soto, & T. Bahri (Eds.) *Climate change implications for fisheries and aquaculture: overview of current scientific knowledge* (FAO Fisheries and Aquaculture Technical Paper. No. 530), pp. 107–150. Rome, FAO. https://agris.fao.org/agris-search/search.do?recordID=XF2006441945

Delicado, A., Schmidt, L., Guerreiro, S., & Gomes, C. (2012). Pescadores, conhecimento local e mudanças costeiras no litoral Português. *Revista de Gestão Costeira Integrada, 12*(4), 437–451.

Descola, P., & Pálsson, G. (2002). *Nature and society – Anthropological perspectives*. Abingdon: Routledge.

Dessai, S., & Trigo, R. (2001). A ciência das alterações climáticas. *Finisterra, 36*(71), 117–132. https://doi.org/10.18055/Finis1651

DGPM. (2020). Relatório de Monitorização da Estratégia Nacional para o Mar 2013–2020. Lisboa: Direção-Geral de Política do Mar. https://www.dgpm.mm.gov.pt/_files/ugd/eb00d2_fa13a5edddd2437aaaadf79cf0682765.pdf

Engels, F., & Marx, K., (1972). *La première critique de l'èconomie politique. Ecrits de 1843–1844*. Paris: UGE-10/18.

ENM. (2021). Estratégia Nacional para o Mar 2021–2030. Lisboa: Direção-Geral de Política do Mar. www.dgpm.mm.gov.pt/_files/ugd/eb00d2_69ba72534a2840c0895ca5483d13df30.pdf

Ferreira, D., & Santos, J. P. (2020). *O Bairro do Troino: Contributos para a sua História*. Setúbal: Estuário História.

Freitas, J. G. D., Bastos, M. R., & Dias, J. A. (2018). Traditional ecological knowledge as a contribution to climate change mitigation and adaptation: The case of the Portuguese coastal populations. In Leal Filho, W., Manolas, E., Azul, A., Azeiteiro, U., & McGhie, H. (Eds.) *Handbook of climate change communication: vol. 3. Climate change management* (pp. 257–269). Cham: Springer. https://doi.org/10.1007/978-3-319-70479-1_23

Garrido, Á. (2018). *As Pescas em Portugal*. Lisboa: Fundação Francisco Manuel dos Santos.

Gillis, J. (2012). *The human shore. Seacoasts in history*. Chicago University Press.

Griffith, D. (1999). *The Estuary's gift: An atlantic coast cultural biography*. Pennsylvania State University Press.

Haas, B., Mackay, M., Novaglio, C., Fullbrook, L., Murunga, M., Sbrocchi, C., . . . & Haward, M. (2022). The future of ocean governance. *Reviews in Fish Biology and Fisheries, 32*, 253–270. https://doi.org/10.1007/s11160-020-09631-x

Hadjimichael, M. (2018). A call for a blue degrowth: Unravelling the European Union's fisheries and maritime policies. *Marine Policy, 94*, 158–164. https://doi.org/10.1016/j.marpol.2018.05.007

Hardin, G. (1975). The tragedy of the commons. *Journal of Natural Resources Policy Research, 1*(3), 243–253.
Heyl, B. S. (2001). Ethnographic interviewing. In Atkinson, P., Coffey, A., Delamont, S., Lofland, J., & Lofland, L. (Eds.) *Handbook of ethnography* (pp. 369–383). London: Sage Publications. https://dx.doi.org/10.4135/9781848608337.n25
Houde, N. (2007). The six faces of traditional ecological knowledge: Challenges and opportunities for Canadian co-management arrangements. *Ecology and Society, 12*(2), 34. http://www.ecologyandsociety.org/vol12/iss2/art34/
Howard, P. M. (2017). *Environment, labour and capitalism at sea. 'Working the ground' in Scotland*. Manchester University Press.
Ingold, T. (2004). *The perception of the environment: Essays on livelihood, dwelling and skill*. Abingdon: Routledge.
Ingold, T. (2011). *Being alive: Essays on movement, knowledge and description*. Abingdon: Routledge.
IPCC. (2019). Technical summary. In Pörtner, H.-O., Roberts, D.C., Masson-Delmotte, V., Zhai, P., Tignor, M., Poloczanska, E. . . . & Weyer, N.M. (Eds.) *IPCC special report on the ocean and cryosphere in a changing climate* (pp. 39–69), Cambridge University Press. doi:10.1017/9781009157964.002
IPCC. (2022). *Climate change 2022: Impacts, adaptation, and vulnerability*. Contribution of Working Group II to the Sixth Assessment Report of the Intergovernmental Panel on Climate Change. Cambridge University Press.
Jentoft, S. (2000a). The community: A mishing link of fisheries management. *Marine Policy, 24*(1), 53–60. https://doi.org/10.1016/S0308-597X(99)00009-3
Jentoft, S. (2000b). Legitimacy and disappointment in fisheries management. *Marine Policy, 24*(2), 141–148. https://doi.org/10.1016/S0308-597X(99)00025-1
Jentoft, S. (2020a). From I to we in small-scale fisheries communities. *Maritime Studies, 19*(4), 413–417. https://doi.org/10.1007/s40152-020-00204-z
Jentoft, S. (2020b). Life above water: Small-scale fisheries as a human experience. *Maritime Studies, 19*(4), 389–397. https://doi.org/10.1007/s40152-020-00203-0
Jentoft, S., & Chuenpagdee, R. (2009). Fisheries and coastal governance as a wicked problem. *Marine Policy, 33*(4), 553–560. https://doi.org/10.1016/j.marpol.2008.12.002
Lam, D. P., Hinz, E., Lang, D., Tengö, M., Wehrden, H., & Martín-López, B. (2020). Indigenous and local knowledge in sustainability transformations research: A literature review. *Ecology and Society, 25*(1), 3. https://doi.org/10.5751/ES-11305-250103
Lopes, J., & Pereira, A. (2015). *A Indústria das Conservas de Peixe em Setúbal*. Setúbal: Estuário.
Madureira, N., & Amorim, I. (2001). *História do Trabalho e das Ocupações – Vol. II: As Pescas*. Lisboa: Celta Editora.
Martins, L. (1999). Mares electrónicos em fundos sem peixe: Um estudo de caso na Póvoa de Varzim e nas Caxinas. *Etnográfica, 3*(2), 235–270. https://doi.org/10.4000/etnografica.3039
Martins, L. (Ed.) (2013). *Mares de Sesimbra: História, Memória e Gestão de uma Frente Marítima*. Lisboa: Âncora Editora.
Martins, R., Carneiro, M., & Rebordão, F. R., (2005). *Contribuição para o Conhecimento das Artes de Pesca utilizadas no Rio Sado*. Lisboa: IPIMAR.
McClanahan, T. R., Castilla, J. C., White, A. T., & Defeo, O. (2009). Healing small-scale fisheries by facilitating complex socio-ecological systems. *Reviews in Fish Biology and Fisheries, 19*(1), 33–47. https://doi.org/10.1007/s11160-008-9088-8
Murray, G., Bavington, D., & Neis, B. (2005). Local ecological knowledge, science, participation and fisheries governance in Newfoundland and Labrador: A complex, contested and

changing relationship. In Gray, T. S. (Ed.) *Participation in fisheries governance* (pp. 269–290). Dordrecht: Springer. https://doi.org/10.1007/1-4020-3778-3_16

Oneto Nunes, F. (Coord.). (2008). *Culturas Marítimas em Portugal*. Lisboa: Âncora Editora.

Ostrom, E. (1990). *Governing the commons: The evolution of institutions for collective action*. Cambridge University Press.

Pauly, D. (2019). *Vanishing fish: Shifting baselines and the future of global fisheries*. Vancouver: Greystone Books.

Peralta, E (2006). Portugal e o Mar: significações culturais e discursos identitários. In Garrido, A. (Ed.) *A Economia Marítima Existe*. Lisboa: Âncora Editora.

Pina-Cabral, J. (2017). *World: An anthropological examination*. University of Chicago Press.

Pita, C. Roumbedakis, K., Fonseca, T., & Castelo, D. (2020). *Briefing Projecto COVID-PESCA Impacto da Pandemia de Covid-19 nos sectores da pesca e aquicultura em Portugal*. Aveiro: CESAM. www.cesam.ua.pt/files/Briefing_COVIDPESCA_Portugal.pdf

Pitta e Cunha, T. (2011). *Portugal e o Mar: À Redescoberta da Geografia*. Lisboa: Fundação Francisco Manuel dos Santos.

Pussetti, C. (2016). Quando o campo são emoções e sentidos. Apontamentos de etnografia sensorial. In Mendes, P., & Martins, H. (Eds.) *Trabalho de campo: envolvimento e experiências em antropologia* (pp. 39–56). Lisboa: Imprensa de Ciências Sociais.

Quintas, M. C. (1998). *Setúbal: Economia, Sociedade e Cultura Operária 1880–1930*. Lisboa: Livros Horizonte.

Resolução do Conselho de Ministros no 120/2021 de 1 de setembro. Diário da República: I série, No 170 (2021). https://dre.pt/dre/detalhe/resolucao-conselho-ministros/120-2021-170591677

Santos, F. D., & Miranda, P. (2006). *Alterações Climáticas em Portugal: Cenários, Impactos e Medidas de Adaptação. Projecto SIAM II*. Lisboa: Gradiva.

Schmidt, L., Santos, F. D., Prista, P., Saraiva, T., & Gomes, C. (2012). Alterações climáticas, sociais e políticas em Portugal: processos de governança num litoral em risco. *Ambiente & Sociedade, 15*(1), 23–40.

Shagrir, L. (2017). *Journey to ethnographic research*. Cham: Springer. https://doi.org/10.1007/978-3-319-47112-9

Silvano, R. A., & Valbo-Jørgensen, J. (2008). Beyond fishermen's tales: contributions of fishers' local ecological knowledge to fish ecology and fisheries management. *Environment, Development and Sustainability, 10*(5), 657–675. https://doi.org/10.1007/s10668-008-9149-0

Spradley, J. P. (1979). *The ethnographic interview*. Long Grove, IL: Waveland Press.

Stratoudakis, Y., Fernández, F., Henriques, M., Martins, J., & Martins, R. (2015). Situação ecológica, socioeconómica e de governança após a implementação do primeiro plano de ordenamento no Parque Marinho Professor Luiz Saldanha (Arrábida, Portugal): I-informações e opiniões dos pescadores. *Revista de Gestão Costeira Integrada, 15*(2), 153–166.

Tanaka, S (2011). The notion of embodied knowledge. In Stenner, P., Cromby, J, Motzkau, J, Yen, J. & Haosheng (Eds.) *Theoretical phychology: Global transformations and challenges*. Concord, ON: Captus Press.

TBTI. (2016). *European small-scale fisheries: A regional synthesis*. (Too Big to Ignore Research Report Number R-04/2016). St. John's, NL: Too Big to Ignore.

Thornton, T., & Scheer, A. (2012). Collaborative engagement of local and traditional knowledge and science in marine environments: A review. *Ecology and Society, 17*(3), 8. http://dx.doi.org/10.5751/ES-04714-170308

van Ginkel, R. (2008). *Coastal cultures: An anthropology of fishing and whaling*. Amsterdam: Het Spinhuis.

Vasconcelos, L., Coelho, M., Pereira, M. J. R., Sár, R., & Costa, M. H. (2011). MARGov – Governância Colaborativa de Áreas Marinhas Protegidas – O Diálogo Eco-Social na Capacitação de Agentes de Mudança para a Sustentabilidade dos Oceanos: o Caso de Estudo do Parque Marinho Professor Luiz Saldanha.

Viveiros de Castro, E. (2012). Transformação" na antropologia, transformação da "antropologia. *Mana*, *18*(1), 151–171. https://doi.org/10.1590/S0104-93132012000100006

Wilson, J. A., Acheson, J. M., & Johnson, T. R. (2013). The cost of useful knowledge and collective action in three fisheries. *Ecological Economics*, *96*, 165–172. https://doi.org/10.1016/j.ecolecon.2013.09.012

10

CHARACTERIZATION AND VULNERABILITIES OF FISHERIES WITHIN A COASTAL LAGOON IN URUGUAY

A participatory approach

Maira Ramos, Rodolfo Reboulaz, Germán Taveira, Ximena Lagos, Hugo Inda, and Leandro Bergamino

Introduction

Small-scale artisanal fisheries play a critical role in contributing to poverty alleviation in coastal communities by providing food security (FAO, 2014). However, artisanal fisheries and their sustainability are often threatened by human activities, including overfishing and habitat degradation resulting in a considerable fish stock decline (McClanahan et al., 2015). Furthermore, within coastal areas, conflicts can occur among the different coastal resource users because of overlapping activities such as tourism and fishing. In this sense, conservation initiatives are important endeavours when it comes to preserving small-scale artisanal fisheries.

In complement with scientific data, fishers' local ecological knowledge (LEK) plays an important role in successful management and conservation since it is strongly linked to the current level of target species and natural resources (Brook and McLachlan, 2008). Local ecological knowledge is transmitted orally through generations representing a cumulative body of knowledge regarding the relationships between living beings including humans and their environment (Berkes, 2012). It can thus could contribute important information where knowledge about the environment is scarce or absent, including fish feeding behaviour, fish spawning aggregation, environmental changes and fish abundance trends (Uprety et al., 2012; Martins et al., 2018). Furthermore, the utilization of spatial LEK incorporated into a geographic information system (GIS) for resource mapping could generate important information that can be used in spatial prioritization, which is essential for reaching sustainable management goals (Aswani and Lauer, 2006). Local participation in the fisheries management process and the generation of bottom-up processes of empowerment became crucial for a successful management in the long term (Salas et al., 2011).

DOI: 10.4324/9781003311171-13

In Uruguay, artisanal fisheries constitute a significant economic activity for several coastal communities relevant to job creation with 46% of the fishers involved in artisanal fishing (Defeo et al., 2011). Artisanal fisheries traditionally operate gear manually or using small vessels (<10 gross registers tonnage) and products are mostly sold at the local market. Furthermore, coastal lagoons support important fisheries of the Uruguay's Atlantic coast associated with the breaching of a sandbar and the consequent entrance of larvae and juvenile from the sea (Fabiano and Santana, 2006). Despite their importance, small-scale fisheries have been undergoing a socio-ecological crisis due to declining trends in the catches and increasing fishing effort (Trimble, 2013). Previous LEK studies in Uruguay working in small-scale fisheries showed that participatory approaches to research and management increased compliance with socio-ecological sustainability and help to identify important factors for conservation and management policies (Mellado et al., 2014).

The study presented here took place within a coastal lagoon located in the Atlantic coast of Uruguay called Garzón lagoon, which has a small fishing community in the process of designing management plans by national governmental institutions including the National Environmental Agency (DINAMA). Furthermore, Garzón lagoon belongs to the National System of Protected Areas (SNAP) since 2014 under the category of "Protected area with species and habitat management", and has undergone an important socio-environmental transformation due to an increasing tourism and urban development. The fishing community of Garzón belongs to the traditional artisanal fisheries in coastal lagoons of Uruguay where both women and men are strongly involved in fishing. Furthermore, this protected area shows an important density of ichthyoplankton, thus representing an important nursery area for fish species with economic importance including silverside *Odonthestes* spp., the whitemouth croacker *Micropogonias furnieri*, blue crab *Callinectes* sapidus, and pink shrimps *Farfantepenaeus paulensis* (Defeo et al., 2009). In Garzón, fish catches are dominated in numbers by silverside *Odontheses* spp., flatfish *Paralichthys orbingnyaunus*, and mullets *Mugil* spp., with high importance for subsistence activities (Santana and Fabiano, 1999). This protected area can be understood as a socio-ecological system with multiple subsystems that interact, including resources, users, and governance (Ostrom, 2009). In this context, the implementation of LEK into management processes constitutes a key step for the area. The purpose of this study was to characterize fishing activities within the lagoon. The approach used in this study and its results will help to identify factors that minimize the vulnerabilities of the fishing communities and provide management strategies in Garzón lagoon and in other coastal lagoons with similar problems worldwide.

Material and methods

Study area

Garzón Lagoon (34°47'02.9"S, 54°33'51.8"W) is located at the south-eastern coast of Uruguay (Figure 10.1). This lagoon is intermittently connected to the Atlantic Ocean and fed by three streams, of which Arroyo Garzón has the greatest inflow

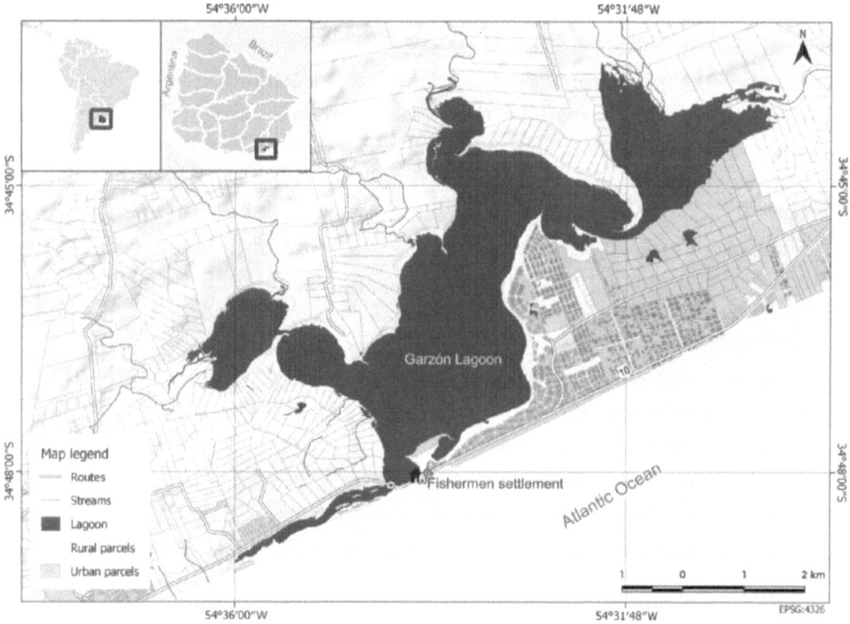

FIGURE 10.1 Location of Garzón Lagoon in the south-western Atlantic, Uruguay.

in the northern zone. The average depth of the lagoon is 0.5 m and it has a surface area of 18 km^2 and a watershed area of 695 km^2. The shallow lagoon encloses natural and cultural values since it forms part of the MAB-Biosphere Reserve Bañados del Este and belongs to the National System of Protected Areas. In 2014, it was included in the National System of Protected Areas (SNAP) of Uruguay with the category of "Protected area with species and habitat management". Furthermore, presently there is a lack of a management plan for the protected area. However, a Local Advisory Committee has been installed formally in 2020 following the Protected Areas National Law and including local and national institution, fishers, farmers, and other local stakeholders. During these meetings, diverse environmental problems have been discussed and presented to the national authorities in order to promote a Management Plan to be submitted to the Ministry of the Environment for legal approval.

In Garzón Lagoon, including the catchment area, there is an estimated population of 546 inhabitants (Rodríguez-Gallego et al., 2017). The fishers' settlement located seaward includes around eight families. A distinct trait of their activities is that women are actively involved in every stage of fishing, from sailing to catching fishes to the commercialization. Small boats typically have a range size between 4 and 10 m length, are constructed of wood, and propelled by outboard (4–6 HP) gasoline engines. Fishing boats are operated by one or two fishers.

Documentation of local ecological knowledge

Data for this study was collected from August 2018 to September 2020 after a preliminary visit to Garzón Lagoon in order to meet local leaders, present the research objectives and obtain verbal consent for the research. Interviewed fishers participated voluntary in the activity. Semi-structured interviews allowed the researchers to guide the discussion and, at the same time, fostered flexible responses that gave room to deeper discussions (Huntington, 1998). The interviews were based on a standardized semi-structured questionnaire on fishing gear, fisheries, and behaviour of economically important fish species and habitat type. Interviews were performed including young and veteran fishers of both genders, representing roughly one individual per family of the whole community that live permanently in Garzón Lagoon. The interviews included women and men, aged between 40 and 80 years with experience in fishing of between 20 and 60 years.

Transcribed interviews were analysed by means of ATLAS.ti 8 (http://atlasti.com/), which is a qualitative data analysis software package that has been used for a variety of research fields including fisheries. The transcribed interviews were analysed for code frequency and code co-occurrence (Daliri et al., 2016).

Spatial analysis

In order to obtain spatial information, we used the participatory mapping technique based on a qualitative approach that collects spatial information from the viewpoint of the actors involved (Dunn, 2007). During the workshop, held in May 2020 and conducted in the fishing community of Garzón, fishers were asked to indicate their responses with points, lines, and polygons on base maps printed at 1: 40,000 scale (Close and Hall, 2006; Mellado et al., 2014). Before conducting the interviews, the thematic components to be surveyed were: (1) draft preference zones; (2) their knowledge about the lagoon ecosystem in terms of morphology and dynamics; (3) their assessment and identification of relevant areas according to their ecological role in the system; (4) the threats to which these areas are exposed; and (5) the main distribution of fishing activities, and other concurrent ones in Garzón Lagoon. Each individual map was digitized and processed in QGIS 3.10 software (QGIS, 2022). The resulting maps are presented in thematic cartography that summarizes the results obtained. The basic geographic information was obtained from official sources (IDEUY, 2020).

Results: fishers' local knowledge in the Garzón Lagoon

The results of the survey on the characteristics of the fishing gear indicated similarities (Table 10.1) with gillnets encircle (remolino) and set (calado) as the main fishing gear used in Garzón Lagoon with mesh sizes that can vary according to the depth of the water and the target species. Furthermore, mullets *Mugil*

TABLE 10.1 Fishing methods characteristics for each target species mentioned by fishers as commercially relevant.

	Mugil *spp. (Mullet)*	Paralichtys *spp. (flatfish)*	Odontesthes *spp. (silverside)*
Fishing methods			
Fishing gear	Set/encircle gillnet	Set gillnet	Set/encircle gillnet
Mesh Size (cm)	9/10/11/12	18–20	4/5
Length (m)	800–1000	500–1000	500–1200
Other species capture	*Brevoortia* spp./Juvenile white croaker/*Rhamdia quelen*	white croaker/ *Brevoortia* spp.	*Oligosarcus jenynsis*/*Brevoortia* spp.
Fishing periods	Annual	Annual	Annual

spp., silverside *Odontesthes argentinensis*, together with flatfish *Paralichtys* spp. were identified as the main target species in interviews and are caught year-round. Silverside seems to be mostly caught in cold waters in the main channel during the day and in shallow water during night, whereas mullets are caught in the centre of the lagoon.

Biological and social information according to fisher's knowledge

The search for an alternative job is mentioned frequently among interviewers. In this context, fishers motivate their children to look for other jobs due to low income from fishing, thus threatening the artisanal fishing tradition as there is no generational renewal. This low economic condition creates the space for dialogue between fishers and the local government necessary for future decision-making that is sustainable over time. In this context, four fishers gave the following statements:

1 "Living from catfish and wolffish fishing is not enough for a living since a very low number of people consume these species." (interview quote, translated)
2 "Because of the low number of fishes I decided to look for a job outside fishing. I need to bring food to my family. Here in Garzón the fishing is being very poor since two years ago." (interview quote, translated)
3 "I always advise my children to look for alternative jobs since in the lagoon fishing is hard work with very low income nowadays, and the future is not promising." (interview quote, translated)
4 "We do not sell the fish directly to the restaurant, we sell it to a middleman that contacts the restaurant. If the restaurant pays well the middleman pays a little more, but they do not improve the payment a lot. I can say that since three years ago they have started to improve the payment because I told them if you don't pay more for the fish, I won't go fishing to look for the fish." (interview quote, translated)

In Garzón Lagoon, factors that limited the succeed of management included the lack of sandbar opening policies and zoning of fishing territory, as can be summarized with the following:

1 "The bar is poorly managed, I cannot say that there is no fish because the bar is opening badly, because in others lagoons at this time there is hardly any fish. But I'm sure that the sandbar opening affects fish abundance. The more days the lagoon it is open, the more possibilities there are for fish to enter." (interview quote, translated)
2 "In the mouth of the lagoon you cannot use any mesh. However, sport fishers do not respect that and they fish at the mouth of the lagoon and do not allow the fish to enter the lagoon." (interview quote, translated)
3 "We have five kite schools. Check out there are 300 sails every day in 20 cm of water column, you see a seagull like me. I have a photo, for example, in the sandbar with seagulls standing in the bar at 9 o'clock the morning and if I pass at 12, the sand isn't there anymore because they leave and they don't come back until the next day. It is a serious problem and difficult to overcome. They have keels down and they cut your mesh, they cut them". (interview quote, translated) (Figure 10.2).

In Garzón, fisher's knowledge is focused on water column depth and water turbidity as proxies to determine a good day for fishing. The fishers also mention the presence of carps (*Cyprinus carpio*) in the lagoon as a major concern that affects the fishing resource, since it predates the eggs of all native fish species.

1 "The wind affects the suspension of sand. When the water is dark, you will probably catch more because if the water is more transparent, the fish can see the mesh and avoid the mesh by going deeper into the water." (interview quote, translated)
2 "The silverside, at night it gets close to the shore and during the day it goes deep, because the silverside prefer cold water." (interview quote, translated)
3 "Nowadays, we have a problem with carps, there are a lot of them in the lagoon and they eat fish eggs." (interview quote, translated)

A conceptual diagram illustrating factors related to the fishing activities derived from the description of the survey, which showed a link between sandbar opening that impacts on the hydrodynamic condition of the lagoon and fish community composition (Figure 10.3). Furthermore, different actors including local authorities, cattle ranching, and fishers were related with the sandbar opening.

Participatory mapping

The spatial analysis using LEK allowed the identification of six important fishing areas within Garzón Lagoon (Figure 10.4), including typical fish caught in each

FIGURE 10.2 Recreation activity of Kite surfing in co-occurrence with shrimp fishery within Garzón Lagoon.

Source: © Hugo Inda

area. It should be noted that "cachuleta" is a vernacular word used by fishers as synonym of "lagoon pockets". Most fishers considered mullet (*Mugil* spp.), silverside (*Odontesthes* spp.), and flatfish (*Paralichtys* spp.) as relevant species for monetary income. The resulting maps from workshops with fishers showed fishing areas for flatfish, silverside, and mullets usually linking to the upstream area associated with

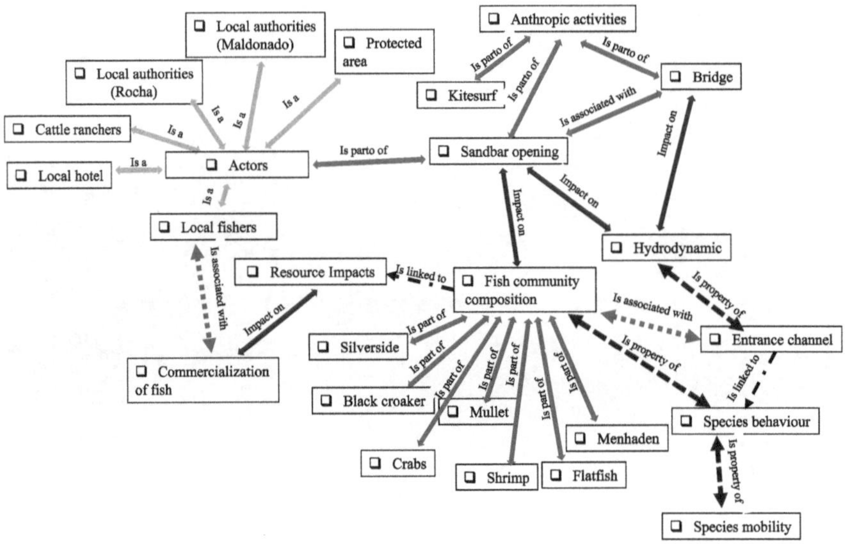

FIGURE 10.3 Network of the fishing activity characterization within the Garzón Lagoon, elaborated by ATLAS.TI using the description of the interviews.

banks and the freshwater inflow (Figure 10.4). Furthermore, the distribution of fishes is influenced by seasonality and water temperature. For instance, silverside abundance was usually associated with cold water according to the interviewees. Fishers clearly identify banks and also the main canal within the lagoon.

Discussion

The evaluation of local uses in protected areas is crucial to assure an effective management with the smallest impact (Lopes et al., 2013). This research based on a collaborative framework allowed the characterization of the fishing activities together with the identification of factors that influence the vulnerabilities of the fishing community affecting current and long-term subsistence of fishers within the protected area of Garzón lagoon. In this sense, in Garzón lagoon fishers deploy gillnets as the most common fishing gear throughout the year with varied mesh sizes to catch different species, as also has been reported in similar areas of Uruguay (Mellado et al., 2014; Trimble, 2013). Fishers showed a perception of decreasing catches over the year, with mullets, silverside, and flatfish species targeted yearlong, as also has been found in similar areas (Mellado et al., 2014). This study promotes transdisciplinary research approaches, combining academic disciplines and community involvement. This approach represents a powerful tool of co-producing knowledge and supporting governance agencies with relevant information for fisheries management issues and solving social challenges (e.g. Bower et al., 2019).

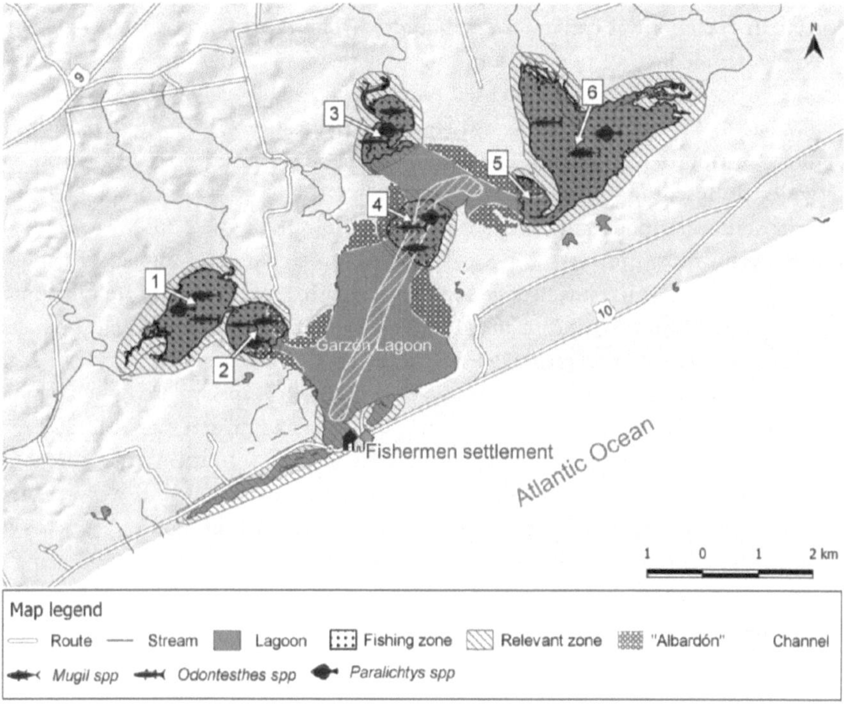

FIGURE 10.4 Map resulting from workshop together with fishermen showing environmental (canal, banks) and fishing areas for silverside, flatfish and mullets. Area 1: Shallow area; fishing area for mullet, flatfish, and silverside; Area 2: Shallow area; nursery area for silverside (mostly in winter); fishing area for silverside, flatfish, and mullet; Area 3: shallow area with hard substrate; nursery area for silverside; Area 4: mid-depth area; important fishing area during winter; Area 5: shallow area; fishing area for flatfish, mullet, and silverside; Area 6: mid-depth area important fishing area during winter.

Fishers identified (a) the lack of job alternatives and (b) a protocol regarding sandbar opening together with the restriction of nautical sports to specific areas inside the lagoon as the main threats affecting fishing activities. Furthermore, the need for additional income sources has also been highlighted previously in coastal fisheries of Uruguay (Trimble and Johnson, 2013). Congruently, previous researchers in Patos Lagoon identified that employment benefits may represent important income in order to decrease the vulnerability of small-scale fisheries, together with a strong institutional presence in resource governance (Kalikoski et al., 2010). A potential solution in order to minimize their vulnerability has been proposed including self-organization with clarification of the rights and local responsibilities, internal rules, and a legitimisation of local participation in co-management (Defeo and Castilla, 2012).

In Garzón Lagoon, fishers ages ranged from 40 to 80 and their mean age for starting this activity was early in their childhood. This fact evidenced a current lack of young people involvement with fishing activities. This aging pattern of the fishery activity has also been documented previously in small-scale fisheries (Mourão et al., 2020). Previous work suggested that the education of young people after school contributes to reducing the number of people getting involved in fishing activities (Johnsen and Vik, 2013). Furthermore, the scarcity of complementary income sources to put food on the table during low fishing seasons also emerged as a powerful reason for young people to look somewhere else instead of fishing as a way of living. It has been suggested that increasing levels of economic income together with higher significant market access by adding value to fishing products and by-products could revert this trend and revert the willingness to exit artisanal fisheries (Cinner et al., 2008). In Garzón lagoon, the fish species are usually commercialized locally via middlemen, and this lack of access to relevant markets creates a scenario of high vulnerability of the fishing system. In this sense, fishers in Garzón lagoon are extremely vulnerable to external factors and thus trapped in a poverty cycle. The development of innovative strategies including the certification of sustainable fishing, eco labelling and community-managed tourism linked to traditional fishing as heritage, as well as building connections between fishers and final consumers, could improve monetary income and, at the same time, facilitate the adaptation of fishers to a dwindling resource.

Management strategies incorporating the local history of resource use together with the social and economic status may reduce conflicts with environmental agencies (Lopes et al., 2013). In this study, using interviews and participatory mapping, we identified that the intensity in fishing activities varied within the lagoon and relevant areas for fishing are associated with shallow waters located toward the mouths of tributaries. These findings clearly illustrate the deep knowledge about ecological dynamics of the lagoon held by fishers, which is not restricted to fish species behaviour but includes several other aspects and is a direct result of a know-how originated lifestyle nourished by a daily and sustained interaction with the environment and its inherent dynamics. Furthermore, the loss of natural grassland and the increase of farmland areas may have significantly contributed to the fish decrease with nutrient loads in coastal lagoons and has been linked to changes in water quality indicators (Redriguez-Gallego et al., 2017). In this context, upstream areas within the Garzón Lagoon may represent risk areas for eutrophication process and consequently there is a need for defining policies integrating the entire catchment area.

Contribution of LEK to fisheries assessment and management proposal

Overall, interviews revealed that use of remolino (active fishing) and calada (passive fishing) are the most used methods for fishing. Fishers that prefer to use the method of calada agree that locations should be considered in the future management plan prohibiting this method near the mouth of the lagoon, and also ask for the protection of the cachuletas representing important shallow areas for fishing. Concurrently, previous work in a similar lagoon found that in general, fish tend to go towards upstream where freshwater inflow causes increased turbidity and higher amounts of food (Mellado et al., 2014). This idea of spatial planning within the lagoon defining geographic boundaries has been identified as an important contribution to co-management success (Douvere, 2008). The identification of these sensitive areas constituted an important first step towards resource management and conservation. Furthermore, according to fishers, the presence of exotic predatory fish species, including carps in the upstream zone is a major concern that poses several threats not only for future management plans but also for the very subsistence of native fish species.

Regarding environmental variables, temperature, wind patterns, and the water level within the lagoon are important factors in the fishing activities. In addition, the fishers identified that the presence of some species depends to a large extent on the state of the sandbar. Generally, the rupture of the sandbar due to the discharge of freshwater, the reduction of the water level followed by marine intrusion creates salinity gradients within the lagoon with an adverse effect on the presence of freshwater fish (Griffiths, 1999). In addition, in Rocha Lagoon, previous work evidenced the positive effect of the opening of the sandbar on the fisheries, promoting estuarine conditions and consequently ideal conditions for feeding and reproduction (Fabiano and Santana, 2006). According to the interviews, the fishers identified a sharp reduction in the fish assemblage with a strong reduction in marine species, including white croaker (*Micropogonias furnieri*) and flatfish (*Paralichtys* spp.). It should be noted that white croaker (*M. furnieri*) has not been mentioned by fishers in Garzon and previous work, which shows that this species was identified as having overfishing status leading to a reduction of juveniles in coastal waters between the 1980s and 1990s (Haimovici, 1998). Congruently, previous estimations of catch per unit effort (CPUE) using gillnet in Garzón for the period January 1992 to August 1994 indicated a total mean of 72.9 kg with the highest catches of *Brevoortia* spp. (41 kg) followed by *Odontesthes* spp. (15 kg) (Santana and Fabiano, 1999). Furthermore, this estimation of fish biomass seems to be low compared with similar coastal lagoons in Uruguay (Santana and Fabiano, 1999). These analyses suggest that a clear long-term monitoring of the fish biomass level is needed in order to estimate trends and suggest conservation management decisions.

Considering that Garzón Lagoon does not have a protocol that regulates the opening of the sandbar, this scenario generates conflicts and uncertainty. A critical point in discussion represents the water level and the location of the sandbar opening during flood periods, since the fishers propose a water level and a location to open the lagoon artificially, however, these conditions are usually not respected since local cattle ranchers press authorities for the decision to opening the bar. The fishers' community expresses the need for a protocol of artificial sandbar opening that considers not only the water level but also wind patterns. In this context, future studies need to develop a protocol for the artificial opening including the active participation of locals and neighbouring fishers (Conde et al., 2015). In this sense, conflicts between fishers and governance agencies indicate a need for institutional arrangement involving multiple stakeholders (Trimble and Berkes, 2015).

Participatory research initiatives involving fishing communities may overcome some of the barriers to co-management, helping to reduce conflict by representing opportunities to achieve sustainable fisheries (e.g. Trimble, 2013; Lopes et al., 2013; Trimble and Berkes, 2015; Mourao et al., 2020). The frequency of meetings represents an important factor that facilitated the participatory research. Also, organized fishers and leadership have also been described as an important condition that may facilitate participation in management and governance (Dias et al., 2020). Particularly in Uruguay, a previous case study showed how social cohesion and leadership may promote the incorporation of fishers' input in monitoring programmes based on a long-term partnership between researchers, the government, and local fishers (Gianelli et al., 2018). Furthermore, the organization of a Festival may be an important opportunity for building trust among fishery stakeholders and, at the same time, attract tourism and government agencies showing significant interest in artisanal fisheries.

Conclusion

Small fisheries of Garzón Lagoon are experiencing severe threats as is the case for small fisheries worldwide, thus compromising their viability even in the short term. The urgent need for the development and enforcement of a sandbar opening protocol became apparent in this case study as well as the need for participatory strategies including the whole set of actors on the lagoon and its basin. Furthermore, this study identifies the need of increasing government programmes to assist fishers within the lagoon and promote their organization and participation in the management process empowering local communities. Participatory process may represent an important step towards addressing economic and management issues.

As a final remark, it should be highlighted that making the fishers' LEK visible could lead to the empowerment of the local communities as a whole, not only in terms of self-recognising as a community and holders of valuable and useful knowledge but also beyond the community as key actors in the Protected Area. This fact is particularly relevant in Uruguay, where fishing communities have historically been neglected and have had little to no representation in the political decision-making spheres. Hopefully, this is also an opportunity to put in value the heritage linked to artisanal fisheries, including both the fishing knowledge and the way to interact with and inhabit the territory. This work began as a contribution focused on fishers' knowledge about fish species, but along the participatory research process it turned into a synergetic venture between the academy, the researchers and a project linking local communities with the productive sector. We would like to emphasize the role of LEK as a tool for generating fruitful transdisciplinary interactions and with a high potential for providing technical knowledge as a means for local community development. Furthermore, in order to understand complex social systems such as artisanal fisheries, transdisciplinarity with its collaborative and participatory approaches is essential for allowing the integration of divers inputs and considering the interest of local people (e.g. Espinoza-Tenorio et al., 2013). Future participatory research approaches considering a sustainable development context should include active participation of the local communities, especially vulnerable groups, from the start of the project and the definition of the central problem (e.g. Jacobi et al., 2022).

Acknowledgements

This work was possible thanks to the participation of the fishers from the communities of Garzón Lagoon, including Angel "El Sacapoco" (and also his family), "El Chico", "Beto", "Ñieca", "Milton", "Rosario", and "Querido". In more than one way, they should be considered as co-authors of this contribution because the knowledge they held makes made this article possible. This work was funded by programa de Vinculación Universidad – Sociedad y Producción (project number 32: Generando lineamientos para el desarrollo de un plan de gestión ambiental en laguna garzón).

References

Aswani, S., and Lauer, M. (2006). Benthic mapping using local aerial photo interpretation and resident taxa inventories for designing marine protected areas. *Environmental Conservation*, 33, 263–273. https://doi.org/10.1017/S0376892906003183

Berkes, F. (2012). Implementing ecosystem based management: Evolution or revolution? *Fish and Fisheries*, 13, 465–476. https://doi.org/10.1111/j.1467-2979.2011.00452.x

Bower, S.D., Song, A.M., Onyango, P., Cooke, S.J., and Kolding, J. (2019). Using transdisciplinary research solutions to support governance in inland fisheries. In Chuenpagdee, R., and Jentoft, S. (Eds.) *Transdisciplinarity for Small-Scale Fisheries Governance*. MARE Publication Series, vol 21. Springer, Cham.

Brook, R.K., and McLachlan, S.M. (2008). Trends and prospects for local knowledge in ecological and conservation research and monitoring. *Biodiversity and Conservation*, 17, 3501–3512. https://doi.org/10.1007/s10531-008-9445-x.

Cinner, J.E., Daw, T., and McClanahan, T. R. (2008). Socioeconomic factors that affect artisanal fishers' readiness to exit a declining fishery. *Conservation Biology*, 23, 124–130. https://doi.org/10.1111/j.1523-1739.2008.01041.

Close, C.H., and Hall, G.B. (2006). A GIS-based protocol for the collection and use of local knowledge in fisheries management planning. *Journal of Environmental Management*, 78(4), 341–393 52. https://doi.org/10.1016/j.jenvman.2005.04.027

Conde, D., Rodríguez-Gallego, L., de Álava, D., Verrastro, N., Chreties, C., Lagos, X., Solari, S., Piñeiro, G., Teixeira, L., Seijo, L., Vitancurt, J., Caymaris, H., and Panario, D. (2015). Solutions for sustainable coastal lagoon management: From conflict to the implementation of a consensual decision tree for artificial opening. In: Baztan, J., Chouinard, O., Jorgensen, B., Tett, P., Vanderlinden, J.-P., and Vasseur, L. (Eds.) *Coastal Zones: Solutions for the 21st Century*. Elsevier, Amsterdam, pp. 217–25

Daliri, E.M., Kamrani, E., Jentoft, S., and Paighambari, S.Y. (2016). Why is illegal fishing occurring in the Persian Gulf? A case study from the Hormozgan province of Iran. *Ocean & Coastal Management*, 120, 127–34. http://dx.doi.org/10.1016/j.ocecoaman.2015.11.020

Defeo, O., and Castilla, J.C. (2012). Governance and governability of coastal shellfisheries in Latin America and the Caribbean: Multi-scale emerging models and effects of globalization and climate change. *Current Opinion in Environmental Sustainability*, 4, 344–350.

Defeo, O., Horta, S., Carranza, A., Lercari, D., Álava, A., Gómez, J., Martínez, G., Lozoya J.P., and Celentano, E. (2009). Hacia un manejo ecosistémico de pesquerías. Áreas marinas protegidas en Uruguay. Facultad de Ciencias-DINARA, Montevideo, 122 p

Defeo, O., Puig, P., Horta, S., and Álava, A. (2011). Coastal fisheries of Uruguay. In: Salas, S., Chuenpagdee, R., Seijo, J., and Charles, A. (Eds.) *Coastal fisheries in Latin America and the Caribbean: An interdisciplinary perspective. FAO Fish. Tech. Pap.*, 544, 357–384.

Dias A.C.E., Cinti A., Parma A.P., and Seixas S.C. (2020). Participatory monitoring of small-scale coastal fisheries in South America: Use of fishers' knowledge and factors affecting participation. *Reviews in Fish Biology and Fisheries*, 30, 313–333. doi:10.1007/s11160-020-09602-2

Douvere, F. (2008). The importance of marine spatial planning in advancing ecosystem-based sea use management. *Marine Policy*, 32, 762–771.

Dunn, C.E. (2007). Participatory GIS – A people's GIS?. *Progress in Human Geography*, 31(5), 616–637. https://doi.org/10.1177/0309132507081493

Espinoza-Tenorio, A., Wolff, M., Espejel, I., and Montaño-Moctezuma, G. (2013). Using traditional ecological knowledge to improve holistic fisheries management: Transdisciplinary modeling of a lagoon ecosystem of southern Mexico. *Ecology and Society*, 18(2), 6. http://dx.doi.org/10.5751/ES-05369-180206

Fabiano, G., and Santana, O. (2006). Las pesquerias en las lagunas salobres de Uruguay. In: Menafra, R., et al. (Eds.) *Bases para la Conservación y el Manejo de la Costa 441 Uruguaya*. Vida Silvestre Uruguay, Montevideo, pp. 557–565

FAO. (2014). *The State of World Fisheries and Aquaculture: Opportunities and Challenges*. Food and Agriculture Organization of the United Nations, Rome

Gianelli, I., Horta, S., Martínez, G., de la Rosa, A., and Defeo, O. (2018). Operationalizing an ecosystem approach to small-scale fisheries in developing countries: The case of Uruguay. *Marine Policy*, 95, 180–188. https://doi.org/10.1016/j.marpol.2018.03.020

Griffiths, S.P. (1999). Consequences of artificially opening coastal lagoons on their fish assemblages. *International Journal of Salt Lake Research*, 8, 307–327

Haimovici, M. (1998). Present state and perspectives for the southern Brazil shelf demersal fisheries. *Fisheries Management and Ecology*, 5(4), 277–289. http://dx.doi.org/10.1046/j.1365–2400.1998.540277.x.

Huntington, H.P. (1998). Observations on the utility of the semi-directive interview for documenting traditional ecological knowledge. *Arctic*, 51(3), 237–242.

IDEUY. n.d. Infraestructura de Datos Espaciales de Uruguay. Accessed September 19, 2020. www.gub.uy/infraestructura-datos-espaciales/

Jacobi, J., Llanque, A., Mukhovi, S.M., Birachi, E, von Groote, P., Eschen, R., Hilber-Schöb, I., Kiba, D.I., Frossard, E., and Robledo-Abad C. (2022). Transdisciplinary co-creation increases the utilization of knowledge from sustainable development research. *Environmental Science and Policy*, 129, 107–115.

Johnsen, J.P., and Vik, J. (2013). Pushed or pulled? Understanding fishery exit in a welfare society context. *Maritime Studies*, 12, 1–20.

Kalikoski, D.C., Neto, P.Q., and Almudi, T. (2010). Building adaptive capacity to climate variability: The case of artisanal fisheries in the estuary of the Patos Lagoon, Brazil. *Marine Policy*, 34, 742–751.

Lopes, P.F.M., Rosa, E.M., Salyvonchyk S., Nora, V., and Begossi, A. (2013). Suggestions for fixing top-down coastal fisheries management through participatory approaches. *Marine Policy*, 40, 100–110. https://doi.org/10.1016/j.marpol.2012.12.033

Martins, I.M., Medeiros, R.P., Di Domenico, M., and Hanazaki, N. (2018). What fishers' local ecological knowledge can reveal about the changes in exploited fish catches. *Fisheries Research*, 198, 109–116. https://doi.org/10.1016/j.fishres.2017.10.008

McClanahan, T, Allison, E.H., and Cinner, J.E. (2015). Managing fisheries for human and food security. *Fish and Fisheries*, 16, 78–103. http://dx.doi.org/10.1111/faf.12045

Mellado, T., Brochier, T., Timor, J., and Vitancurt, J. (2014). Use of local knowledge in marine protected area management. *Marine Policy*, 44, 390–396. https://doi.org/10.1016/j.marpol.2013.10.004

Mourão, J.S., Baracho, R.L., Martel, G., Barboza, R.R.D., and Lopes S.F. (2020). Local ecological knowledge of shellfish collectors in an extractivist reserve, Northeast Brazil: Implications for co-management. *Hydrobiologia*, 847, 1977–1997. https://doi.org/10.1007/s10750-020-04226-w

Ostrom, E. (2009). A general framework for analyzing sustainability of social-ecological systems. *Science*, 325, 419–422.

QGIS (2022). QGIS – A Free and Open Source Geographic Information System. Official website. Available online at www.qgis.org/en/site/index.html.

Rodríguez-Gallego, L., Achkar, M., Defeo, O., Vidal, L., Meerhoff, E., and Conde, D. (2017). Effects of land use changes on eutrophication indicators in five coastal lagoons of the Southwestern Atlantic Ocean. *Estuarine, Coastal Shelf Science*, 188, 116–126. https://doi.org/10.1016/j.ecss.2017.02.010

Salas, S., Chuenpagdee, R., Charles, A., and Seijo, J.C. (2011). Coastal fisheries of Latin America and the Caribbean, FAO Fish. Tech. Pap. No. 544. Rome: FAO.

Santana, O., and Fabiano G. (1999). Medidas y mecanismos de administración de los 515 recursos de las lagunas costeras del litoral atlántico del Uruguay (Lagunas José Ignacio, Garzón de Rocha y de Castillos) En: Amestoy, R. M. F. & Arena, G. (Eds.) *Plan de investigación Pesquera*. INAPE-PNUD URU/92/003 165

Trimble, M. (2013). *Towards Adaptive Co-Management of Artisanal Fisheries in Coastal Uruguay: Analysis of Barriers and Opportunities, With Comparisons to Paraty (Brazil)*. Winnipeg: Doctoral dissertation, University of Manitoba. http://umanitoba.ca/institutes/natural_resources/canadaresearchchair/thesis/trimble_micaela.pdf.

Trimble, M., and Berkes, F. (2015). Towards adaptive co-manage- ment of small-scale fisheries in Uruguay and Brazil: Lessons from using Ostrom's design principles. *Maritime Studies*, 14, 1–20.

Trimble, M, and Johnson, D. (2013). Artisanal fishing as an undesirable way of life? The implications for governance of fishers' wellbeing aspirations in coastal Uruguay and southeastern Brazil. *Marine Policy*, 37, 37–44. https://doi.org/10.1016/j.marpol.2012.04.002

Uprety, U., Asselin H., Bergeron, Y., Doyon, F., and Boucher, J.-F. (2012). Contribution of traditional knowledge to ecological restoration: Practices and applications. *Ecoscience*, 19, 225–237. https://doi.org/10.2980/19-3–3530.

11

DIALOGUE OF KNOWLEDGE FOR THE ASSESSMENT OF THE IMPACTS OF THE OIL SPILL DISASTER ON THE BRAZILIAN COAST IN 2019

Louise Oliveira Ramos Machado, Luize da Silva Rezende da Mota, Cristina Larrea-Killinger, Priscilla Andrea Orsi, Josilan da Silva Nascimento, Amanda Laura Northcross, and Rita de Cássia Franco Rêgo

Introduction

The current economic development model, which relies on the use of fossil fuels as an important source for the world energy matrix, leads to an increase in environmental disasters with potential economic, environmental, and social consequences. It is estimated that the annual amount of oil spilled into the oceans reaches 4.5 million tons. Tanker disasters are responsible for about 10% of the global marine oil pollution. Approximately 35% of these disasters originate from regular transport operations, this includes oil released during incidents involving all types of ships, as well as oil from illegal tank cleaning. The increase in the occurrence of oil disasters at sea contributes to the increasing degradation of the oceans (World Ocean Review, 2014). The United Nations (UN) declared the Decade of Oceans Science for Sustainable Development (2021–2030), aiming to lead countries to plan sustainable management and preservation of biodiversity, and to develop knowledge about the oceans. The Decade of the Oceans presupposes a better understanding of humanity's relationship with the ocean. This includes traditional and local knowledge, which contribute to highlighting the multiplicity of cultural value for the ocean. Therefore, principles such as equity, inclusion, respect, justice, and scientific integrity are fundamental to the Decade of the Oceans (United Nations Educational, Scientific and Cultural Organization [UNESCO], 2020). We understand that a dialogue that encourages active cooperation between scientific and popular knowledge is essential in this process (Hersch-Martínez, 2013; Menéndez, 2008; Haro, 2010), which is reached from a sociocultural epidemiology perspective. In this chapter, we focus on the experience of knowledge dialogue used in a Brazilian study that aims to assess the impacts of a major disaster/crime crude oil spill that occurred in 2019 along the northeast coast of Brazil. The spill had significant impacts on the life and health of artisanal fishermen.

DOI: 10.4324/9781003311171-14

Between 2019 and 2020, the Brazilian coast experienced its largest marine environmental disaster with crude oil/oil spill, in terms of both extension and duration (Soares et al., 2020). This disaster/crime is considered the biggest crude oil spill in the country's history and affected eleven states, 130 municipalities and 1009 locations. The Brazilian Navy classifies the disaster as an environmental crime and an "unprecedented event in the history of combating pollution at sea", due to the extent and time of oil spreading, and the affected population (Marinha do Brasil, 2019). Until February 12, 2020, from a total of 159 occurrences of oiled fauna, 67 (42%) were in the State of Bahia, and among these, 42 (62.7%) were found dead. Of the 67 fauna species affected in Bahia, 25 (37%) were birds, 34 (51%) were sea turtles and 8 (12%) were other species. Analyses performed from January 29 to February 12, 2020, revealed high levels of nine Polycyclic Aromatic Hydrocarbons (PAHs) for oyster and crab species in some localities (Bahia Pesca, 2020).

Artisanal fishers represent about 1 out of 200 Brazilians (Rêgo et al., 2018) and were the most affected group by the disaster/crime, both economically and psychologically. These populations were acutely and chronically exposed to the contact with crude oil through dermal, ingestion and respiratory routes. Artisanal fishers were the first to enter the sea, estuaries, and mangroves as they attempted to contain and remove the crude oil trying to save the ecosystems, but without adequate personal protective equipment. Due to lack of food options, fishermen from affected communities consumed oiled fish and shellfish.

The disaster had direct repercussions on environmental health, affecting the fauna and flora, and, consequently, posing a threat to the work and health of artisanal fishermen. Artisanal fishermen are mostly Afro-descendants and earn on average less than a dollar a day (Rêgo et al., 2018). The reduction in income and survival conditions had serious consequences for communities and fishing territories. The disaster temporarily affected the capture of fish, crustaceans, and mollusks in several areas, and drastically reduced the consumption and sale of these products, as many buyers feared the fish contamination by oil (Araújo et al. , 2020). The COVID-19 pandemic, which occurred shortly thereafter, in March 2020, resulted in greater vulnerability to communities that live off artisanal fishing, compromising food security, income, and health conditions of populations (Magalhães et al., 2021)

Fishing communities report minor accidents, such as paraffin or oil spills that have been managed quickly. However, the 2019 disaster was marked by the absence or failure of immediate actions by the federal government in the face of the environmental and social emergency. This intensified the crisis due to the lack of coordination/monitoring and the cleaning process of affected areas. Many communities took initiative to begin the process of oil removal, often without adequate equipment. From an epidemiological point of view, these individuals are an exposed population, whose health can be potentially affected by the harmful properties of the oil. Literature review study involving five oil spill accidents provides evidence on the association between this exposure to spilled oils and the onset of acute physical, psychological, genotoxic, and endocrine effects in exposed individuals (Aguilera et al.,2010).

Oil spills are complex events that affect various human activities at sea and on land and demand the participation of various actors in decision-making on protection priorities. Research and actions to respond to accidents of this nature involve several interests: environmental (presence of rare or endangered species or the ecological importance of certain habitats), sociocultural (areas occupied by traditional communities and fishermen, where local productive arrangements are still in force and popular knowledge), economic (fishing, fish and crustacean farming or water harvesting areas), and recreational (tourism or sport fishing areas), among other uses of the water surface and the soil.

In the context of this disaster, several investigations were performed, including an epidemiological study to evaluate the impacts on the health of the population affected by the crude oil spill on the coast of Bahia State. The partnership for the research development was carried out with representatives and leaders of artisanal fisheries. Bahia was one of the most affected Brazilian states, in terms of coastline and volume of oil collected (Instituto Brasileiro do Meio Ambiente e dos Recursos Naturais Renováveis [IBAMA], 2019). In the effort to study the different impacts of this fortuitous and complex event, emergency transdisciplinary research was carried out. The project congregates researchers from different areas of knowledge (hard and earth sciences, biological sciences, health sciences, applied social sciences and human sciences) and uses methodologies involving analysis of environmental samples, soil sediment and marine species, environmental epidemiology, and analysis of social, economic, cultural, and subjective aspects experienced by fishermen. Environmental activists and researchers joined the coastal communities affected by the disaster to respond to the emergency demands of containment, identification, monitoring of different impacts, and remediation of the effects caused by the contaminants. The epidemiological research was developed to study the effects on human health using participatory methodology known as Community-Based Participatory Research (CBPR) (Detroit Community-Academic Urban Research Center,2011).

The construction of interpretive models of the health-disease process that integrate quantitative perspectives, such as an epidemiological method, which privileges generalizations, and qualitative, which privileges deepening, may be possible from different strategies. One of these perspectives has been called ethnoepidemiology, which assumes that health-disease phenomena are social, historical, complex, fragmented, conflicting, dependent, ambiguous, and uncertain processes (Almeida-Filho, 1992; Almeida-Filho, 2000).

For Almeida Filho (1992), ethnoepidemiology seeks to explore methodological alternatives for research on social processes and practices related to health, combining qualitative and quantitative approaches in a single ethnoepidemiological strategy, allowing reflection on the potential of an approach that integrates disciplines for the study of health phenomena (Almeida Filho, 1992). The ethnoepidemiological approach assumes that transdisciplinary and community participation at all stages of the process are important components of research and policy development (Larrea-Killinger et al., 2013).

This chapter describes the participatory methodology used in the epidemiological study, with interdisciplinary interfaces, of the impacts of the oil spill on the health of fishing communities. In addition, we share the perspective of community partners in this process. It brings subjective, social, and cultural aspects of the phenomenon, with the support of health anthropology, analysed in the light of the CBPR principles as a fundamental methodological axis for development and execution of a research.

Methodology: epidemiological and ethnoepidemiological studies at the interface of participatory process

Ethnoepidemiology subsumes the epidemiological study by approaching it through a participatory approach, sensitive to subjective aspects that are the object of analysis of the anthropology of health while at the same time allowing economic and environmental concerns to emerge which are the study object of other disciplines. Here they are presented as different methodological approaches.

A two stages method was used: an epidemiological study and an ethnoepidemiological study. The epidemiological study consisted of the following steps: identification of the problem, design of the project, approval by the research ethics committee (REC, IRB, in English) for human beings and the environment (SISBIO), definition of the study areas, definition of the population to be investigated, contacts with leaders of the study areas, elaboration of a questionnaire, validation of the questionnaire using Delphi Methods (Beiderbeck et al., 2021), selection and training of field interviewers, questionnaire application, data analysis and interpretation of results, and publication and dissemination. The results are expected to influence policy making (e.g. emergency response, health care, environmental protection, and protection of Indigenous and Quilombola communities).

Regarding the ethnoepidemiological method, the definition of the type III classification, proposed by Almeida Filho (2020), was adopted. This definition consists of studies that take the scientific practice of epidemiology as an object of investigation, with the application of anthropological concepts and ethnographic methods to the environments, everyday life and institutional cultures in which epidemiological knowledge is produced (Almeida Filho, 2020). Aim is to understand how a "knowledge dialogue" is established within a research team that shares different knowledge and practices on the development of impact studies on environmental disasters.

To carry out the ethnoepidemiological approach, two online discussion groups were organized, formed by three researchers from two universities with two local researchers for each area (North and South Coast of Bahia). Besides, an in-depth interview was carried out by two university researchers with the only local researcher with previous research experience, who participated in three study areas. The script for the online discussion groups and the interview was the same, developed to better understand the dialogue established in the research process, based on three axes: the internal team experience during the project activities (selection process and training, application of the questionnaire, internal dynamics of the teams and monitoring); the individual experiences in these research activities; the expectations about the research (return of results). The discussion groups and the interview were applied in an open and long-drawn way.

The study was approved by the Research Ethics Committee (REC) of the Medical School of the Federal University of Bahia, under advice No. 4.444.084, CAAE: 29570620.3.0000.5577, and in the Biodiversity Information Authorization System (SISBIO) with No. 76691–1.

The Community-Based Participatory Research approach

Community-Based Participatory Research (CBPR) principles and practices enabled the use of different investigative methods associated with the ethnoepidemiological research project for assessing the impacts of the oil spill on the northeast coast of Brazil. In the present investigation, bibliographic and documental research, questionnaire application, semi-structured interview, and participant observation were used. Thus, the cross-sectional epidemiological study articulates with the health anthropology to perform research by using participatory methodology aiming to study a major disaster due to a crude oil spill, which occurred in 2019 on the Brazilian northeast coast, and its repercussion in life and health of artisanal fishermen residing on the coast. The study focuses on artisanal fisheries workers because they were severely affected and possibly had the greatest potential for diverse exposures to crude oil, either during work activity or in the removal of oil residues on beaches or through food ingestion.

The research team's partnership with fishing communities was already present in previous studies (Carvalho et al., 2014; Müller et al., 2022). During the disaster, large mobilizations took place by various groups from universities in support of fishermen, in the press and in public hearings, at the state and federal levels, calling for protection of the coast against the disaster, financial support for those affected and carrying out research. During the project preparation and the execution of the epidemiological research, this partnership was intensified, through contacts with the leaders of artisanal fishing communities in different regions.

Faced with the complexity of the environmental, social, economic, and health problems resulting from the oil spill disaster on the Bahia coast, CBPR was chosen for providing the active involvement of artisanal fishermen, organizational representatives, and researchers in all aspects of the research, targeting to integrate knowledge and action in favour of the involved communities. Table 11.1 presents CBPR principles.

Based on these principles, it is possible to visualize the research-action-integration in communities at different research stages. The involved university researchers recognize the existing inequalities between them and the community participants and, therefore, the experience of the involved partners and the need to share information for decision-making proves to be valuable (Cook, 2008). In this research, the recognition of limitations (little funding, extension of the

TABLE 11.1 CBPR Principles

The core principles of CBPR:

1 CBPR promotes collaborative and equitable partnerships in all research phases and involves an empowering and power-sharing process.
2 CBPR recognizes community as a unit of identity.
3 CBPR builds on strengths and resources within the community.
4 CBPR facilitates co-learning and capacity building among all partners.
5 CBPR for health focuses on problems of relevance to the local community using an ecological approach that attends to multiple determinants of health and disease.
6 CBPR balances research and action for the mutual benefit of all partners.
7 CBPR disseminates findings and knowledge gained to the broader community and involves all partners in the dissemination process.
8 CBPR promotes a long-term process and commitment to sustainability.

Source: www.detroiturc.org/about-cbpr/community-based-participatory-research-principles

territory, difficulty in obtaining data about fishing communities) enabled a collaborative construction with researchers from other areas and counted on the knowledge of fishing leaders about the territory and their particularities in the decision-making of areas to be studied and how to select local researchers to participate in data collection.

Starting from the recognition of the available resources per territory (infrastructure to carry out data collection and team lodging, ability to mobilize and communicate with possible research participants, identification, and contact of artisanal fishermen), the project team could organize the work and build networks of integrated social relationships, contributing to reinforce and/or increase the sense of community in the collaborative research process.

CBPR makes it possible for researchers to learn about the history, culture, social context, and forms of internal organization of communities, in the same way that it makes it possible for community members to develop or improve skills in areas such as the organization and production of research or preparing fundraising proposals. The community comes to recognize that the knowledge produced by the research will be useful for the development of action plans that benefit the community, and the elaboration of the results is debated among the members before the wide dissemination. The final product can be elaborated collaboratively, as in the construction of this chapter.

CBPR also emphasizes that the commitments go beyond the project, the presence of researchers in the field or funding period, as the participatory research takes place in a way that strengthens the bond and continuous partnership between the involved institutions (academy) and the community.

It is a premise that the disaster in question is a "complex synthetic, non-linear, multiple, plural and emergent object" (Almeida-Filho, 2005, p. 38). The impacts of disasters can be long-term, requiring prevention and preparation actions for the possibility of future disasters, which challenges scientists to open borders, to consider the contexts.

This interdisciplinarity transdisciplinary thinking is an important condition for proposing and executing projects, involving the disciplinary fields of epidemiology and anthropology. The definition of interdisciplinarity adopted in the context of this project is understood as structural to the project, providing for reciprocity, mutual enrichment, with a tendency to horizontalize power relations between the fields involved. The realization of interdisciplinarity requires the identification of a common problem, a platform for joint work, the adoption of common fundamental principles and concepts, and the search for mutual learning, which is not carried out by simple addition or mixing, but by a recombination of internal elements (Almeida Filho, 1997).

The attentive look of the research team, the constant dialogue and the collaborative construction corroborate the research effectiveness. We cannot fail to mention that reality is not static, during the process of construction and development of the research, the COVID-19 pandemic occurred, the overlap of disasters brought new strategies and investigation demands to be incorporated, that is: incorporation of health education actions and vaccination incentives, and changes in the research instrument to deal with the investigation of the pandemic impacts on fishing communities.

Fishing communities have historically suffered from environmental racism, defined as "inequities in exposure to toxicants and the resultant health effects in communities that are most vulnerable to environmental threats: low-income communities and inhabited by people of color" (Bullard and Wright, 1993), land conflicts, and historical injustices, as the denial of access to rights guaranteed by the Brazilian constitution and that provide basic living conditions such as access to health, education, transport, sanitation, and housing, among others. These injustices result in the loss of fishing territory, environmental devastation, and the precariousness of artisanal fishing as the main source of subsistence (Barros et al., 2021).

Among the recommendations proposed to face environmental injustice, Bullard proposes the participation of those investigated in all stages of epidemiological research (planning, development, dissemination, and communication) (Bullard and Wright, 1993). The research context is, without a doubt, a political choice, while prioritizing the needs and demands of territories and populations. The integration between community members, researchers, and the search for dialogue with public policy managers, seeks to bring the research results closer to the planning of disaster risk management actions in the territories, in a culturally appropriate way, reflecting the perspective of fishing communities.

The continuous involvement of community members acting as researchers and participants has been essential in all phases of this research, from project design, data collection, analysis, and interpretation, and later in the sharing of results (Harper et al., 2012). As the project intends to accompany this population for a few years, the search for strengthening communication channels and carrying out joint actions has been a challenge for the project team, and this includes the need to build monitoring indicators that can measure the success of the partnership and the project (Brush et al., 2020).

It is worth mentioning that the CBPR framework has been successfully used in other studies involving oil disasters, such as the one that occurred in the Gulf of Mexico in 2010. The authors report that the project measured exposure to petrogenic polycyclic aromatic hydrocarbons, researched the toxicity of these polycyclic aromatic hydrocarbons, and communicated project findings and seafood consumption guidelines (Sullivan et al.,2018).

Project activities and data collection

Table 11.2 shows the synthesis of the phases of the research process relating to the actors involved and the methods used.

Identification and selection of research areas

Between the end of 2019 and March 2020, some meetings were organized involving fishing community representatives and researchers from different areas of knowledge, aiming to identify and prioritize the areas that should be the object of the study. Given that funding and research teams were insufficient to analyse the impacts in all 1,009 affected locations, in February 2020, a meeting was held involving researchers developing projects on the oil spill in Bahia and representatives of the movements linked to fishing for discussing criteria to be used in the selection of locations. It was decided to prioritize areas that received a higher volume of oil and had a greater population dependent on fishing for survival. In all states, the document Environmental Sensitivity to Oil Charts was consulted. In Bahia, besides consulting documents such as the Atlas of Environmental Sensitivity to Oil in the Sea Basins of Bahia (Dominguez, 2012), records of sightings or strandings of oil slicks were integrated, thus defining areas of vulnerability, considering the geomorphological, biological, and socioeconomic characteristics in an integrated way. All these decisions were made in common agreement with the representative entities of fishing populations and with other researchers who are part of the project named National Institute of Science and Technology (INCT) AmbTropic Phase II, made up of geographers, oceanographers, biologists, chemists, and epidemiologists. In Bahia, the municipalities selected for the research were: Prado (four locations), Canavieiras (five locations), Belmonte, Conde (four locations), and Cairu (one location).

Data collection instrument – development and validation

The elaboration of the data collection instrument, a structured epidemiological questionnaire, involved dialogues with international researchers (epidemiologists and an environmental engineer) with experience in investigating health impacts in communities affected by oil or with groups of workers directly involved in the process of cleaning oil residues.

TABLE 11.2 Research and participatory process stages

Steps	*Goals*	*Strategies and Method*	*Subjects/Actors Involved*
Identification and selection of research areas	Identify and prioritize the areas and communities that should be the object of the study.	Bibliographic and documentary consultation; Consultation of documents resulting from public hearings in the legislative assembly where the affected areas were reported by the fishermen; Meeting of researchers from different areas of knowledge	Representative entities of fishing populations, geographers, oceanographers, biologists, chemists, and epidemiologists
Data collection instrument – Development and Validation	Elaborate and validate the epidemiological questionnaire	Bibliographic and documentary consultation; Validation by consensus-building Delphi approach.	Research from public health, toxicology, or another field related to environmental health. All had significant work experience in this field.
Selection and training of interviewers and Data collection	Select and train local interviewers. Apply the epidemiological questionnaire and semi-structured interview.	Preparation of selection criteria and public notice for selection of interviewers carried out together with researchers and members of the fishing community; Meetings for the elaboration of strategies to mobilize the interviewees to answer the epidemiological questionnaire; Application of semi-structured qualitative interviews, with script prepared by researchers from different areas of knowledge.	Research from public health, anthropologist, toxicology, or another field related to environmental health. Representative entities of fishing populations.
Mobilization and dissemination of the research.	Publicize the survey and invite participants to fill in questionnaire; Publication of articles and participation in conferences; Delivery of reports to each community.	Meetings with the local team (leaders and interviewer); Publicity in local media; Groups for analysis and discussion of results and definitions of important themes.	Representative entities of fishing populations; Research from public health, anthropologist, toxicology, or another field related to environmental health.

Source: own table.

The questionnaire elaboration and validation process done by the research team took place between March and December 2020. It was based on the Delphi method and further reviewed/validated by experts (researchers in the artisanal fisheries area and representative of the Pastoral Council of Fishermen). The questionnaire was reviewed using the consensus-building Delphi approach. Our review committee included 16 experts, all of whom held doctorates in public health, toxicology, or another field related to environmental health.

The epidemiological questionnaire consists of 317 questions, distributed in 13 blocks (Identification and control; General Information; Socioeconomic and housing characterization; Work history and work organization; Oil spill; Organization of fishing/shell fishing work; Exposure associated with oil stain removal; Fish consumption; Perception of the oil spill impact; Clinical measures; Other health information; Lifestyle; Quality of Life; and COVID-19).

Selection and training of interviewers and data collection

The project team presented the general lines of the study in meetings with representatives and presidents of the Fishermen's Associations and/or Colonies located in the chosen municipalities, (North and South Coasts of Bahia State). During the meetings, the research dissemination strategies were designed together with the members of the fishermen's associations, according to the reality of each community.

Assuming that all fishermen residing in the territory could have been affected by the disaster, a sample was not chosen. Individual invitations were delivered to fishermen's domiciles, and interviews were scheduled through messaging apps. Videos to publicize the research were broadcast on social networks (WhatsApp groups) and the research coordinator gave interviews to the local media (radio and newspaper) for promoting mobilization and ensuring the participation of artisanal fishermen in the study.

Local interviewers were chosen through a selection process, whose criteria were: Ability to communicate well; Knowing and having a good relationship with the population of the territory, especially with artisanal fishermen and fisherwomen; Familiarity with electronic devices – tablets; Affinity with the health area; Have carried out research in the region; Being or coming from a family of fisherman/shellfish gatherers; Participate in fishermen/shellfish associations; Participate in organized groups representing fishermen/shellfish gatherers; To have been involved in the collective activities developed during the oil spill. A score was assigned to each criterion and the candidates with the highest scores were integrated into the team. The number of vacancies for local interviewers was proportional to the number of potential artisanal fishermen participating in the study.

In total, 16 people were selected as "local interviewers", 10 women and 6 men, aged between 16 and 48 years, distributed in five locations: Cumuruxatiba (2), Belmonte (2), Canavieiras (7), Conde (4), and Cairu (1). Regarding schooling, four had incomplete high school, five had completed high school, five had incomplete higher education and two had completed higher education. All received a grant of

R$400.00/month (US$80.00). The working period varied between one and four months.

The local interviewers were trained by the researchers regarding the application of the questionnaire and ethical aspects involving the research, they collaborated in the mobilization and application of the questionnaires. This community engagement strategy was fundamental to the success of the research in the communities. Partnership ties with local interviewers remain even after the data collection stage is completed, through messaging app groups.

In the areas of the North Coast and Lower South Coast, an anthropologist joined the research team. In these regions, participant observation of the communities where the questionnaires were applied was carried out. In these places, semi-structured interviews were also applied, which allowed for a deeper understanding of the experience of artisanal fishermen who were most affected by the disaster. Some interviews were also carried out with residents of the community who participated in the management and collecting of oil during the disaster. Most of these interviews were carried out individually and a few with the participation of members of the research team, near the colonies or in their domiciles.

Mobilization and dissemination of the research

In the Marine Extractive Reserve (RESEX) areas, the invitation for artisanal fishermen to participate in the study was based on the list of beneficiaries provided by the RESEX management. In the other areas, the register of beneficiaries of the emergency aid granted to fishermen during the disaster, provided by Bahia Pesca, the Bahia State agency responsible for coordinating fishing actions, was used. These lists were updated in the field with the help of local fishermen association leaders and local interviewers.

Meetings were held with the local team (leaders and interviewers) to outline the best strategies for publicizing and presenting the project, such as choosing the radio stations most listened to by fishermen in the region, using social networks and sometimes contacting health workers who usually carry out home visits. For both mobilization and dissemination, the participation of field interviewers was essential.

Several successful experiences have been reported with CBPR, especially in the academic-community partnership, but there is a relative dearth of community voices addressing these topics (Caldwell et al., 2015). In the following topic, the interviews portraying community voices, particularly evidenced in this research, will be described, and analysed.

Lessons learned from CBPR in an epidemiological research

The narratives included originate from native researchers who participated in the discussion and interview groups. To preserve anonymity, the referenced names have been changed. Three axes are approached to analyse the experience of local interviewers from North and Lower South Coasts of Bahia on the application of the CBPR methodology, once the questionnaire application phase has been completed. First, the experience during the project activities (selection and training process, application of the questionnaire, internal dynamics of the teams and monitoring); second, individual experiences in research activities; and third, expectations from the research (return of results).

The results of this experience are organized into two parts, the first related to the experience of local, collective, and individual interviewers during the project, and the second to the future of the research.

"It was a very cool experience": local interviewers during project activities

As already indicated, there were significant differences between communities during the process of selecting local interviewers. In the Lower South Coast in Bahia, where the RESEX of Canavieiras brings together several entities with great socioeconomic research activity and monitoring of marine biota species, the selected local interviewers (two men and five women) had previous experience, which facilitated the training. One of them continued to work with the research team in the North and Lower South Coasts. The fellows were experienced people, and some of them had already worked together in previous research.

> As I had a scholarship . . . Professor M. from the University always came and we followed up . . . simple things.
>
> *(João, interview January 6, 2022, Lower South Coast)*

> it was something I had already worked on, with a tablet and everything, doing questionnaire and other stuff, so for me it was super easy.
>
> *(Maria, discussion group December 16, 2021, Lower South Coast)*

On the contrary, the local interviewers on the North Coast were indicated by the fishermen's colony, maintaining family relationships with members of its board, very young and inexperienced, not directly linked to the fishing activity. Since the beginning, they related the experience of participating in the research as a work opportunity and considered that they contributed to improve communication with the community and to create a conviviality/friendship with the research team. The emotional dimension was fundamental for strengthening the bonds of friendship that led the interviewers to participate with motivation and pleasure.

> For me it was very nice because it even helped me a little . . . then I'm a little shy, I'm not used to talking to people and on the days, I was there alone it helped me a lot. I managed to feel more comfortable and able to talk to people, explain things.
>
> *(Rose, North Coast discussion group, December 15, 2021)*

> I worked feeling like without the intention to work to receive something, the way you treat people, I don't know, it's very good, it was the real conviviality.
>
> *(Diana, discussion group interview December 15, 2021, North Coast)*

> We kind of created a kind of friendship, it involved so many really cool things, there were lots of jokes.
>
> *(Diana, discussion group interview, December 15, 2021, North Coast)*

The interaction during the data collection stage was intense, the team made meals together, shared dormitories, leisure, and social spaces outside the research application facilities, which contributed to the strengthening of bonds and horizontality between researchers.

This human dimension in the conviviality of all members of the research team was mainly reflected by the women and M. from South Bahia also mentioned that it was "a very good experience, because I met you, some excellent wonderful people . . . and until today I miss that. Experiences that made it easy to meet different people and other different learnings" (Alice, discussion group from the Lower South Coast, December 16, 2021). The dialogue of knowledge was established in a climate of trust and empathy.

These experiences that emotionally bond team members were also shared by men, as observed in interviewee Joao. By the way, these expressions that come after the practice of acquiring knowledge:

> Knowledge. Knowledge, experiences, experiences, experiences of other communities, we see from a totally different point of view, the difference from what we live and the experience of other people, knowledge.
>
> *(João, interview January 6, 2022, Lower South Coast)*

This knowledge brought by the participation in the research team and the coexistence in the communities stimulated the comparison and a differentiated look between them. For example, Joao signalled economic differences between communities by looking at people's needs and suffering. He highlighted that this look focused on needs is easier to see outside the own community.

> Many times, we are here in our locality, we do not say "I live wonderfully", we live not as well as we expected, but from my point of view, seeing the needs of those people there, I was very moved.
>
> *(João, interview January 6, 2022, Lower South Coast)*

Differences were centred on the observation of the number of shellfish collected and the value achieved by selling, and Joao realizes that compared to his community, which has abundance of shellfish, and all are united to control prices, there are others who are more needy and disunited. In fact, in other narrative of political nature, Joao recognized that if, on the one hand, belonging to an extractive reserve supposes "a totally more advanced level", in the other, the fishing colony leaves people "a little more forgotten", favouring only the ones who live closer to the headquarters.

> Because we have the extractive reserve here, we still have someone who looks after us, someone who is always together with us and we ourselves, who are the reserve, we have to make our reserve and we give the RESEX our hands and run everything together, always protecting our biodiversity, those things. And they don't have that [compared to communities on the North Coast].
>
> *(João, interview, January 6, 2022, Lower South Coast)*

These needs that Joao is able to see more clearly from the comparison between communities different from his own, are also recognized by other local interviewers. A. recognizes having seen the suffering in her own family when the oil spill happened, "because there were people who suffered a lot, including myself, I have a brother at home who had some problems because of the oil" (João, interview, January 6, 2022, Lower South Coast).

The fact of collaborating in a university survey is seen by local interviewers as an opportunity to work, learn and meet other people. Several times the local interviewers adopted an active attitude, indicating people to be interviewed by the anthropologist who participated in the project, after the questionnaire was applied. Sometimes they even followed her closely, and when that was not possible, they were interested in knowing from the anthropologist how the interview had gone. They felt satisfied for having contributed to the fact that these people's statements were listened to more carefully.

When asked about the activity of training and application of the questionnaire, all responded that they did not observe any difficulties. There were mixed opinions about the format of the questionnaire. Most considered that it was simple to apply, good and useful, but there were also those who thought it was too time-consuming because it contained repetitive questions. In this case, they learned to clarify doubts and be cautious with filling. Despite having to interview relatives, most of them did it without fear, and relatives looked for them to know about the research.

"The questions in the questionnaire, I think were super necessary. I felt that some people felt it was difficult, but at least for me, the few interviews I did, I didn't feel any difficulty in applying" (Rose discussion group, December 15, 2021, North Coast).

The much more experienced local interviewers from Lower South Coast were able to better mobilize people in the community to participate in the research, talking, "running after", "doing a task force", "mouth to mouth", and "knocking on the door." Alice acknowledged that "some people were too lazy to go . . . there was even one that I got at home "let's do it, it is important, let's do it".

"People asked if you would come back with the survey": the future return of results

Local interviewers are asked by the community for the return of the results, as participants are interested in whether the spill has had any impact on their own health. On the other hand, many of them were asked when a future meeting with the research team would take place in the community, then they see the research as an opportunity to claim rights regarding financial compensation for the disaster effects. The delay in getting the results of the research puts pressure on local interviewers, but they manage to get around it because now they know better the procedure of scientific practice.

> Yes, always questions. Questions about the research, about the results, about the day it will arrive, the day researchers will come with the results. But I always have to protect myself, so I talk about what happened, how it works, and that it takes time. I say you have to wait, and I say they'll come. I know the result will come. I'm one who is waiting too.
>
> *(João, interview January 6, 2022, Lower South Coast)*

The main wish is that the results arrive as soon as possible and that all this has an effect on the government for compensating those affected, both financially and in health. Local researchers and participants desire the results to contribute to a future improvement in health as well. "I hope something comes along to help people who have been harmed . . . health issues, then until today there are people who are sick from it . . . the disaster that happened" (Maria, discussion group, interview December 16, 2021, Lower South Coast).

"There are also people who ask about values "when will this oil stuff come, the question of this payment that the oil was going to make". Then I said:

> Ah, it is not that we are going to give money, we are more in the logistics of examination and see how the health of the fishermen is doing. Regarding this value stuff, what day it shall come, you have to ask the fishers' colony and you will know that this money comes there. This is the information I give.
>
> *(João, interview, December 16, 2021, Lower South Coast)*

All community members are of the opinion that the return of the research results should be done in the form of a seminar or meeting in the community's associations. They consider that the mobilization should be the same as that used to engage the community at the beginning of the research. They are interested in keeping linked with the research team in the longer term, despite the short financial support having ended, and they are willing to help in the mobilization for the return of the results and in future research. The human and emotional bond between all the researchers continues today through the same messaging app used during the field research, which is a space where other disasters are also shared, such as the flood that happened in Lower South Coast in December 2021. And the way they see their communities is a learning curve for future disasters.

Final considerations

The CBPR methodological choice enabled a dialogue of knowledge with different actors involved in the research process. Without a doubt the inclusion of community members and direct communication with associations and entities of the fishing populations contributed to the adhesion of the population, as well as to establishing a link between universities and fishing communities. In this chapter, we sought to describe this experience, which was based on the principles of CBPR. This study reflected on the voices of communities in the research process, a topic that is relatively scarce in research involving CBPR (Caldwell et al., 2015). However, this research experience can be defined, strictly, by some authors as community engagement. Therefore, it becomes necessary further critical analysis of this experience with a focus on the principles of CBPR, which will be carried out in future studies.

The project's objective is to highlight the various impacts of environmental disasters on fishing communities in Brazil. However, understanding this process is only possible when considering the following phrase: "Nothing about us, without us". This means, do not talk about us without having actively participated in this process. In the perspective proposed by the Ocean's Decade, we understand that the CBPR, based on collaborative research principles, constitutes an important tool in the study of environmental disasters, which are complex and transdisciplinary. Previous experience of this interdisciplinary collaboration was developed by the team in epidemiological impact assessment projects (Larrea-Killinger et al., 2013). This chapter constitutes a transdisciplinary collaboration between epidemiology and anthropology, involving communities.

Community involvement, co-creation, and transdisciplinary are increasingly valued as essential ingredients to implement policies (Roura, 2021). In this context, participatory research represents a promising path for stakeholders to collaborate and seek to solve problems in an innovative way (Roura, 2021) However, it is important to admit that there is a complex relationship between the partners of the community and the academy, and that this, in itself, constitutes a complex issue in the dynamics of research (Minkler, 2005).

Furthermore, the social ecology of power in participatory research, a conceptual theoretical model (Roura, 2021), helps to understand the relationships established in the present investigation, which involves the complexity of the disaster and the need for a dialogue of knowledge within and outside the university. In this publication, the author provides a list of monitoring questions to guide the assessment of power dynamics at each of the socio-ecological levels exacerbating inequities when disasters happen (Roura, 2021). Future reflection on this process, in the light of this or other conceptual theoretical models that emerge from participatory processes, can contribute to a better understanding of the research results and favour greater integration of the knowledge acquired with intervention actions, generating more co-participatory and co-designed solutions.

Extensive literature has reported successful experiences of academic-community partnerships with CBPR; however, there is a lack of consensus in the field on what defines partnership success and how to measure the factors that contribute to this success in short- and long-term CBPR partnerships (Brush et al., 2020) There are experiences from specific studies involving partnerships in relation to disasters (Sullivan et al., 2019) that deserve further attention in the future. It is expected that the reflection of the experience of using this methodology can be useful in other contexts: in the processes of monitoring populations affected by disasters and in the interventions in the affected territories; collaborating in the development of public policies to protect fishing communities; encouraging the dialogue of knowledge through social control and the production of knowledge, strengthening the culture of ocean preservation, and recognising the role of traditional fishing communities in this process.

References

Aguilera, F., Méndez, J., Pásaro, E., & Laffon, B. (2010). Review on the effects of exposure to spilled oils on human health. *Journal of Applied Toxicology: JAT*, 30(4), 291–301. https://doi.org/10.1002/jat.1521

Almeida Filho, N. (1992). Hacia una etnoepidemiología (Esbozo de un nuevo paradigma epidemiológico). *Revista de la Escuela de Salud Pública*, 3(1), 33–40.

Almeida-Filho, N. (1997). Transdisciplinaridade e Saúde Coletiva. *Ciência & Saúde Coletiva*, 2(1–2), 5–20. https://doi.org/10.1590/1413–812319972101702014

Almeida Filho, N. (2000). *La ciência tímida*. Ensayos de desconstruccíon de la epidemiologia Lugar Editoria.

Almeida Filho, N. (2005). Transdisciplinaridade e o paradigma pós-disciplinar na saúde. *Saúde e Sociedade [Online]*, 14(3), 30–50. https://doi.org/10.1590/S0104-12902005000300004

Almeida Filho, N. (2020). Etnoepidemiologia y salud mental: perspectivas desde América Latina. *Salud Colectiva*, 11, https://doi.org/10.18294/sc.2020.2786

Araújo, M.E., Ramalho, C.W.N., & Melo, P.W. (2020). Pescadores artesanais, consumidores e meio ambiente: consequências imediatas do derramamento de óleo em Pernambuco, Nordeste do Brasil; Cad. *Saúde Pública*, 36(1), e00230319. www.scielo.br/j/csp/a/66t7BVfM6X4pBBCJwLcqmjf/?format=pdf&lang=pt

Bahia Pesca. Technical advice. (February 10, 2020). www.bahiapesca.ba.gov.br/arquivos/file/parecertecnicolitoralnortefinal.pdf.

Barros, S., Medeiros, A., & Gomes, E. B (2021). Conflitos socioambientais e violações de direitos humanos em comunidades tradicionais pesqueiras no Brasil: relatório. Conselho Pastoral dos Pescadores. www.cppnacional.org.br/publicacao/relat%C3%B3rio-dos-conflitos-socioambientais-e-viola%C3%A7%C3%B5es-de-direitos-humanos%C2%A0em%C2%A0comunidades

Beiderbeck, D., Frevel, N., von der Gracht, H.A., Schmidt, S.L., & Schweitzer, V.M. (2021). Preparing, conducting, and analyzing Delphi surveys: Cross-disciplinary practices, new directions, and advancements. *MethodsX*, 8, 101401. https://doi.org/10.1016/j.mex.2021.101401

Brush, B.L., Mentz, G., Jensen, M., Jacobs, B., Saylor, K.M., Rowe, Z., Israel, B.A., & Lachance, L. (2020). Success in long-standing community-based participatory research

(CBPR) partnerships: A scoping literature review. *Health Education & Behavior: The Official Publication of the Society for Public Health Education*, 47(4), 556–568. https://doi.org/10.1177/1090198119882989

Bullard, R.D., & Wright, B.H. (1993). Environmental justice for all: Community perspectives on health and research needs. *Toxicology and Industrial Health*, 9(5), 821–841. https://doi.org/10.1177/074823379300900508

Caldwell, W.B., Reyes, A.G., Rowe, Z., Weinert, J., & Israel, B.A. (2015). Community partner perspectives on benefits, challenges, facilitating factors, and lessons learned from community-based participatory research partnerships in Detroit. *Progress in Community Health Partnerships: Research, Education, and Action*, 9(2), 299–311. https://doi.org/10.1353/cpr.2015.0031

Carvalho, I.G.S., Rêgo, R.C.F., Larrea-Killinger, C., Rocha, J.C.A., Pena, P.G.L., & Machado, L.O.R. (2014). Por um diálogo de saberes entre pescadores artesanais, marisqueiras e o direito ambiental do trabalho. *Ciênc. saúde coletiva*, 19(10). https://doi.org/10.1590/1413-812320141910.09432014.

Cook, W.K. (2008). Integrating research and action: a systematic review of community-based participatory research to address health disparities in environmental and occupational health in the USA. *Journal of Epidemiology and Community Health*, 62(8), 668–676. https://doi.org/10.1136/jech.2007.067645

Detroit Community-Academic Urban Research Center. (2011). Community-based participatory research principles. The Detroit Community-Academic Urban Research Center. www.detroiturc.org/about-cbpr/community-based-participatory-research-principles

Dominguez, J.M.L. (Orgs). (2012). *Atlas de sensibilidade ambiental ao óleo das bacias marítimas da Bahia*. Brasília: MMA.

Haro, E.J.A., (2010). *Epidemiología sociocultural: Un diálogo en torno a su sentido, métodos y alcances*. Sonora y Buenos Aires: El Colegio de Sonora - Lugar Editorial.

Harper, S.L., Edge, V.L., Cunsolo Willox, A., & Rigolet Inuit Community Government (2012). 'Changing climate, changing health, changing stories' profile: using an EcoHealth approach to explore impacts of climate change on inuit health. *EcoHealth*, 9(1), 89–101. https://doi.org/10.1007/s10393-012-0762-x

Hersch-Martínez, Paul. (2013). Epidemiología sociocultural: una perspectiva necesaria. *Salud Pública de México*, 55(5), 512–518. Recuperado en 20 de mayo de 2022, de www.scielo.org.mx/scielo.php?script=sci_arttext&pid=S0036-36342013000700009&lng=es&tlng=es

Instituto Brasileiro do Meio Ambiente e dos Recursos Naturais Renováveis. (2019). Monitoring in the management of the environmental emergency related to the oil that hit the beaches of the Northeast. Affected locations – maps. www.ibama.gov.br/manchasdeoleo#localidades

Larrea-Killinger, C., Rego, R.F., Strina, A., & Barreto, M.L. (2013). Epidemiologists working together with anthropologists: Lessons from a study to evaluate the epidemiological impact of a city-wide sanitation program. *Cad Saude Publica*, 29(3): 461–474, www.scielo.br/j/csp/a/k8XTzLWtYftjThnL8HjDvws/?lang=en

Magalhães, K.M., Barros, K., Lima, M., Rocha-Barreira, C.A., Rosa Filho, J.S., & Soares, M.O. (2021). Oil spill + COVID-19: A disastrous year for Brazilian seagrass conservation. *The Science of the Total Environment*, 764, 142872. https://doi.org/10.1016/j.scitotenv.2020.142872

Marinha do Brasil. Press release (December 9, 2019). www.marinha.mil.br/sites/default/files/nota_a_imprensa_09dez_comunidade_cientifica.pdf

Menéndez, E.L. (2008). Epidemiologia sociocultural: propostas e possibilidades. *Região e Sociedade*, 20(2). https://doi.org/10.22198/rys.2008.2.a526

Minkler, M. (2005). Community-based research partnerships: Challenges and opportunities. *Journal of Urban Health: Bulletin of the New York Academy of Medicine*, 82(Suppl. 2), ii3–ii12. https://doi.org/10.1093/jurban/jti034

Müller, J., da Silva, E.M., & Franco Rego, R. (2022). Prevalence of musculoskeletal disorders and self-reported pain in Artisanal Fishermen from a Traditional Community in Todos-os-Santos Bay, Bahia, Brazil. *International Journal of Environmental Research and Public Health*, 19(2), 908. https://doi.org/10.3390/ijerph19020908

Rêgo, R.F, Muller, J.S, Falcão I.R, & Pena, P.G.L. (2018). Vigilância em saúde do trabalhador da pesca artesanal na Baía de Todos os Santos: da invisibilidade à proposição de políticas públicas para o Sistema Único de Saúde (SUS). *Revista Brasileira de Saúde Ocupacional*, 43(Suppl. 1), 1–9. https://doi.org/10.1590/2317-6369000003618.

Roura, M. (2021). The social ecology of power in participatory health research. *Qualitative Health Research*, 31(4), 778–788. https://doi.org/10.1177/1049732320979187

Soares, M.O., Teixeira, C.E.P., Bezerra, L.E.A., Paiva, S.V., Tavares, T.C.L., Garcia, T.M., Araújo, J.T., Campos, C.C., Ferreira, S.M.C., Matthews-Cascon, H., Frota, A., Mont'Alverne, T.C.F., Silva, S.T., Rabelo, E.F., Barroso, C.X., Freitas, J.E.P., Melo Júnior, M., Campelo, R.P.S., Santana, C.S., Carneiro, P.B.M., Meirelles, A.J., Santos, B.A, Oliveira, A.H.B., Horta, P., Cavalcante, R.M. Oil Spill in South Atlantic (Brazil) (2020). Environmental and governmental disaster. *Marine Policy*, 115, 103879. https://doi.org/10.1016/j.marpol.2020.103879.

Sullivan, J., Croisant, S., Howarth, M., Rowe, G.T., Fernando, H., Phillips-Savoy, A., Jackson, D., Prochaska, J., Ansari, G., Penning, T.M., Elferink, C., & Community Partner Authors: Louisiana Environmental Action Network, United Houma Nation, Bayou Interfaith Shared Community Organizing, Dustin Nguyen-Vietnamese Community Partner, Center for Environmental & Economic Justice, and Alabama Fisheries Cooperative Project Community Scientist Author: Wilma Subra (2018). Building and maintaining a citizen science network with fishermen and fishing communities post deepwater horizon oil disaster using a CBPR approach. *New Solutions: A Journal of Environmental and Occupational Health Policy: NS*, 28(3), 416–447. https://doi.org/10.1177/1048291118795156

Sullivan, J., Croisant, S., Howarth, M., Subra, W., Orr, M., & Elferink, C. (2019). Implications of the GC-HARMS fishermen's citizen science network: Issues raised, lessons learned, and next steps for the network and citizen science. *New Solutions: A Journal of Environmental and Occupational Health Policy: NS*, 28(4), 570–598. https://doi.org/10.1177/1048291118810871

United Nations Educational, Scientific and Cultural Organization -Intergovernmental Oceanographic Commission. (2020). *The United Nations Decade of Ocean Science for Sustainable Development (2021–2030) Implementation Plan*. UNESCO, Paris (IOC Ocean Decade Series, 20), p. 56.

World Ocean Review. Marine Resources (2014) Opportunities and risks. https://worldoceanreview.com/wp-content/downloads/wor3/WOR3_en.pdf

12

THE MARINE RESERVE OF FISHING INTEREST AT CAPE ROCHE (CONIL, SPAIN)

Transdisciplinarity and academic challenges of a conflictive process

David Florido-del-Corral and Mar Abbot-Jiménez

Introduction

A still unfinished activation process for the declaration of a Marine Reserve of Fishing Interest (MRFI, one of the possible legal protections of marine spaces in Spanish legislation) has been ongoing from 2009 to the present day, in the area around the port of Conil de la Frontera. In territorial terms, the proposed area lies between Sancti-Petri islet (Hercules Castle) and the Cape of Trafalgar and is 18 nautical miles offshore (Figures 12.1 and 12.2). The area corresponds to the historical fishing grounds of the Conil and Sancti-Petri (Chiclana de la Frontera, Cadiz province) small-scale fleet, although artisanal boats from Barbate, Rota (Cadiz province), and the city of Cadiz (around 100 vessels in total) flock to the area, as well as longer-range fleets: trawlers (from the province of Huelva to the northwest) and purse seiners (from Barbate to the southeast). A tuna trap fishery is also located in the proposed area for the marine reserve. With the initial involvement of environmental organizations, academics, and the fisheries sector but with the incorporation of other actors throughout the process, the difficulty of securing a shared discussion and decision-making space to approve the proposal has been brought to light. The factors that hinder this include the size of the potentially affected area; the wide range of actors with diverse uses, expectations, and experiences in the territory; the conflicts that have emerged among users, in particular, those that pit the trawlers against the small-scale fishers, and the latter against a specific segment of recreational fishers, the underwater fishers; and the lack of a history of collaboration between the fisheries administration and recreational fishers. Also, directly related to the theme of this book, there are other underlying issues, such as the wide range of different perspectives and legitimation frameworks that set small-scale fishers against underwater fishers. And lastly, there is a discrepancy between the marine biology-based scientific knowledge – which

DOI: 10.4324/9781003311171-15

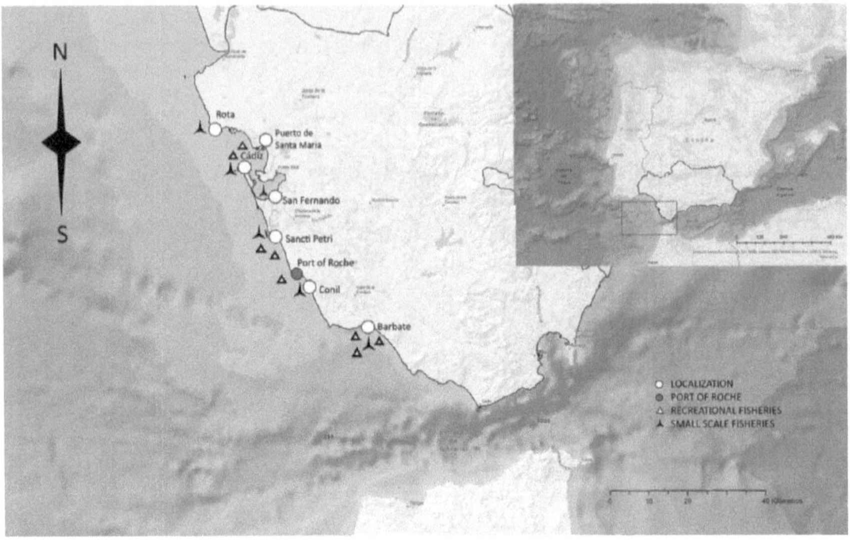

FIGURE 12.1 Location map.

Source: © Helena Pérez Gámuz

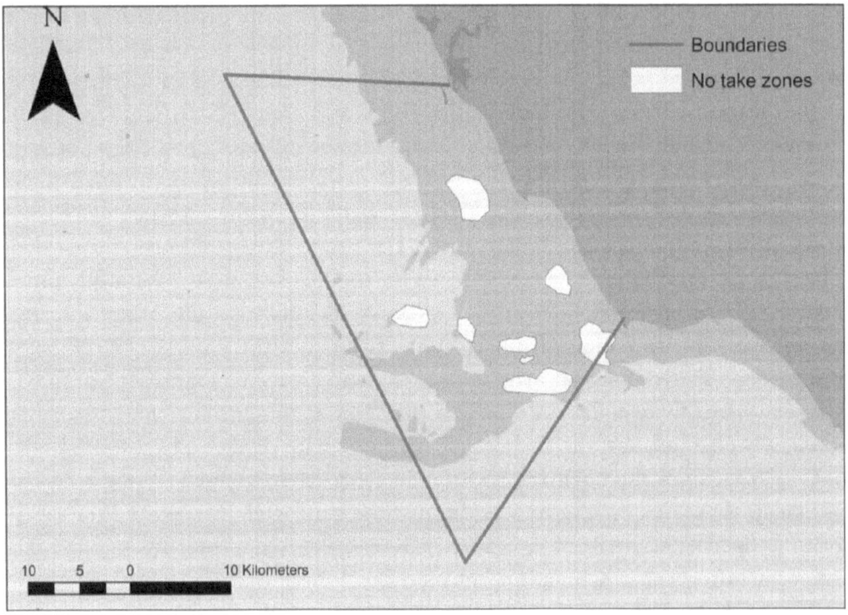

FIGURE 12.2 Boundaries and no take zones of the MRFI, as proposed by promoters.

Source: © Francesc Sobrado

validates the decisions adopted by the state administration in marine affairs through an oceanographic research institute – and the knowledge of the actors and users who, through a variety of both extractive and discursive strategies, appropriate the coastal strip under discussion in both material and symbolic terms.

In this chapter, procedures will be described through which the knowledge of different actors was assimilated to identify uses and their intensity in the area, the delimitation of the area's boundaries, and the MRFI's proposals for no-take zones in the various phases of a long-drawn-out and complex political process. The methodology was based on the use of participatory techniques coordinated by a promoter group led by environmentalist non-governmental organizations (NGOs) and academic staff (social anthropology) in close collaboration with professional fishers. Senior representatives from the administration and other stakeholders were also involved in this process until both the atmosphere and the tone of the discussions changed drastically from 2021 on: the social networks were turned into an echo chamber for recreational fishers' hostility towards the reserve, the opposition took to the streets and the media with a public demonstration calling for its rejection (March 2021).

In the Spanish case,[1] MRFI are legal marine governance concepts that depend on the central government in political terms. There is a small team that is very active in direct talks with fisheries representatives and environmentalist NGOs taking part in both formal and informal meetings, in workshops, and organized events, on their own behalf or for other entities. The MRFI are innovative institutional tools that explicitly plan to create a governance style based on collaboration between actors and the attainment of multi-level sustainability to guarantee both the preservation of ecological elements and dynamics in a delimited area of the marine environment and, at the same time, the social reproduction of small scale fisheries and their cultural practices and other traditional uses in the area. From our point of view, MRFI represent a new relationship model between social actors and the marine environment and its ecological dynamics, and a new form of interaction between actors based on knowledge of alterity, and the ability to define common rules and assume local wisdom-based management that adapts to the changing circumstances as a result of permanent scientific monitoring. In this sense, this management arrangement represents an innovative concept of interactive, transparent, and embedded democratic governance based on the notion of public property and the appreciation of the socio-ecosystem and cultural services provided by artisanal fleets and traditional fishing uses, with their knowledge and practices, and other types of users.

We will discuss this conflict process and the role that anthropology can play in participatory techniques in scenarios of this type. Our theoretical perspective assumes the importance of achieving knowledge transfer between social and institutional actors and between the latter themselves. Not only schemes with practical recipes and media representation models will be taken into account, but also an exchange of evaluation schemes and legitimacy discourses on the role of one or other of the actors. We start from the hypothesis that the conflicts triggered around marine reserves do not only represent a clash of interests but also struggles between legitimation schemes and the acceptance of otherness, which requires a knowledge and governance model quite different from the established model.

Methodology and institutional process

In 2009, an unplanned research process got underway (Table 12.1) that combined different strategies: focus groups, workshops, participatory forums, in-depth interviews, and questionnaires were combined in a framework perspective (Maya-Jariego et al., 2016), that conformed to the so-called stakeholder analysis (Quesada et al., 2019) and participatory methodologies (Florido, 2021). At that time, the role of MRFI was already attracting attention in fisheries anthropology (Jentoft et al., 2011; Pascual and De la Cruz, 2011), so one of the authors, David Florido, began to attend the focus groups that the environmental NGO Ecologists in Action was organizing with the support of the local fishers guild at the latter's headquarters. This was the beginning of what could be termed the process's pre-critical phase (2009–2011) (Figure 12.3). These weekly meetings went on for two months to discuss among local artisanal fishers whether a MRFI was an appropriate management tool to cope with the fishing crisis. At the same time, the meaning of this legal concept was being explained to them and questionnaires were being passed around at the guild's headquarters to identify the extent of the fishers' knowledge and their degree of internal agreement and support to this initiative.[2] In short, an initial proposal was defined for a boundary and some possible no-take zones within the reserve. With this information, the fishers' guild sent a formal written proposal to the General Secretariat of Maritime Fisheries (central government) and the Directorate General of Fisheries and Aquaculture (regional government) to support the marine reserve.

The financial crisis in the public administration, which was particularly severe in 2010 and the following years, halted the process, which subsequently acquired some new actors and some new techniques. This was the beginning of an expansive phase (2015–2017) based on the World Wide Foundation for Nature (WWF) initiative and supported by a local NGO (Soldecocos, the Coastal Community Development Partnership), both of which are environmentalist groups. At that time, the proposal for the reserve was being capitalized by the Conil Fish Market Artisanal Producers Organization (Fish Producer Organization-PO 72), an organization that had emerged out of the local fisheries sector in 2010. This was partly the outcome of the guild crisis of the previous years and a response from the field of associationism to adapt to the European Fisheries Policy. During this second period, conservationist proposals made some headway: not only was the proposal for the MRFI project taken up again but also, in parallel, the declaration of the Nature 2000 network was triggered based on some passages written into the Habitat Directive also found in the OSPAR Convention. These principles have outstanding environmental value due to the associated marine biodiversity, including species of fisheries interest and other species that attract larger-size ichthyofauna

TABLE 12.1 Methodological process: phases, activities, and main stakeholders

	Promoters	*Research techniques*	*Stakeholders involved*	*Main results*
PRE-CRITICAL PHASE (2009–2011)	Conil Fishers' Guild *Ecologistas en Acción* – Conil	Questionnaires (boat owners and crew) Focus groups with Conil and Sancti-Petri professional fishers.	Conil and Sancti-Petri professional fishers Environmentalist NGO (*Ecologistas en acción*).	Explanation of project to local sector to obtain its support. 1st proposal of boundaries and possible no-take zones incorporating TEK of local fishers.
EXPANSIVE PHASE (2015–2017)	PO-72 Conil, WWF-Soldecocos, supported by David Florido (University of Seville).	Focus group with professional fishers: Conil, Barbate, Sancti-Petri, S. Fernando, and Cadiz. Stakeholder workshop in Conil In-depth interviews given to Conil professional fishers to input their TEK	Representatives of fishers from area. Recreational Fishing Clubs from the area. Representatives from local, regional, and central gov. (marine reserves). Environmental organizations, local associations.	Presentation of project to everyone affected. A new means of promoting the declaration of the area is opened up in the area of the Nature 2000 network habitats. Interest of central government representatives captured.

(*Continued*)

TABLE 12.1 (Continued)

	Promoters	*Research techniques*	*Stakeholders involved*	*Main results*
PROJECTIVE INSTITUTIONAL PHASE (2018–2019)	PO-72 Conil, WWF-Soldecocos, supported by David Florido (University of Seville).	Workshops with recreational fishers Questionnaires (recreational fishers) Briefings to explain the results of the Sea of Gades project to oceanographic research centre members.	Professional fishers from the area. Political representatives of Marine Reserves. Special emphasis on communication with recreational fishers. Personnel from the Spanish Institute of Oceanography	Sea of Gades project. Presenting project to recreational fishers and their perspective added into documents. Estimating recreational sector extractive activity: zones, species, seasonality. Economic importance and social diversity of recreational sector. Creation of an opposition platform to the reserve: Recreational Fishers "Responsible Fishing" Association.
INSTITUTIONAL CONFLICT PHASE (2020–)		Virtual meetings with stakeholders In-depth interview of research centre technician. Discourse analysis of declarations of opposition to the marine reserve on social networks and in media.		Publication of public consultation on declaration of the MRFI Social expansion of protest, especially among the recreational sector and trawler fishers.

FIGURE 12.3 Focus Group (Conil) in the initial phase. Fishers and promoters exchange visions and knowledge.

Source: © David Florido

such as the loggerhead turtle, and mammals such as the bottlenose dolphin, the killer whale, and the porpoise.[3] A new focus group was held at the Conil fish market in June 2016 with artisanal fishers from Cadiz, San Fernando, and Barbate (Cadiz province). But the most important act at this time was a workshop organized in April 2017 to which new stakeholders were invited; in particular, recreational fishers; central, regional, and local government representatives; representatives of fishers from other areas and other modes of fishing that had traditionally worked the area, and also representatives from other parts of Spain with experience in marine reserves. The aim was to disseminate the project socially, to spread knowledge of the proposal, and to garner support from these other bodies. The presence of government marine reserve representatives was expected to secure their commitment to political support and they fulfilled their role of setting out the legal models that the projected MRFI in Cape Roche should adhere to. The representatives of the regional government explained an artisanal surveillance and monitoring tool that had been applied in the area since 2007 and had resulted in the comprehensive mapping of the fisheries grounds visited, the extracted biomass in each area by quantity and species, and fishing and sailing times.[4] At the workshop, the new promoters (WWF, Soldecocos, PO 72, advised by the University of

Seville's Social Anthropology department) committed to requesting and executing a project within the Biodiversity Foundation's "Pleamar Program" framework.

This research was developed during 2018–2019 and generated the projective institutional phase in which the promoters executed the Sea of Gades project with three studies designed to give an empirical basis to the proposal: bionomic cartography, a report on professional fishing in the area, and another report on recreational fishing.[5] What was new at the time was the incorporation of the information and positioning of the recreational fishers, which also meant that a "discordant wing" simultaneously emerged in the marine reserve project. In addition, a new recreational fishing association was founded, built around a core of fishers in nearby areas but not confined to any particular area territorially: The Recreational Fishers "Responsible Fishing" Association, which combined four types of fishing: sea fishing from boats, underwater fishing, fishing from kayaks, and surfcasting, all of which were done in the coastal area of the proposed reserve.

The research activities undertaken at the time were: in-depth interviews of professional fishers to gather traditional ecological knowledge (TEK) on sea beds and fisheries areas, and several focus groups, meetings, and workshops with recreational fishers at El Puerto de Santa María (1), Sancti Petri (1), Conil (1) and Barbate (1) (Cadiz province), which are points on the coast where there are yachting clubs with recreational fishing from boats, and Medina Sidonia (1) and Chiclana (1) (Cadiz province), with underwater fishers. At the same time, questionnaires were given out at the previously mentioned workshops and focus groups to habitual boat-owning recreational fishers in the framework of the Sea of Gades project, albeit with little success.[6] Fishers at the Media Sidonia workshop were given the opportunity to take part in the information generation process by aggregating data on fishing trips, with the identification of species, capture areas, and fishing trips on a mobile phone APP called Pesgadir, but this initiative generated very few results. This proposal was presented by biologists from an oceanographic research centre who had already had experience with other reserves in Andalusia. The Sea of Gades report results were presented at the oceanographic research centre headquarters in Cadiz to keep the marine biology organization updated on the collected information, as, by law, it would have to endorse any declaration of a reserve.

The pandemic affected the tasks executed in 2020 and 2021, when the institutional conflict phase began. On the one hand, the promoter group relaunched the meetings with professional fishers, albeit virtually, to confirm the areas that could be defined as no-take zones and map them. In parallel, a formal request was made for a video conference with the Secretariat General of Maritime Fisheries to begin the legal process for the reserve to be declared. The outcome of this meeting was the publication in the State Gazette (February 16, 2021)[7] of the public consultation concerning the processing of the order by which a marine reserve of fisheries interest is established at Conil de la Frontera (Cadiz province), which opened up new prospects. A reserve effect was triggered, not in the usual biological sense, however, but in a socio-political sense: an opposition platform to the reserve was set up both on the social networks and through a variety of actions, including a demonstration in Cadiz to voice rejection to the project. This gave rise to a few days of unease among the promoters, who agreed to new virtual meetings with the recreational fisher representatives that were also attended by other actors (such as the Chiclana de la Frontera town council, which had spent some time expressing its opposition to the project). Attempts were made on social networks to respond to the stereotyped and incorrect information and opinions regarding some of the project's technical aspects and the promoters' intentions.

Arguments were heard opposing the Secretariat General of Maritime Fisheries' declaration of intentions to rubber-stamp the MRFI and again led to the people in charge of marine reserves and the promoter group agreeing to explain the technical details of the proposal to the various stakeholders involved in yet more virtual meetings: one meeting with representatives of the professional fishers and another with representatives of the recreational fishers, called and hosted by the administration between October and November 2021.

As might be deduced from the described process, the research activities have been recalibrated and redesigned as new circumstances have affected the process's socio-political conditions, which is plausible in the framework of qualitative methodologies. In addition, two consolidated theoretical principles in the political theory of marine reserves were once more fulfilled: (i) the proviso for the so-called step zero in the initial phase requiring that all the interested actors should be informed and involved in the institutional process – otherwise they will be angered, block the project and thus delegitimize it (Chuenpagdee et al., 2013); and (ii) the need to secure the various actors' images of the reserve – their knowledge, ideas, expectations, and associated values have to be determined, as they will be decisive in the inception process and, if approved, in the governance mode that is instituted (Jentoft et al., 2012).

Results: transferring knowledge, perspectives, and goals

Setting participatory processes in motion to put legal protections in place such as this MRFI requires a precise grasp of the modes of knowledge that the various collectives possess and deploy in their actions and discourse. Usually, consideration is almost exclusively given to the modes of knowledge of technical operations; that is, the lore and the know-how that, incorporated into and articulated with different devices as a kind of prosthesis (fishing gear, technological devices), enable fishers of whatever type to catch species in the intended way. However, it is usually forgotten that the practical know-how has other dimensions: on the one hand, related to epistemic modes (metaphors that include the reasoning that social actors use to imagine the relationships between living species and non-biotic elements – from diverse types of oceanographic factors to meteorological factors); and on the other hand, the practical knowledge that is built up over a fisher's life course, which, handed down from one generation to the next, can be translated into practical formulae to point the way for the regulatory measures of the socio-ecosystemic relationships or values that legitimize the position of people and collectives; that is, modes of governance. Techne, episteme, and phronesis (practice, knowledge, and wisdom, terms that refer to the Aristotelian notion of knowledge) are the modes of symbolic appropriation (nested in material and political appropriation) that experienced fishers use in their relationship within the socio-ecosystems that they themselves inform. Next, we shall address the interpenetration and conflict relationships that we have collated in the Cape Roche marine reserve institutionalization process and the roles of the different academic disciplines in securing collaboration mechanisms.

Diversity of goals and legitimization discourses

At the beginning of the research process, the main effort was directed at obtaining information that could be turned into maps in order to propose the boundaries of the reserve and the no-take zones. The habitual professional fishers' accurate knowledge served to plot the perimeter of the rocky sea beds supposedly inhabited by the greatest variety of biodiversity, the basis for the most complex food chains (Aswani and Lauer, 2006). These are the areas that play a major ecological and economic role as they are the fishing grounds where commercial species are found. The habitual professional fishers (from Conil and Sancti-Petri) felt that they were the right people for this operation by virtue of the intergenerational history of artisanal fishing activities in the area (Figure 12.5). Their TEK legacy was passed on to the technicians at the environmental NGO to be entered geographic information systems that could be submitted to the administration for the reserve proposal. In the focus groups, members of the environmental NGOs (trained in marine biology) and fishers from other areas with experience in marine reserves took charge of conveying or transferring the rationality of MRFI to fishers concerned by the perceived fall in their catches on their fishing trips. There was, however, another dimension to this exercise, and it was the express wish of the artisanal fleet to prevent the presence of the trawler fleet and the growing underwater fishing collective operating in some of the selected zones in such an environmentally-sensitive area (Gallego and Reyes, 2015). In other words, the arguments that are the object of marine biology and underpin the philosophy of the definition of MRFI – that the administration should act as a guarantor of the environmental values of some specific micro-ecosystems through the declaration of legally protected zones such as reserves, where certain human activities considered incompatible with these environmental values would be banned – were accepted by a local fisheries collective as the guarantee of social and economic sustainability (Figure 12.5).

Two types of actors were identified as the cause of the fall in catches: external and internal. The external actors were the trawler vessels, which were accused of unauthorized fishing with banned gear in waters and rocky sea beds with a high environmental value. Moreover, some of the underwater fishers, who were beginning to proliferate and not only become ecological but economic competitors, were selling their catches (despite it being illegal) (Maya et al., 2021). So the professional fishers' perspective combined objective ecological arguments taken from marine biology and applied to marine reserves by the administration with arguments to delegitimize other actors that they hoped to exclude (Figure 12.5).

However, there were also internal actors in the spotlight, such as professional fishers who use monofilament single trammel entangling nets (known locally as "tripe gear"). Since the 1990s, strong tensions have arisen between this collective and fishers who use longlines and other hook tackle and nylon gillnets deploying similar environmental arguments. In fact, since the 1980s, the fishers' guild had opted for regulation measures that reflected this practical wisdom in their grounds, which had sparked conflicts both internally and with the trawlermen: the positioning of artificial reefs to make it difficult to trawl; the setting of close seasons and minimum weights for some species, and the definition of minimum hook and mesh sizes that were even more restrictive than those in the state regulatory framework. Over time, this tradition of norm-setting created a feeling of a legitimate link with the territory and its ecosystem that was now translated into the expectation that a marine reserve (with its well-defined boundaries, no-take zones, regulatory measures, limitations on use, and surveillance system) would strengthen its conservation measures (Figure 12.5).

However, the image of what was for the local fishers, or most of them, a legitimation mechanism became a simple territorial exclusion strategy for other actors. During the critical and conflictive expansion phases, an idea was repeated in the various focus groups and workshops held with recreational fishers and also on the social networks and in the media, whose successful propagation enabled opposition to the proposed reserve: that the reserve complied with the fishers' objective of turning it "into a private fishing ground for their own interests" (Figure 12.5). This image was not only directed at the professional fishers with the intention of delegitimizing them but also directly at the state authorities, who have to watch over the public nature of the marine environment's ecosystem goods. As a result, the authorities established a clear differentiation between MRFI and the Marine Protected Areas (MPAs) whose intended purpose is environmental. Each has its legal instruments[8] and specific objectives,[9] and everyone was reminded that MRFI cannot be instruments to contend with Illegal, Unreported, and Unregulated Fishing (IUU).

As far as the promoters are concerned, during the expansive phase, they deemed it necessary to include recreational fishers through participatory means, especially the underwater fishers. So, in one of the workshops organized to convey information about the proposal for the reserve to the Recreational Fishers "Responsible Fishing" Association, they were invited to provide information about catch areas, species, and frequency of visits for the map.

This information was eventually added to the final Sea of Gades project document on recreational fishing (Figure 12.4), as had been done with the information provided by the professionals. Recreational fishers believe that they have better knowledge of sea beds than other collectives, as they actually see them, the densities of biomass, the locations of super-reproducers, and they state that there is some

FIGURE 12.4 Focus Group (Medina Sidonia, Cádiz). Recreational fishers contribute geo-located information on their fishing activities to the promoters.

Source: © David Florido

fishing gear on the sea beds that have the effect of ghost fishing. They do not only contend that their inclusion in the participatory processes and their contribution to improving the quality of the information on the ecosystems are needed through these images but they also delegitimize the professional fishers and the administration for not having duly taken them into consideration (Figure 12.5). This is an

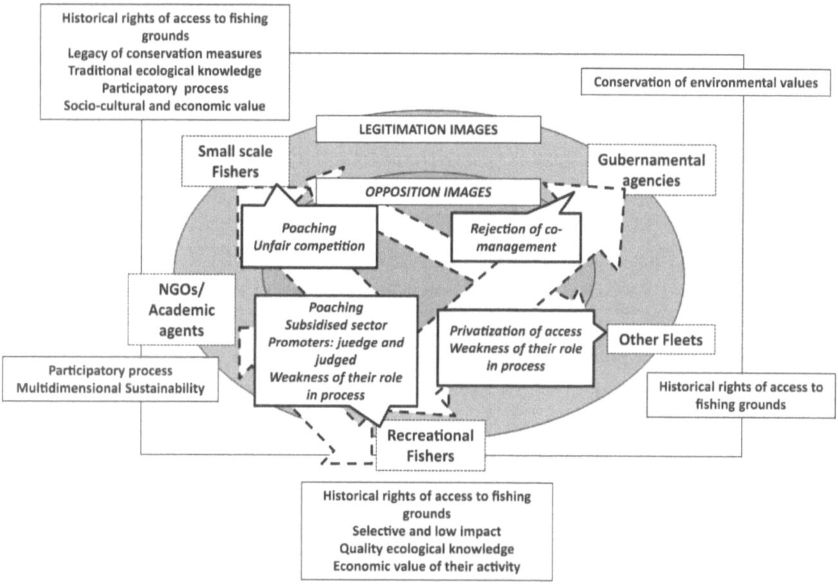

FIGURE 12.5 Stakeholders' legitimation and opposition images during an MRFI inception process at Cape Roche.

Source: Own image

especially tricky subject in MRFI declaration processes, where underwater fishers and the trawler fleet are systematically excluded for environmental reasons. While these fishers argue that their activity is very selective, some marine biologists believes that they have an environmental impact on reserves that restricts the anticipated reserve effect through their targeting of older and bigger super-reproductive specimens (Coll et al., 2004).

The recreational fishers also delegitimize the professional fishers who, they say, are responsible for the growing fishing effort through the application of sophisticated fishing equipment and gear that cannot be compared to theirs. Mention is also made of the decreasing performance of professional fishing, which contrasts with the high added value that all recreational activities (sailing, recreational fishing, recreational activities) generate for local economies. Furthermore, they claim that maritime territory and its ecosystems are public property, so any measure taken by the authorities that restricts their participation is contested and considered unjustified (Figure 12.5). They are convinced that their activity does not generate any appreciable impact on the ecosystem, with the argument that their fishing trips are sporadic, their catches are limited by law, and they are never intensive. The promoter team insisted in the workshops that the relationship that has existed between the state and recreational fishing is under review and that a new legal framework backed by Europe that imposes the principle of no data, no fish is on the horizon and that it will also be applied to recreational fishing.[10] In fact, since 2019, the oceanographic research centre has been running a project funded by the European Maritime Fisheries Fund for the study of recreational fishing in Andalusia,[11] in the framework of which there have been collaborations between the reserve's promoters and biologists from the oceanographic research centre.

The group of recreational fishers in the area is very heterogeneous and growing as part of the fishing sector crisis is being reabsorbed by retraining unemployed and retired sailors who go out on fishing boats, thanks to the accumulated knowledge on the activity and the handling of fishing gear, even though they are at a subsistence level (Martínez and Florido, 2018). Their activity fully falls into the IUU category and has a very sensitive social profile for political leaders. Locally they are known as retirees or recreational fishermen, although they are well-differentiated from other groups of recreational fishers and claim the right to fishing that they are afforded by their past career, their knowledge, and the fact that they have received the culture from their forebears.

Participation as a controversial discourse

The participation process, its planning, and progressive rollout have been another point of friction between collectives' images and representations. Whilst the fishers and the promoters perceived that they had followed a praxis that guaranteed access to the different stakeholders, especially as the vision of collectives such as the recreational fishers and the intensive fishing fleet has been very different since 2015 when the expansion phase began. The promoter group undertook a self-correcting effort after reflecting on the process. As a result, as advances were made in the institutional projection and conflict stages, efforts were made to communicate directly with the less-informed stakeholders, such as the leaders of the purse-seine fleet and, above all, boat recreational fishing clubs (Figure 12.5).

The participatory and dialog techniques are especially appropriate for identifying the social actors' images of an instituting process such as the Cape Roche MRFI. In-depth interviews given to different types of experienced and recognized fishers in the collective also generated knowledge of tides and currents, meteorological factors, knowledge of sea beds (types of gear, catches), fishing grounds, main species and their feeding patterns, associated avifauna, and location systems (traditional and with new technological equipment), among other things. Not only were qualitative data on biotic and abiotic aspects collected in these interviews but also on fishing strategies and the possible effects of the restrictions on the different sectors, with a nautical chart used to geolocate the data. In the framework of the Sea of Gades project, the role of the professional fishers' TEK was decisive in the survey of the bionomic map of sea beds in the reserve area, as the coordinates that they gave enabled to carry out both the sampling dives and the ROV dives to take photographs of the most environmentally-sensitive areas. Such were the available time and budget available for this underwater mapping that without this knowledge input from the fishers, the scientific work would have been seriously limited.[12] Later, in the last phase, another virtual meeting was held with the professional fishers to gain their commitment to an agreement on no-take zones that could be presented to the representatives of the administration. All this conferred legitimacy on the process for the artisanal fishers collective.

From the beginning, both the NGOs involved and the fishers spoke of a co-management model. They put their trust in effective participation in a monitoring committee to be able to discuss methods to regulate the activity, census regulation and accompanying the scientific monitoring policy, as is usual in the regulation of MRFI. However, the heads of the Marine Reserves of the General Secretariat of Maritime Fisheries made a clear statement: "there is no place for co-management, understood as co-responsibility in decision-making, in the Spanish constitutional framework. Any explicit reference to co-management will be deleted from any and every document".[13] In the same line, they argued that any decisions regarding the public domain are the responsibility of the central State administration, as is demanded by the state lawyers in the regulatory projects for all the approved reserves. Given that this is not possible, they argued, their presence at all the forums in which they are guests and their good relationship with the stakeholders throughout the entire process illustrate that they do believe in what they call participatory management (Figure 12.5), which in the theory of natural resource management could be identified as a collaborative, informed management model (Thompson et al., 2019).

The effective participation of the recreational fishers came later, from the expansive phase on (from 2017 on), once the promoters accepted that the selection of stakeholders demanded their inclusion, given the importance of their activity in this coastal zone. Their late inclusion affected the legitimacy of the process as the recreational fishers repeatedly claimed that the promoters had already made the decision on the reserve's boundaries and permitted uses, despite neither the professional fishers nor the NGOs involved having any political authority to define any aspect of regulations. Any management plan for MRFI is the responsibility of the General Secretariat of Maritime Fisheries, with the endorsement of marine biology provided by the oceanographic research centre. Despite this, at different meetings and workshops between 2017 and 2018, the recreational and underwater fishers' knowledge of frequent fishing areas, species, and mean sizes was included in an initial, but incomplete approach to estimate their importance for extraction in the area. The promoter group attempted to convey the objectives, goals, and philosophy of the proposal for a reserve. However, the recreational fishers believed that their role in the participatory process was always late in coming and not on an equal political footing: this image is lodged in their cognitive framework and has hardly changed (Figure 12.5).

Arlinghaus (2008) has drawn attention to the problem that ignoring the impact of recreational fishing on marine ecosystems entails while also outlining different factors of this problem: the lack of focus of the administration and, ultimately, of academic disciplines; the weakness of cooperative institutional links between the recreational fishers and the other stakeholders; the absence of a common framework of rules and values (the phonetic dimension to which we allude) (Linke and Jentoft,

2014) for the control of stereotyped images, and the lack of self-critical reflection by the recreational fishers themselves. As one of the oceanographic research centre biologists stated to us in an interview:

> *[Our collaboration with] recreational fishing is only just beginning, a very short time ago. But it has a potential for information with important knowledge, of course it does, because, i.e. the underwater lot knows just where some certain creatures are. So, you can identify the habitats of these species, the habitats where they feed, the habitats where they lay, where the recruitment is concentrated . . . Naturally, this offers huge potential, but the administration needs to make some big advances here.*

Concluding remarks: lessons learnt

The first consequence of comparing images that we documented is the emergence of a social space disputed between recreational fishers and small-scale fleets, a political arena in which very stereotyped discourses and imageries are brought into play and seriously compromise the good governance of the MPAs (Gómez et al., 2021). Their different perceptions of the rights of access to the territory and public goods, the sense of community, social norms, and the way of thinking and practising fishing have been brought to the surface in the participatory process laid out in this chapter. The professionals stand by their history and their cultural value, a dimension that retired fishers and those driven out by the crisis in the sector also assert. Meanwhile, the recreational fishers invoke legal criteria (right of access to the public domain) and their economic supremacy. Both one and the other blamed the crisis in catches on illegal sales practices and poaching, so once again we encounter the scenario described by Folchi (2001): conflicts that go beyond the environment but in which environmental arguments are used. Despite the efforts of the promoter group, the gap between recreational and professional fishers has not been closed. The agreements on governance must be based on a broad ethical framework in which the different actors can acknowledge each other.

We are convinced that anthropology is in the best place to conduct this transfer of ethical frameworks, images, and feelings that the various ontologies of the social groups organize in relation to their conflicts. The images are interspersed with goals and expectations, modes of legitimization, and reasoning. Identifying which of these images are incompatible with the institutionalization of an innovation concept such as a marine reserve is extremely important, and this is not usually a task commended to biology or marine ecology, which is more focused on defining the ecosystemic relationships between non-human elements and communities. Is not, after all, the specialty of anthropology to convey the meaning of social worlds in all their diversity, under the assumption that within these created worlds people and collectives live with practical formulae, aspirations, and different interests? Over all these years, we have been able to transfer the technical know-how of professional and recreational fishers to different documents to form a practical knowledge base that has enabled the seabed to be mapped and the no-take zones to be proposed, which might serve for making management decisions. And, in the same way, we have also been able to benefit collaboration between biological and sociological know-how.

Nevertheless, we have also encountered the counter positions of a scientific episteme represented in this case especially by biology and ecology, and another, vernacular or folk episteme. Whereas the state grants the former qualities such as objectivity, being based on protocols, and universal methodological requisites that validate management measures, the latter, incorporated, personal, strictly experimental in nature, and closely adapted to local conditions, is not recognized socially (beyond the communities that personify this knowledge) or politically. The epistemological suppositions of standard science (clear separation between social and natural processes, a subject/object dichotomy, a lack of trust in knowledge born out of a historical connection between human communities and their surroundings) make this cross-disciplinarity difficult (Huntington, 2000); as a power relationship always exists between the protocolized procedure of science and the experiential and personalized forms embodied in the feelings of fishers.

However, the role that TEK can contribute to conservation biology has already been affirmed (Zalles, 2017), as has been demonstrated: territorial distribution of species, feeding patterns, relations between species, associations between animals (ichthyofauna-avifauna), and other meteorological elements. Those who live in the ecosystem are able to perceive it as a community of multiple connections that include the agency of humans and their technological artefacts) (Florido del Corral, 2020). And this is a cognitive heritage that can enrich the heritage of the scientific community. Both epistemes, science's and the users', need to give each other feedback to contribute to mutual enlightenment. The epistemological breach persists that refers to a world of feeling (cognitive and practical), an ontology, different. A researcher from the oceanographic research centre with wide experience in the relationship with professional fishers for different fishing plans and an assiduous collaborator in some of the workshops as a tool of the social sciences states:

> *There really is a big difference between what they [the fishers] tell you and* ***what really is.*** *They say something and they concoct their idea . . . And what I do is listen, I always try to learn. . . You're always very humble. Being with them year after year and, little by little, you learn how to put it over and get them to take it in. . . There's more and more feedback. However, sometimes I think they go too far. Here's an example: they dare question the sampling methodology that's been scientifically validated. They questioned my sampling system for the baby clam. I try to explain to them: "I have to randomly sample all of them, the places where there are lots of baby clams and other places where there are none, but then there's extrapolation, and that's math, statistics".*
>
> *(emphasis added by authors)*

These technicians are taking an increasing part in meetings where research results are presented to fishers who collaborate in management plans, something that was quite unthinkable 20 years ago, when biologists and the agents of the administration were fishers' main enemies in the dominant imaginary (Herrera et al., 2015; Truchet et al., 2022).

Despite an effort being made in the workshops and focus groups to mediate and transfer the interpretative frameworks and interests of the various stakeholders, the promoters do not possess the effective power to transfer the collected proposals to the legal level. This leads to the stakeholders losing confidence in the promoters, who they see as judge and judged in the process. The objective is to persuade all the actors to consider themselves agents that inform the social ecosystem and who can enrich each other reciprocally. However, this desideratum can only be put into practice through a step decided by the administration to boost dialog processes in the indicated direction. Post-normal science has already been proposed as a paradigm for achieving transdisciplinarity (Zalles, 2017), which we can understand as a path for interculturality between the objective world recreated by standard science and which the majority of conservationists cling on to, and the subjectivized world that professional and recreational fishers carry within themselves. Socio-ecosystem relationships are not reduced to those of biotic and abiotic elements but must include the plural and interconnected modes of knowledge of humans who compile and contribute data and contexts on this socio-ecosystem, informing it from within. Transdisciplinarity and the translation of modes of knowledge are sympathetic to a new way of politically addressing socio-environmental challenges that are, at the same time, ontological.

Acknowledgements

This work would not have been possible without the active participation of scientists from the Spanish Institute of Oceanography, and without the collaboration of the Conil Fishermen's Guild, the Conil Producers' Organisation 72, The Recreational Fishers "Responsible Fishing" Association and the various organisations and of recreational fishing. We would also like to thank Xisco Sobrado and Helena Pérez-Gámuz for their collaboration in the elaboration of the maps.

Notes

1 Subdirectorate General for the Protection of Fisheries Resources of the General Secretariat of Maritime Fisheries.
2 The questionnaire was handed out to 75 fishers (34% sample of the universe); 48 had been skippers (64% of the sample) and 27 crew (36% of the sample).
3 Specifically, types of habitats: 1,170 (reefs) and 1,110 (sandbanks permanently covered by shallow seawater), and for the type of habitat 8,330 (submerged or semi-submerged sea caves), between the north-western point of the proposed reserve and its south-eastern vertex. Seventy species of coralline algae, anthozoans, bryozoans, calcareous red algae, echinoderms, and annelids were detected.
4 We refer to the Andalusian Fishing Vessel Locator and Monitoring System, SLSEPA.
5 SEA OF GADES Project: Protection and Governance of the Cape Roche Sea (Cape of Trafalgar-Sancti Petri Islet) during 2018–2019 funded by the Pleamar Program of the Biodiversity Foundation of the Ministry of Ecological Transition and the Demographic

Challenge <www.programapleamar.es/proyectos/gades-proteccion-y-gobernanza-del-mar-de-cabo-roche-cabo-de-trafalgar-islote-de-sancti>.

6 Thirty-five questionnaires were given out to an estimated universe of 1,670 sport fishers (not necessarily habitual). The reason for the sample's low level of representation is this segment's progressive resistance to collaboration, in keeping with the Recreational Fishers "Responsible Fishing" Association capitalizing the political representation of the collective.

7 www.mapa.gob.es/es/pesca/participacion-publica/pconsultapreviaconil2021.aspx

8 MRFI are regulated by Law 3/2001 concerning State Maritime Fishing and Law 33/2014, which modifies State Maritime Fishing Law in the European framework of the Common Fisheries Policy (CFP); whereas Marine Protected Areas (MPAs) are environmental and regulated by Law 42/2007 concerning Natural Heritage and Biodiversity, in the European framework of the Habitats Directive (92/43/EEC) and Directive 79/409/CEE on the conservation of wild birds.

9 MRFI are aimed at the conservation of the fisheries resource and, ultimately, of traditional activities, especially artisanal activities, as expressed in three-dimensional sustainability: environmental-social-economic, whereas the environmental MPAs are aimed at the conservation of species and habitats with special ecological value, which can result in general regulations and more restrictive use and management plans.

10 A public information process started in March 2021 regarding the Draft Royal Decree concerning Maritime Recreational Fishing in External Waters. This alarmed the recreational fishing sector as meant greater control over their activity. It was seen as an attempt to enforce the general conditions for undertaking their activity and for collecting data on the activity that ensure the sustainability of fisheries resources pursuant to Art. 25 of the Regulation (EU) no. 1380/2013 on the CFP. This new norm will override Royal Decree 347/2011, of 11th March, concerning the regulation of Maritime recreational fishing in external Waters.

11 Study of recreational fishing in Andalusia: FEMP_AND_Project 03/08. (Spanish Federation of Municipalities and Provinces Project).

12 The report, its methodology, and samples of the photographs taken can be examined via the following link: www.programapleamar.es/sites/default/files/informe_final_biono.pdf

13 This quote is taken from the April 2017 forum at Conil and corresponds to one of the technicians from the Marine Reserves of the General Secretariat of Maritime Fisheries. Similar statements have been made at subsequent meetings and workshops.

References

Arlinghaus, R. (2008). Social Barriers to Sustainable Recreational Fisheries Management Under Quasi Common Property Fishing Rights Regime. *American Fisheries Society Symposium*, 49, 105.

Aswani, S., & Lauer, M. (2006). Incorporating Fishermen's Local Knowledge and Behavior into Geographical Information Systems (GIS) for Designing Marine Protected Areas in Oceania. *Human Organization*, 65, 81–102.

Chuenpagdee, R., Pascual-Fernández, J.J., Szelianszky, E., Alegret, J.L., Fraga, J.M., & Jentoft, S. (2013). Marine Protected Areas: Re-thinking their Inception. *Marine Policy*, 39, 234–240. https://doi.org/10.1016/j.marpol.2012.10.016.

Coll, J., Linde, M., García-Rubies, A., Riera, F., & Grau, A.M. (2004). Spear Fishing in the Balearic Islands (West Central Mediterranean): Species Affected and Catch Evolution During the Period 1975–2001. *Fisheries Research*, 70, 97–111.

Florido del Corral, D.F. (2020). Hibridaciones de saberes y lógicas culturales en la pesca: Vivir de la mar y en la mar en Andalucía (España) y Chiloé (Chile) en el contexto contemporáneo. *Estudios Atacameños*, 65, 21–45. https://doi.org/10.22199/ISSN.0718-1043-2020-0019

Florido del Corral, D. (2021). Todas las voces. La elaboración de mapas de actores para la salvaguarda del patrimonio cultural inmaterial. En G. Carrera Díaz (Ed.), *La salvaguardia del patrimonio inmaterial como acuerdo social*. Sevilla: Instituto Andaluz del Patrimonio Histórico, Consejería de Cultura y Patrimonio Histórico. Junta de Andalucía, pp. 317–335. https://www.juntadeandalucia.es/servicios/publicaciones/detalle/80362.html

Folchi Donoso, M. (2001).Conflictos de contenido ambiental y ecologismo de los pobres: no siempre pobres, ni siempre ecologistas. *Ecología política*, 22, 79–100. https://repositorio.uchile.cl/handle/2250/122793

Gallego, M.A., & Reyes, O.M. (2015). La población de meros Epinephelus marginatus en el litoral andaluz (España). *Chronica Naturae*, 5, 68–80.

Gómez, S., Carreño, A., & Lloret, J. (2021). Cultural Heritage and Environmental Ethical Values in Governance Models: Conflicts Between Recreational Fisheries and Other Maritime Activities in Mediterranean Marine Protected Areas. *Marine Policy*, 129, 104529.

Herrera-Racionero, P., Lizcano-Fernández, E., & Miret-Pastor, L. (2015). Us and Them. Fishermen From Gandía and the Loss of Institutional Legitimacy. *Marine Policy*, 54, 130–136.

Huntington, H.P. (2000). Using Traditional Ecological Knowledge in Science: Methods and Applications. *Ecological Applications*, 10, 1270–1274.

Jentoft, S., Chuenpagdee, R., & Pascual-Fernández, J.J. (2011). What Are MPAs for: On Goal Formation and Displacement. *Ocean & Coastal Management*, 54, 75–83. https://doi.org/10.1016/J.OCECOAMAN.2010.10.024.

Jentoft, S., Pascual-Fernández, J.J., De la Cruz Modino, R., González-Ramallal, M., & Chuenpagdee, R. (2012). What Stakeholders Think About Marine Protected Areas: Case Studies From Spain. *Human Ecology*, 40, 185–197. https://doi.org/10.1007/s10745-012-9459-6.

Linke, S., & Jentoft, S. (2014). Exploring the Phronetic Dimension of Stakeholders' Knowledge in EU Fisheries Governance. *Marine Policy*, 47, 153–161.

Martínez Alba I., & Florido D. (2018). Las estrategias económicas de la pesca marítima de recreo en el entorno de Trafalgar (Cádiz). *I Simposio Internacional sobre Pesca Marítima Recreativa*, Vigo, España. Not published.

Maya-Jariego, I., Florido, D., Holgado, D., & Hernández, J. (2016). Network Analysis and Stakeholder Analysis in Mixed-Methods Research. In Jason, L., & Glenwick, D. (Eds.) *Handbook of Methodological Approaches to Community-Based Research: Qualitative, Quantitative, and Mixed Methods*. New York: Oxford University Press, p. 325–334.

Maya-Jariego, I., Martínez-Alba, I., & Alieva, D. (2021). "Plenty of black money": Netnography of Illegal Recreational Underwater Fishing in Southern Spain. *Marine Policy*, 126, 104411. https://doi.org/10.1016/j.marpol.2021.104411

Pascual-Fernandez, J.J., & De la Cruz Modino, R. (2011). Conflicting Gears, Contested Territories: MPAs as a Solution? In Chuenpagdee, R. (Ed.), *World Small-Scale Fisheries Contemporary Visions*. Delft: Eburon, p. 205–220

Quesada-Silva, M., Iglesias-Campos, A., Turra, A., & Suárez-de Vivero, J.L. (2019). Stakeholder Participation Assessment Framework (SPAF): A Theory-based Strategy to Plan and Evaluate Marine Spatial Planning Participatory Processes. *Marine Policy*, 108. https://doi.org/10.1016/j.marpol.2019.103619

Thompson, S., Stephenson, R.L., Rose, G.A., & Paul, S.D. (2019). Collaborative Fisheries Research: The Canadian Fisheries Research Network Experience. *Canadian Journal of Fisheries and Aquatic Sciences*, 76(5), 671–681. https://doi.org/10.1139/cjfas-2018-0450

Truchet, D.M., Noceti, B.M., Villagran, D.M., & Truchet, R.M. (2022). Alternative Conservation Paradigms and Ecological Knowledge of Small-Scale Artisanal Fishers in a Changing Marine Scenario in Argentina. *Human Ecology*, 50, 209–225. https://doi.org/10.1007/s10745-022-00309-5

Zalles, J. (2017). Conocimiento ecológico local y conservación biológica: la ciencia post-normal como campo de interculturalidad. *Íconos, Revista de Ciencias Sociales*, 59, 205–224.

PART IV

Ways forward for transdisciplinary ocean science and management

Against the background of conceptual considerations, methodological approaches and empirical case studies in the past 12 chapters, what can we learn for the future of marine transdisciplinarity? In Chapter 13, Steins and colleagues reflect on the status quo of transdisciplinary marine research and, based on their own longstanding experiences of collaborating with the fishing sector, debate ways forward for this newly emerging discipline. To reach the next level of transdisciplinary marine research, they argue, what we need is not simply more science but a different way of generating scientific knowledge through participatory approaches and improving the science-policy interface. The authors outline several important steps in this direction:

1 an integration of the notion of marine *socio*-ecological systems as a key thought into marine research;
2 a focus on enabling transdisciplinary approaches to science and management;
3 explicit reflection on the ontological and epistemological perspectives represented in participatory processes;
4 an expansion of statutory requirements for data collection to include socio-economic information; and
5 a regime-shift of existing governance systems towards integrating stakeholders' local, traditional, observational and practical knowledge.

Formats that enable such steps are the setting-up of multi-actor partnerships, the appointment of boundary spanners or knowledge brokers for bridging different ontologies and epistemologies, and a rethinking and re-evaluation of different ways of knowing and living the marine.

DOI: 10.4324/9781003311171-16

13

AIMING FOR THE NEXT LEVEL OF TRANSDISCIPLINARY MARINE RESEARCH

Nathalie A. Steins, Susan de Koning, and Marloes Kraan

Introduction

The future of our oceans is challenged by an increasing number of direct and indirect anthropogenic factors. Traditional and new users, such as fisheries, mariculture, oil and gas production, mining, recreation and renewable energy production, as well as human pressure through land-sea interactions, such as land reclamation, human migration, effluent discharges, agricultural runoff and climate change, impact marine ecosystems around the globe. Addressing the associated complex governance questions requires dealing with different and high stakes with a multitude of underlying values. This implies that there are no objective or definite solutions or answers and that trade-offs must be made (Funtowicz and Ravetz, 1993; Hessels and van Lente, 2008). In fact, sustainable ocean governance is a so-called wicked problem and therefore requires active involvement of those affected by it (Jentoft and Chuenpagdee, 2009). For a long time, such participation was limited to various forms of participation in the management process (Reed, 2008), but there is increasing recognition that stakeholders must also be involved in the science process in support of management (Dickey-Collas and Ballesteros, 2021; Röckmann et al., 2015). This requires transdisciplinary science approaches (Hessels and van Lente, 2008), which are characterised by dynamic interactions *between* one or multiple scientific discipline(s) *and* practitioners or stakeholders in co-producing knowledge.

Co-production of knowledge for sustainable management of oceans is necessary for two main reasons. Firstly, ecological and social systems are interdependent; they form a social-ecological system (Berkes and Folke, 1998). Whereas scientists have a suite of methods for examining the state of marine social-ecological systems to provide advice to managers on potential management actions, user groups possess a wealth of contextual knowledge and sensitivity about these systems. This so-called

DOI: 10.4324/9781003311171-17

experiential knowledge (Stephenson et al., 2016) is a result of their experiences from working in that system, often over many generations, and its associated socio-economic, cultural, technological, physical or other changes (Hind, 2015; Johannes and Neis, 2007; Said and Trouillet, 2020; Steins et al., n.d.; Stephenson et al., 2016). Secondly, the experiential knowledge of users often is the 'best available information' that scientists could make use of, for example, in exploratory research or when changes in the social-ecological system occur. Thirdly, the behaviour of fishers and other users of the system is what in fact is being managed. Having a thorough understanding of their activities, their perceptions and the drivers for the choices they make in exploiting the resource is thus essential (Steins et al., 2020).

Fisheries have a relatively long history of knowledge research and its integration in management (Hind, 2015; Stephenson et al., 2016) compared to stakeholder participation in other marine sciences. In the past decade, particularly in governance systems with well-developed scientific advisory frameworks, research collaboration with fishers and the fishing industry has gained increasing attention. This was driven by growing willingness and clear interest from both parties in working together to provide the best available information to assessment and management problems (Steins et al., n.d.). These well-developed scientific advisory systems witnessed a growing number of research collaboration initiatives, such as the European GAP2 project (Holm et al., 2020), the Dutch OSW programme (De Boois et al., 2021; Röckmann et al., 2015; Steins et al., 2020), the Norwegian reference fleet (Bjørkan, 2011), the Canadian Fisheries Research Network (Thompson et al., 2019) and Trident Systems in New Zealand (Mackinson and Middleton, 2018) to name just a few. If done well, such co-production brings many advantages, including cost-effective, improved data collection, increasing transparency and dialogue, capacity building of fishers, improved societal relevance of research and a collective rationale for durable solutions (Holm et al., 2020; Johnson and Van Densen, 2007; Kraan et al., 2013; Mangi et al., 2018; Steins et al., 2020; Stephenson et al., 2016; Thompson et al., 2019; Verweij et al., 2010). However, there is still significant room for improvement.

First, applied fisheries and marine research is still dominated by natural scientists and economists, resulting in a disciplinary understanding of subsets of the marine social-ecological system, rather than the required integrated social, economic and ecological knowledge base for addressing ocean governance problems. Second, using experiential knowledge from stakeholders in science in an appropriate way requires specific expertise; many scientists have not been trained in collecting such information (interviewing, mapping, participant observation) and analysing it (coding, interpretation, linking to theory) (Macher et al., 2021). Third, most receiving scientific and management systems are not (yet) equipped for integrating experiential information from fishers and other users and qualitative information from social scientists (Holm et al., 2020; Steins et al., n.d.). A fourth, related issue is that natural scientists, social scientists and users have different ways of understanding and studying or observing the world (Moon et al., 2021). This is a barrier in both interdisciplinary science and transdisciplinary knowledge co-production. Related to this is a fifth aspect that needs improvement: (natural) scientists sometimes find it difficult to view knowledge of fishers as best available information and rather call it 'anecdotal' or want to disregard it as they consider it as an attempt to influence science through vested interests (Steins et al., n.d.). Sixth, participation of fishers in governance systems is often still hardly developed, many governance systems have a top-down mode of governance with at best pockets of participation (collaborative arrangements) within the broader structure of 'multi-level governance' (Piattoni, 2009). But what often is lacking is a shared explicit consideration of the underlying goals and values of participation (Reed, 2008). Yet in knowledge co-production for management, there is a strong relation between the level of legitimacy in science and in management that needs to be understood.

In this chapter, we address these critical issues, using examples of knowledge co-production in fisheries. We narrow our focus on well-developed scientific advisory systems, as this is where issues about the transition in participatory approaches in marine science and governance are matters of debate rather than necessity (Macher et al., 2021; Owen et al., 2012; Steins et al., n.d.). While our focus is on fisheries, our discussion is also relevant for knowledge co-production in other ocean science fields. We will first present a framework for assessing the integration of fishers' knowledge in science and management and then discuss different types of knowledge and approaches for integration of these different types of knowledge. This is followed by a discussion of the links between governance and science and how this is relevant for the co-production of knowledge, we will end with the barriers to integration and a discussion on what is needed to reach the next level of transdisciplinary marine science.

Integrating fishers' knowledge in science and management: a framework

In the co-production of knowledge in fisheries science, the traditional focus has been on collecting measurable observational data, where fishing vessels are used as platforms to collect biological data, with or without active assistance of fishers. From the mid-2000s onwards, guidelines for organising this type of science-industry research collaboration were developed (Johnson and Van Densen, 2007; Mangi et al., 2018; Steins et al., 2020). However, 'fishers' knowledge includes, but is much greater than, basic biological fishery information. It includes ecological, economic, social, and institutional knowledge, as well as experience and critical analysis of experiential knowledge' (Stephenson et al., 2016). This observational and experiential knowledge can be integrated in different degrees in both fisheries assessment and management: from fishers just providing biological data to a fully participatory governance regime that demands use of fishers' information and knowledge in shaping management (Stephenson et al., 2016). The Fishers' Knowledge Research (FKR) spectrum (Figure 13.1) provides a useful framework for our discussion of the critical issues around the co-production of knowledge; it provides a typology of fishers' knowledge: (1) (measurable) observations and (2) experiential knowledge,

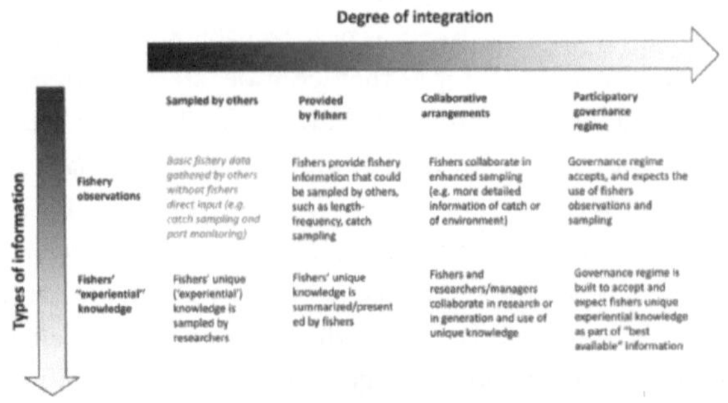

From: Integrating fishers' knowledge research in science and management
ICES J Mar Sci. 2016;73(6):1459-1465. doi:10.1093/icesjms/fsw025

FIGURE 13.1 FKR spectrum.

Source: Stephenson et al. (2016). All but the italicized cell are considered examples of Fishers' Knowledge Research.

in the context of four levels to which this knowledge is integrated in the governance system (Stephenson et al., 2016). We emphasize that in the co-production of knowledge, the type of fishers' knowledge and the degree to which it is integrated is not a static, but a dynamic process (Steins et al., 2020).

Through the course of knowledge co-production, degrees of integration can become stronger, for instance, when mutual trust and trust in the quality of data provided by fishers increases. One example is the discards self-sampling programme in Dutch demersal fisheries, which started out as a voluntary initiative by the fishing industry to improve data limitations in stock assessments, and evolved to become part of the European statutory scientific data collection framework (Kraan et al., 2013; Steins et al., 2020).

When research collaboration with fishers becomes part of the established modus operandi, we also see that scientists become more open to using fishers' experiential knowledge, for example, in seeking their knowledge about historical developments in stock abundance and landings, spawning grounds, fish migration patterns, dietary preferences of fish, fishing strategies, documentation of new and invasive species, mapping of benthic habitats, and performance assessment, rigging and maintenance of fish stock survey gear (Azzurro et al., 2019; Bentley et al., 2019; Bjørkan, 2011; De Boois et al., 2021; Holm et al., 2020; Johannes and Neis, 2007; Murray et al., 2008). The dynamics in the level of integration and use of different types of knowledge in the FKR spectrum are, however, not always progressive. There are documented cases where co-production eroded, often related to trust issues and perceptions about the integrity of data provided by parties with a vested interest, such as fishers. One example is the demise of a successful collaboration between scientists, fishers and gear manufacturers in the US Trawl Survey Panel. Much improved survey nets and gear had been jointly developed, but when the Science Centre unilateraly decided which trawl doors to order because in 'something as important as a resource survey [they could not] allow themselves to be seen as fully cooperative with the industry', the fishers and manufacturers left the panel (Johnson and McCay, 2012). The FKR spectrum is a useful framework to assess how and which types of fishers' knowledge, or indeed that of other users, is embedded in the marine governance system, and how this may change over time.

Taking a different perspective on knowledge for marine management

Established science systems in support of marine management have a strong preference for quantitative information. This preference makes no sense when recognising that true transdisciplinary science is needed to address ocean governance problems. Observational knowledge and experiential knowledge all have value. Experiential knowledge indeed has demonstrably contributed towards an improved knowledge base for management (Azzurro et al., 2019; Bentley et al., 2019; Johannes and Neis, 2007; Murray et al., 2008; Palsson, 2000; Silvano and Valbo-Jørgensen, 2008; Steins et al., n.d.). Yet, this type of knowledge is often dismissed as being anecdotal, unusable or interest-driven. This critical issue needs to be addressed and is linked to ontological and epistemological questions about what truth or knowledge is and how we study and understand the world (as social or natural scientists, as a fisher or a scientist etc.).

The strong preference for quantitative information is rooted in the paradigm of positivism. Positivism assumes that knowledge can only be obtained through empirical observations following the standards of 'the scientific method'. This directly excludes non-scientists from being able to contribute to knowledge production, and disregards processes that are not readily observable, such as values and attitudes. Science that follows a positivist line of thinking assumes there is one reality out there. Many social scientists, however, assume that there are multiple, socially constructed realities (Moon and Blackman, 2014). Studying and including these multiple realities can aid in improving marine resource management practices to become more adapted to different perspectives and contextual factors of the specific social-ecological system, which may lead to more effective management (Moon et al., 2021).

Moreover, viewing the world as consisting of multiple realities also translates into different perspectives on knowledge and how knowledge can be obtained. Different epistemic communities (such as social scientists, fisheries biologists, fishers and conservationists) value and evaluate knowledge differently (Moon et al., 2021). However, academic knowledge is often regarded as the only valuable type of knowledge, thereby marginalizing non-scientific actors (Johannes and Neis, 2007; Moon et al., 2021; Pascual et al., 2021). Fishers' knowledge, for example, is often much more detailed and context-specific than scientifically derived data but is rarely used as an input for marine resource management. If used, the intricate (spatial, seasonal) details of fishers' knowledge are often lost, as their knowledge needs to be integrated into databases based on aggregated biological and economic metrics (Said and Trouillet, 2020). Also, using fishers' knowledge is often also seen as problematic as it is considered to be anecdotal and not objective (Johannes and Neis, 2007; Silvano and Valbo-Jørgensen, 2008). For instance, the Fishers' North Sea Stock Survey collected fisher's perceptions on annual developments in important North Sea fish stocks; this knowledge was, however, barely used as it was regarded as subjective, and neither particularly valid nor reliable from a methodological perspective (Stange, 2017). Issues around validity and reliability can, however, be overcome. Qualitative social scientists who work from constructionist and subjectivist epistemologies have developed a suite of methods and techniques to systematically collect, study and contextualize this experiential information (Guba, 1981; Johannes and Neis, 2007; Moon et al., 2021; Murray et al., 2008; Neis et al., 1999; Palsson, 2000). The idea that fishers' knowledge (or that of other users') is not objective comes from confusing knowledge with vested interests. All actors with a stake in the resource (including scientists) have developed their theoretical or practical understanding through facts, skills or acquaintance, justified by objective or subjective evidence; this knowledge is contextually relevant and both useful and practical in resolving issues or unknown elements of the system (A. Jenkins, 2004). It is when resource use negotiation starts, that interests become prominent. It is therefore important in both science and management, to differentiate between different types of knowledge, which might reveal contrasting perspectives, and differences in values and stakes, which can lead to different interpretations and preferred courses of actions.

Integrating different types of knowledge

From experiences with transdisciplinary knowledge production in the domain of marine science and management, two main lessons emerge. First, it is important to work with partnership approaches to be able to create boundary spaces. A boundary here 'refers to an abstract, shared space between collaborating actors with different knowledge. This "boundary space" is where tacit knowledge becomes explicit and where actors are confronted with, and learn about, each other's interests and perspective' (Stange, 2017). Boundary spaces consist of objects (e.g. fisheries management plans), activities (e.g. workshops) and actors (e.g. fisheries representatives, policy-makers). A study on stakeholder-led transdisciplinary knowledge production in a fisheries management context (Stange, 2017) found that if contested issues arise, it can help to develop new boundary spaces, including new actors and objects. The formation of partnerships can be one way of creating boundary spaces. These partnerships can be limited to scientist and fishers or other users, but can also embrace other stakeholders such as representatives from societal organisations or policy institutions. Involving managers is important as this ensures that the knowledge produced in a transdisciplinary manner is salient (does the information come in time; is it useable?) and considered to be credible by the managing institutions. An experiential review of science-stakeholder-manager partnerships for decision-support in fisheries management (Macher et al., 2021) showed that such partnerships can be formal and institutionalized, but also be very informal. The review indicated that to improve the contribution of transdisciplinary science in decision support for fisheries management, there needs to be: (i) better engagement of decision-makers and managers and more diverse interactions between scientists and managers; (ii) a complete picture of how produced knowledge is taken up in management; (iii) expectations management, including awareness of the (slow) pace of science; and (iv) a clarification of the different roles involved in transdisciplinary approached; and (v) efforts to develop transdisciplinary capacities, places and practices. This study also emphasized that for fostering true transdisciplinary approaches, adaptation of current funding models is needed. Funding models should encourage that non-scientific stakeholders articulate their questions before proposals are submitted rather than becoming involved post-granting, as is the dominant practice (Macher et al., 2021).

The second lesson learnt from experiences with integrating transdisciplinary approaches in marine research is to acknowledge both contributory and interactional expertise (Collins and Evans, 2002). Contributory expertise refers to being an expert who can genuinely contribute to (developments in) this field in a specific field. For example, in fishing gear technology research, contributory expertise can be provided by both fishing gear technologists (scientists) or practitioners (e.g. fishers and net manufacturers). Interactional expertise refers to being able to engage in discussions in a certain field, whilst not being part of this field or being able to put the expertise of this field into practice. It can be acquired through immersion in a specific field of expertise (scientific or practice-oriented, such as marine biology or small-scale fisheries) (Collins and Evans, 2002). Although involving interactional expertise alone is not sufficient for successful transdisciplinary collaborations (see first lesson), it can lead to improved collaboration by improving communication and connecting actors who work from different paradigms (L. D. Jenkins, 2015). In this context, so-called boundary spanners (Johnson, 2011) who recognize the value of experiential knowledge and are able to communicate on both sides of the boundary between scientists' and stakeholders' knowledge can make a positive contribution to transdisciplinary approaches (Johnson, 2011; Steins et al., n.d.).

Studies have shown that where using interactional expertise has been successful, there usually was some form of a (legal) obligation to collaborate (L. D. Jenkins, 2015; Johnson, 2011; Johnson and McCay, 2012). Purely voluntary collaboration disintegrates more easily; actors working in mandatory collaborations need to persevere and may therefore overcome barriers (L. D. Jenkins, 2015). Another factor impacting transdisciplinary collaboration is the role of sentiments in the discernment of expertise. If actors involved do not accept others as contributory or interactional experts, this severely hinders collaboration (L. D. Jenkins, 2015). For example, a fisher who cannot take a scientist seriously because s/he does not know what it is like at sea, or a scientist who does not take fisher serious because s/he does not have an academic education. This links back to our section on different knowledge types: if actors draw on different types of knowledge, this may affect the extent to which they are perceived as experts by other actors who use other types of knowledge (L. D. Jenkins, 2015).

Linking transdisciplinary science and decision-making

Management decisions emerge from interactions between different actors who each play a different role in the decision-making process. In transdisciplinary science, fishers, for example, may have had an active role in generating the knowledge in support of the development of fisheries management plans. However, once the science moves to the decision-making domain, they or their representatives are likely to bring other considerations (interests) forward that may influence the final management decision. The same goes for scientists, who may wear a different hat in a research collaboration setting compared to when advising their government (Dankel et al., 2016). The interactions between scientists, decision-makers and other actors can be assessed in the so-called interaction triangle (Röckmann et al., 2015) (Figure 13.2). The triangle highlights three dimensions of interaction:

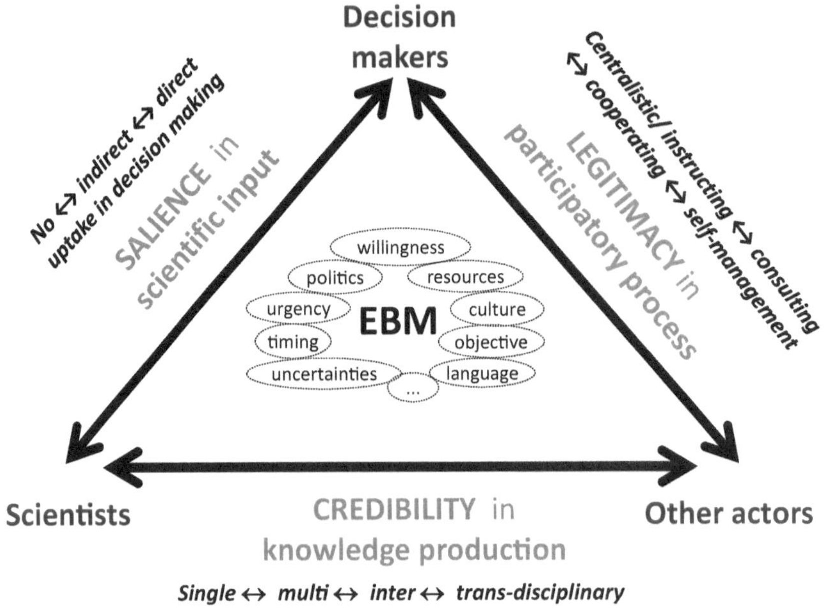

FIGURE 13.2 The EBM triangle of interaction, specifying an interaction spectrum (outside, black) for each of the three dimensions (grey). Encircled inside the triangle, examples of context specific factors.

Source: Röckmann et al. (2015)

(1) credibility in knowledge production; (2) salience in scientific input; and (3) legitimacy in participatory process (Mitchell et al., 2006).

The triangle shows that the three interaction dimensions are related and influence each other. Whilst the credibility of science amongst fishers ('other actors' in the triangle) can be increased through research collaboration between scientists and fishers, the success of that collaboration is also influenced by the uptake of that knowledge by decision-makers. The latter is linked to the saliency of science, but also depends on the political context. Transdisciplinary research projects in fisheries have, for instance, been known to 'hit the management wall' (Holm et al., 2020), that is, the results of collaborate projects was either not in demand by managers at the time it was produced or the receiving science-policy system was simply not equipped to incorporate this knowledge, as became clear in the EU GAP2-project (Holm et al., 2020). The political context can also interfere with the saliency and uptake of science, and the legitimacy in participatory process. One example is the pulse trawl, developed as a more sustainable alternative for the North Sea beam-trawl fishery for Dover sole (Delaney et al., 2022). Its development has been marked by controversies, including the development of a strong anti-pulse lobby by a variety of stakeholders. Even while European fisheries policy demands management decisions being made on the basis of best available knowledge, and a multi-annual ecological impact study (involving fishers) was nearing completion, the science was overtaken by political developments. A decision was made to completely ban this gear innovation in the European Union, even with near-complete scientific results pointing to ecological and environmental benefits of the gear (Delaney et al., 2022). This decision, in turn, affected the pulse fishers' perceptions of the legitimacy of the participatory process and resulted them becoming less inclined to participate in fisheries innovation research. They questioned the usefulness of science when a gear proven to be more sustainable, was no longer allowed (Steins et al., 2022).

The link between the uptake of (transdisciplinary) science and to the legitimacy of the participatory process is also evident in relation to discontent with the lack of adaptive management in response to new scientific evidence. An example is the Plaice Box, an important nursery ground which, in 1989, was put under a spatial protection regime with support from Dutch fishers. It aimed at increasing the North Sea plaice spawning stocks and, hence fishing opportunities. Despite four scientific evaluations showing that the Plaice Box had not fulfilled this aim (Beare et al., 2013), the management regime stayed in place. This has caused much frustration amongst fishers, who in discussions about Marine Protected Areas (MPAs) to protect vulnerable habitats and species keep using the Plaice Box as an example to oppose MPAs ('once we agree to a closure, this will remain so forever even if it does not work'). Similarly, years later, when Dutch policy officers asked fishers to collaborate in research as part of the implementation of the European landing obligation, fishers recalled 'what happened with the Plaice Box' and questioned the usefulness of cooperation (Kraan and Verweij, 2020). This illustrates that, for fishers, collaborating in science is strongly linked to their perception of management:

do management objectives and regulations make sense to them; can they trust that the information they provide is not collected to be used 'against them'? Our experience is that fishers are generally motivated to collaborate in research because they hope this will influence decision-making, making it more closely linked to their experience, the fishing practice. If the co-produced knowledge is then not taken into account, this can lead to disappointment and hence diminishing support for collaborative research (Steins et al., 2020). Therefore, expectation management is crucial: scientists must make sure they have explained clearly that they can neither guarantee the uptake of science (or positive results) nor decide on how scientific conclusions translate into policy decisions.

One example in transdisciplinary knowledge creation where interactions between credibility, saliency and legitimacy come together is a study into the effect of the aforementioned pulse trawl on direct mortality of marine organisms in the wake of the trawl (Schram et al., 2022). One of the controversies surrounding the pulse trawl were allegations from small-scale fishers about the occurrence of high levels of dead organisms following pulse trawl activity. This so-called graveyard hypothesis was not included in an ongoing multiannual research programme. The Dutch government therefore decided to commission additional research, and asked that small-scale fishers should be involved in the design, execution, and interpretation of the research (Schram et al., 2022). The subsequent interactions between fishers and scientists provided insights into small-scale fishers' perceptions of pulse fishing. Their engagement led to an adaptation in the fieldwork location, with the location corresponding with the areas that, according to these fishers, were most vulnerable to pulse fishing. Through addressing specific concerns of the small-scale fishers, their reasoning of why they opposed pulse fishing changed. While initially they focused on biological impacts, they started to discuss political and management-related concerns, showing the importance of addressing gear transitions as a multi-dimensional problem. This can be explained in two (complementary) ways. On the one hand, the research addressed biological knowledge gaps, thereby shifting the focus to other types of concerns. On the other hand, fishers' concerns turned out to be mainly political and management related from the start, yet biological arguments dominated the pulse trawling debate. This study shows that fishers' participation in policy support research, can increase the credibility and saliency of research while at the same time the nature of the problem can be assessed (Schram et al., 2022). Nevertheless, in an ideal situation, the nature of a problem should be assessed before initiating research.

Whereas science is important to understand the impact of fisheries, conflicts often revolve around other issues, such as conflicting values (de Koning et al., 2020) or a perceived lack of government control (Saavedra-Díaz et al., 2015; Schram et al., 2022). In such cases, a transdisciplinary approach involving policy, science and other stakeholder is required in developing a joint problem definition at the policy-science-society interface, as a foundation for the development of the (transdisciplinary) knowledge base for addressing the issue(s). However, this also requires that the governance system is receptive to integrating knowledge from stakeholders in science and policy and able to address associated barriers.

Barriers to integration

An international review into the integration of fishers' knowledge contributions into well-developed scientific advisory systems found a number of barriers for full integration related to (i) data quality considerations, (ii) dealing with the 'uniqueness' of experiential knowledge, and (iii) perceptions about the integrity and credibility of co-produced science (Steins et al., n.d.). All three are related to the receiving governance system.

Only few fishers' knowledge contributions seem to make it to the highest level in the FKR spectrum (Figure 13.1), where the governance system is expecting this integration as part of best available knowledge practices. In these cases, the knowledge contributions that are integrated into the science pillar of the governance system tend to be dominated by (measurable) observational knowledge, which is more easy to implement in existing data collection, storage and analytical tools. Many fishers' knowledge contributions do not make it beyond the collaborative arrangement mode. Even if such contributions are being used, it makes them vulnerable in terms of continuity, which relates to both funding for and support to maintain engaged in collaborative research. Where the receiving governance system is limited to managing fisheries under national jurisdiction, integrating fishers' knowledge contributions is more likely to take place. One example is the management of the mussel fishery in the Dutch Wadden Sea, an important nature conservation area (de Koning et al., 2020). In allocating areas for the mussel seed fisheries, an 'experiential knowledge map' (Ervaringskaart) is used (van Stralen and Troost, 2021). This map is based on the collective experiential knowledge of retired and active mussel farmers and the field staff of the inspection authority. It assesses for the subtidal parts of the Wadden Sea the chance that mussel seed survives the next winter (predation, storms) and hence has potential to develop into a 'stable' subtidal mussel bed with associated nature conservation values. The mussel seed fishery is based on the premise that fishing during the spring and autumn season first takes place on those seed beds that are highly likely not to survive (van Stralen and Troost, 2021). While aspects of the mussel fishery are contested by environmental NGOs (de Koning et al., 2020), the co-produced knowledge in this map and its use in management is considered credible and legitimate by all involved stakeholders. Where there is a direct, short link between science for advice and implementation, transdisciplinary knowledge seems to be easier to integrate (see also (Holm et al., 2020)). However, especially in multi-level governance systems (Piattoni, 2009), such as the European Union or Regional Fisheries Management Organisations, the integration of fishers' knowledge contributions becomes more difficult. In these cases, unique national co-production of knowledge efforts must be integrated in the umbrella system, passing different levels of management, and this is more easily said than done.

Multi-level governance systems are characterised by the involvement of a large and changing number of actors, policy process and interactions between them (Piattoni, 2009). They are generally characterised by a strong top-down, prescriptive, input-based management approach. In such systems, identifying all relevant actors (e.g. managers, user groups and NGOs), setting up dialogue and aligning the different stakes with the implementation of co-produced knowledge is not an easy task; particularly In situations with unequal power relations, where some stakeholders are more organised, resourceful and articulate than others (Turnhout et al., 2020). In EU research programmes, a focus on Responsible Research and Innovation has become increasingly prominent (Owen et al., 2012). This demands a stakeholder-inclusive approach and use of the best available knowledge. However, the EU has so far done little to facilitate its existing scientific advisory system to incorporate fishers' knowledge contributions in its science; the data collection framework regulation, which outlines the data that need to be supplied by Member States for fisheries and ecosystem assessment, and its associated funding continue to be directed at collecting basic biological information from fisheries-independent surveys or observational fisheries-dependent data. Changing this situation to open up for transdisciplinary information may even require difficult policy reforms (Holm et al., 2020). But if the system fails to be adaptive in including transdisciplinary knowledge and collaborative research projects keeps hitting the management wall (Holm et al., 2020), this may eventually lead to diminishing support for collaborative research, eroding trust and reduced legitimacy of the fisheries governance system (Steins et al., 2020). The EU's main provider of fisheries advice, the International Council for the Exploration of the Sea (ICES) is currently undertaking efforts to open up to fishers' knowledge contribution and stakeholder engagement in the advisory process (Dickey-Collas and Ballesteros, 2021), but the burden of transitioning to a participatory science system should not be placed solely on the science system.

To transform the scientific advisory system, policy needs to create the frameworks to really make transdisciplinary research work, including funding structures. However, enabling knowledge co-production and its use in science and management, goes hand in hand with willingness of all actors involved to accept, and their ability to understand, new modes of knowledge (co)production. Here, dealing with different ontologies and epistemologies about 'what is true knowledge and how do we understand reality' and perceptions about the integrity of scientific advice that includes stakeholder knowledge (i.e. trust issues) play an important role. Particularly the impact of public perception about using knowledge co-produced with those having a vested interest in marine resource management should not be underestimated. The predominant view amongst scientists in favour of working in transdisciplinary approaches seems that, as long there are good quality assurance systems in place, it should not matter who provides the data and information (Steins et al., n.d.). But, even when good quality assurance systems are in place, it can still matter who provides the scientific information. One case that made this clear is the multi-annual impact assessment programme for the flatfish pulse fishery, where the integrity of the collaborative science and nationality of the scientists involved was continuously questioned by those opposing the innovation (Delaney et al., 2022). Another illustrative case is that of Trident Systems in New Zealand.

New Zealand's science support system for fisheries management is based on a partnering approach and operates on a basis of a Standard for Research and Innovation (2011, re-issued 2021) (Mackinson and Middleton, 2018). Its underlying principles are to ensure quality and integrity of the data 'irrespective of the source of that information', to require research providers to meet sufficient data quality standards, and to ensure effective and efficient peer-review processes (MinPI, 2011). In 2012, in a collaboration between the government, government research providers and the industry, 'Trident' was set up as a not-for-profit, industry-led research service provider. It 'focussed on areas where direct involvement of the seafood industry in research had clear benefits, undertaking Research and Development to develop innovative systems and processes and using these to address fisheries management information needs' (Middleton and Guard, 2021). It provided both data collected under statutory requirements and voluntary data contributions from research collaborations, following the aforementioned Standard. Internationally, Trident Systems was considered to be exemplary of how quality controlled use of industry data in science and management could be organised (Mackinson and Middleton, 2018). However, some environmental NGOs publicly questioned its integrity. When Trident became involved in the development of video monitoring systems for fisheries observation, controversy grew. In NGO press releases, Trident's involvement in this video monitoring was confused with involvement in fisheries compliance, a clear government responsibility. Following a conflict of interest review, Trident decided in 2019

> that it was not possible to meet their objectives of improving the efficiency of fisheries data collection and extracting greater value from fisheries data in an environment in which Trident's industry ownership had become a barrier to its participation in Government funded or supported research.
>
> *(Middleton and Guard, 2021)*

As a result, Trident is now in the process of being wound up. This shows that it is somewhat naïve to assume that the presence of quality assurance systems alone can deal with issues around credibility (and hence legitimacy) of transdisciplinary science.

Towards the next level of transdisciplinary marine research

There is a tendency to focus on 'more science' (e.g. more data collection, better or more complex models) to cope with the increasing complexity of marine resource management problems. We argue that the way forward does not lie in spending more public funding on merely generating more (single, multi- or inter-disciplinary) scientific knowledge. A focus on enabling transdisciplinary approaches to science and management is required to deal with knowledge gaps and trade-offs and foster support for sustainable management practices. This calls for 'transformative change' (Visseren-Hamakers et al., 2021) in our marine governance systems, which simultaneously requires integrative, inclusive, adaptive and pluralist modes of governance (Visseren-Hamakers et al., 2021). In the context of building the knowledge base for marine management, this means that the science-policy interplay needs to change. As a part of this change, the notion of marine social-ecological systems needs to become more than a buzz word.

The science base in support of marine management is dominated by natural sciences, followed by economics. This is rooted in the advisory requests by governments and regional management organisations, which have traditionally been focused on 'single issues' and which are driven by a desire for 'numbers', such as sustainable levels of fish stock exploitation. Consequently, funding has mainly been geared towards biological and environmental data collection. With this dominance of natural and economic sciences, a culture of cooperation between scientists from different disciplines is not a given, let alone collaboration with stakeholders. In a culture of cooperation, there will be time to develop interdisciplinary and transdisciplinary concepts, approaches and methods, whilst explicitly reflecting on ontological and epistemological aspects. Although scientific institutions are increasingly opening up towards interdisciplinarity and the inclusion of stakeholders engagement in science and advisory processes, it is governmental bodies that can make the difference. By requesting integrated advice that takes the ecological and the social system into account, science institutions will be compelled to revisit their current disciplinary modes of operandi. Naturally, this will not happen overnight but if the clients of scientific advice do not ask for it, it is unlikely that scientific advice practices will transform.

This also means that governments need to enable scientific disciplines that are currently underrepresented to also play a role in social-ecological advice, for example, by including socio-economic data collection in statutory research obligations. This will enable assessment of the impact of potential fisheries management decisions on the socio-economic side of the social-ecological system, allowing for political decisions with clear and informed trade-offs based on scientific information that is readily available. Furthermore by providing funding opportunities for social scientists to collect and translate stakeholders' experiential knowledge for use in marine science, marine resource management practices become more adapted to different perspectives and contextual factors of the specific social-ecological system. This change towards integrated advice requests must go beyond bringing different scientific disciplines together, and be inclusive of stakeholders' knowledge of the marine social-ecological system as part of using the best available information practice.

Government managers can actively encourage transdisciplinary knowledge production, but this is not enough. The receiving governance system also needs to adapt to a fully participatory regime that expects use of stakeholders' observational and experiential knowledge (Figure 13.1). Participation in knowledge production and governance are tightly linked as we have illustrated with the interaction triangle (Figure 13.2). Adapting the governance system might be a challenge in large multiple-level governance systems such as the European Union. Yet in the end, only having collaborative arrangements as 'pockets of participation' within a broader multi-level merely top-down governance structure may eventually lead to the erosion of trust between the actors involved and delegitimise policy. Tearing down the management wall (Holm et al., 2020) and genuinely opening up for using knowledge co-produced between scientists and stakeholders that meets scientific quality criteria is the way forward and will require explicit reflection on current modes of governance and the level of participation of stakeholders in management decisions. Whilst acknowledging that power relations also influence knowledge co-production and need addressing (see next paragraph) (Turnhout et al., 2020), for the management process, this means that the different actors involved must be able to distinguish between knowledge used in the transdisciplinary process and vested interests that are prominent once knowledge enters the policy arena. Developing trust in the knowledge production system itself and between different stakeholders involved is a crucial element here. In this context, it is important to note the impact of 'the culture of moving jobs' whereby managers and interest groups representatives move positions every few years on participatory processes. This culture of moving in and out of positions significantly influences trust relations and interactional expertise which, as a result, constantly need rebuilding. Having people in the same position for longer periods of time creates historical knowledge and solid relations which is really valuable in such contested and complex topics like fisheries management.

Scientific knowledge has a prominent place in advising marine management and has in general more authority within co-production processes. Apart from this inhibiting the integration of other types of knowledge – often the only type of information available – this also creates inequality between actors coming from an academic and non-academic background. Power imbalances in knowledge co-production can contribute to existing inequalities between stakeholders. This affects decisions in the knowledge production process and the subsequent scientific advice and management actions, leading to disadvantageous or irrelevant outcomes for less powerful stakeholders, such as small-scale fishers. In transdisciplinary knowledge production, it is important to explicitly acknowledge power relations, interests, values and different types of knowledge and how these may impact outcomes before, during and after the process (Turnhout et al., 2020). Even if some groups, for different reasons, cannot participate in knowledge co-production, this explicit acknowledgement will at least identify limitations in approach and outcomes.

Participation of stakeholders in knowledge production and marine management both need 'to be underpinned by a philosophy that emphasises empowerment, equity, trust and learning' (Reed, 2008). Setting up boundary spaces (Stange, 2017) in the form of partnerships that include scientists, users with experiential knowledge, managers and stakeholders who may not have knowledge to contribute but do have an interest to represent, is a way forward towards operationalising and strengthening this philosophy. Actively appointing boundary spanners (Johnson, 2011) who understand different scientific ontologies and epistemologies of different scientific disciplines and non-academic knowledge holders, and how this affects (beliefs about) co-produced knowledge, the knowledge production process itself and its outcomes is an important element of such partnerships. Equally, there should be boundary spanners for the science-policy interface, as credibility, saliency and legitimacy of knowledge are entwined with different stakes and values, policy challenges and management decisions. Meaningful co-production of knowledge for marine resource management requires clear and continued communication to facilitate a culture of cooperation between all actors involved. This includes making different types of knowledge (and contrasting perspectives) explicit and making differences in values and stakes visible as these influence interpretation of results and preferred actions (de Koning et al., 2020). Boundary spanners can play a key role in this process. Fostering a culture of cooperation in marine science and management is ultimately what will make transdisciplinary approaches tick, and for it to become the philosophy of ocean governance for the future.

References

Azzurro, E., Sbragaglia, V., Cerri, J., Bariche, M., Bolognini, L., Ben Souissi, J., Busoni, G., Coco, S., Chryssanthi, A., Fanelli, E., Ghanem, R., Garrabou, J., Gianni, F., Grati, F., Kolitari, J., Guglielmo, L., Lipej, L., Mazzoldi, C., Milone, N., . . . Moschella, P. (2019). Climate change, biological invasions, and the shifting distribution of Mediterranean fishes: A large-scale survey based on local ecological knowledge. *Global Change Biology*, 25(8), 2779–2792. https://doi.org/10.1111/gcb.14670

Beare, D., Rijnsdorp, A. D., Blaesberg, M., Damm, U., Egekvist, J., Fock, H., Kloppmann, M., Röckmann, C., Schroeder, A., Schulze, T., Tulp, I., Ulrich, C., Van Hal, R., Van Kooten, T., & Verweij, M. (2013). Evaluating the effect of fishery closures: Lessons learnt from the Plaice Box. *Journal of Sea Research*, 84, 49–60. https://doi.org/10.1016/j.seares.2013.04.002

Bentley, J. W., Serpetti, N., Fox, C., Heymans, J. J., & Reid, D. G. (2019). Fishers' knowledge improves the accuracy of food web model predictions. *ICES Journal of Marine Science*, 76(4), 897–912. https://doi.org/10.1093/icesjms/fsz003

Berkes, F., & Folke, C. (1998). *Linking social and ecological systems: management practices and social mechanisms for building resilience*. Cambridge University Press.

Bjørkan, M. (2011). *Fishing for advice: The case of the Norwegian reference fleet* (PhD dissertation). UiT The Arctic University of Norway. https://munin.uit.no/handle/10037/3770

Collins, H. M., & Evans, R. (2002). The third wave of science studies: Studies of expertise and experience. *Social Studies of Science*, 32(2), 235–296. https://doi.org/10.1177/0306312702032002003

Dankel, D. J., Stange, K., & Nielsen, K. N. (2016). What hat are you wearing? On the multiple roles of fishery scientists in the ICES community. *ICES Journal of Marine Science*, 73(2), 209–216. https://doi.org/0.1093/icesjms/fsv199

De Boois, I. J., Steins, N. A., Quirijns, F. J., & Kraan, M. (2021). The compatibility of fishers and scientific surveys: increasing legitimacy without jeopardizing credibility. *ICES Journal of Marine Science*, 78(5), 1769–1780. https://doi.org/10.1093/icesjms/fsab079

de Koning, S., Steins, N. A., & Toonen, H. M. (2020). Struggling over shellfish: How diverging perceptions of marine nature distort deliberative governance. *Ocean and Coastal Management*, 198, 105384. https://doi.org/10.1016/j.ocecoaman.2020.105384

Delaney, A. E., Reid, D. G., Zimmermann, C., Kraan, M., Steins, N. A., & Kaiser, M. J. (2022). Socio-technical approaches are needed for innovation in fisheries. *Reviews in Fisheries Science and Aquaculture*, 1–17. https://doi.org/10.1080/23308249.2022.2047886

Dickey-Collas, M., & Ballesteros, M. (2021). The process in ICES of opening up to increased stakeholder engagement (1980–2020). *ICES Cooperative Research Report*, 353. https://doi.org/10.17895/ices.pub.8516

Funtowicz, S. O., & Ravetz, J. R. (1993). Science for the postnormal age. *Futures*, 25, 739–755. https://doi.org/10.1016/0016-3287(93)90022-L

Guba, E. G. (1981). Criteria for assessing the trustworthiness of naturalistic inquiries. *ECTJ*, 29(2), 75–91.

Hessels, L. K., & van Lente, H. (2008). Re-thinking new knowledge production: A literature review and a research agenda. *Research Policy*, 37(4), 740–760. https://doi.org/10.1016/j.respol.2008.01.008

Hind, E. J. (2015). Knowledge research: A challenge to established fisheries science. *ICES Journal of Marine Science*, 72(June), 341–358.

Holm, P., Hadjimichael, M., Linke, S., & Mackinson, S. (2020). *Collaborative research in fisheries: Co-creating knowledge for fisheries governance in Europe*. MARE Publication Series, vol 22. Springer. https://doi.org/10.1007/978-3-030-26784-1

Jenkins, A. (2004). Why define? The case for definitions of knowledge. *Proceedings of the Tenth Americas Conference on Information Systems*. New York, August 2004, 4165–4173. http://aisel.aisnet.org/amcis2004/520

Jenkins, L. D. (2015). From conflict to collaboration: The role of expertise in fisheries management. *Ocean and Coastal Management*, 103, 123–133. https://doi.org/10.1016/j.ocecoaman.2014.10.006

Jentoft, S., & Chuenpagdee, R. (2009). Fisheries and coastal governance as a wicked problem. *Marine Policy*, 33, 553–560. https://doi.org/10.1016/j.marpol.2008.12.002

Johannes, R. E., & Neis, B. (2007). The value of anecdote. In N. Haggan, B. Neis, & I. G. Baird (Eds.), *Fishers' knowledge in fisheries science and management*. UNESCO Publishing.

Johnson, T. R. (2011). Fishermen, scientists, and boundary spanners: Cooperative research in the U.S. Illex squid fishery. *Society and Natural Resources*, 24(3), 242–255. https://doi.org/10.1080/08941920802545800

Johnson, T. R., & McCay, B. J. (2012). Trading expertise: The rise and demise of an industry/government committee on survey trawl design. *Maritime Studies*, 11(1), 1–24. https://doi.org/10.1186/2212-9790-11-14

Johnson, T. R., & Van Densen, W. L. T. (2007). Benefits and organization of cooperative research for fisheries management. *ICES Journal of Marine Science*, 64(4), 834–840. https://doi.org/10.1093/icesjms/fsm014

Kraan, M., Uhlmann, S., Steenbergen, J., Van Helmond, A. T. M., & Van Hoof, L. (2013). The optimal process of self-sampling in fisheries: Lessons learned in the Netherlands. *Journal of Fish Biology*, 83(4), 963–973. https://doi.org/10.1111/jfb.12192

Kraan, M., & Verweij, M. (2020). Implementing the landing obligation: An analysis of the gap between fishers and policy-makers in the Netherlands. In P. Holm, M. Hadjimichael, S. Linke, & S. Mackinson (Eds.), *Collaborative research in fisheries: Co-creating knowledge for fisheries knowledge in Europe*. MARE Publication Series, vol 22. (pp. 231–248). Springer. https://doi.org/10.1007/978-3-030-26784-1_14

Macher, C., Steins, N. A., Ballesteros, M., Kraan, M., Frangoudes, K., Bailly, D., Bertignac, M., Colloca, F., Fitzpatrick, M., Garcia, D., Little, R., Mardle, S., Murillas, A.,

Pawlowski, L., Philippe, M., Prellezo, R., Sabatella, E., Thébaud, O., & Ulrich, C. (2021). Towards transdisciplinary decision-support processes in fisheries: Experiences and recommendations from a multidisciplinary collective of researchers. *Aquatic Living Resources*, 34, 21. https://doi.org/10.1051/alr/2021010

Mackinson, S., & Middleton, D. A. J. (2018). Evolving the ecosystem approach in European fisheries: Transferable lessons from New Zealand's experience in strengthening stakeholder involvement. *Marine Policy*, 90(December 2017), 194–202. https://doi.org/10.1016/j.marpol.2017.12.001

Mangi, S. C., Kupschus, S., Mackinson, S., Rodmell, D., Lee, A., Bourke, E., Rossiter, T., Masters, J., Hetherington, S., Catchpole, T., & Righton, D. (2018). Progress in designing and delivering effective fishing industry – science data collection in the UK. *Fish and Fisheries*, 19(4), 622–642. https://doi.org/10.1111/faf.12279

Middleton, D. A. J., & Guard, D. (2021). Summary and evaluation of the electronic monitoring programmes in the SNA 1 trawl and bottom longline. *Fisheries New Zealand, New Zealand Fisheries Assessment Report 2021/37.*

MinPI. (2011). Research and Information Standard for New Zealand Fisheries (May 2011). https://fs.fish.govt.nz/NR/rdonlyres/D1158D67-505F-4B9D-9A87-13E5DE0A3ABC/0/ResearchandScienceInformationStandard2011.pdf

Mitchell, R. B., Clark, W. C., & Cash, D. W. (2006). Information and influence. In R. B. Mitchell, W. C. Clark, D. W. Cash, & N. M. Dickson (Eds.), *Global environmental assessments: Information and influence* (1st ed., pp. 307–338). The MIT Press.

Moon, K., & Blackman, D. (2014). A Guide to Understanding Social Science Research for Natural Scientists. *Conservation Biology*, 28(5), 1167–1177. https://doi.org/10.1111/cobi.12326

Moon, K., Cvitanovic, C., Blackman, D. A., Scales, I. R., & Browne, N. K. (2021). Five questions to understand epistemology and its influence on integrative marine research. *Frontiers in Marine Science*, 8, 1–9. https://doi.org/10.3389/fmars.2021.574158

Murray, G., Neis, B., Schneider, D. C., Ings, D., Gosse, K., Whalen, J., & Palmer, C. T. (2008). Opening the black box: Methods, procedures and challenges in the historical reconstruction of marine socio-ecological systems. In J. S. Lutz & B. Neis (Eds.), *Making and moving knowledge* (pp. 100–120). McGill-Queen's University Press.

Neis, B., Schneider, D. C., Felt, L., Haedrich, R. L., Fischer, J., & Hutchings, J. A. (1999). Fisheries assessment: What can be learned from interviewing resource users? *Canadian Journal of Fisheries and Aquatic Sciences*, 56(10), 1949–1963. https://doi.org/10.1139/f99-115

Owen, R., Macnaghten, P., & Stilgoe, J. (2012). Responsible research and innovation: From science in society to science for society, with society. *Science and Public Policy*, 39(6), 751–760. https://doi.org/10.1093/scipol/scs093

Palsson, G. (2000). Finding one's sea legs: "Learning, the process of enskilment, and integrating fishers and their knowledge into fisheries science and management". In B. Neis & L. Felt (Eds.), *Finding our sea legs: Linking fishery people and their knowledge with science and Management* (pp. 26–40). ISER Books.

Pascual, U., Adams, W. M., Díaz, S., Lele, S., Mace, G. M., & Turnhout, E. (2021). Biodiversity and the challenge of pluralism. *Nature Sustainability*. https://doi.org/10.1038/s41893-021-00694-7

Piattoni, S. (2009). Multi-level governance: A historical and conceptual analysis. *Journal of European Integration*, 31, 163–180. https://doi.org/10.1080/07036330802642755

Reed, M. S. (2008). Stakeholder participation for environmental management: A literature review. *Biological Conservation*, 141(10), 2417–2431. https://doi.org/10.1016/j.biocon.2008.07.014

Röckmann, C., van Leeuwen, J., Goldsborough, D., Kraan, M., & Piet, G. (2015). The interaction triangle as a tool for understanding stakeholder interactions in marine ecosystem based management. *Marine Policy*, 52, 155–162. https://doi.org/10.1016/j.marpol.2014.10.019

Saavedra-Díaz, L. M., Rosenberg, A. A., & Martín-López, B. (2015). Social perceptions of Colombian small-scale marine fisheries conflicts: Insights for management. *Marine Policy*, 56, 61–70. https://doi.org/10.1016/j.marpol.2014.11.026

Said, A., & Trouillet, B. (2020). Bringing 'deep knowledge' of fisheries into Marine Spatial Planning. *Maritime Studies*, 19(3), 347–357. https://doi.org/10.1007/s40152-020-00178-y

Schram, E., Molenaar, P., de Koning, S., & Rijnsdorp, A. D. (2022). A transdisciplinary approach towards studying direct mortality among demersal fish and benthic invertebrates in the wake of pulse trawling. *Frontiers in Marine Science,* 9, 1–12. doi:10.3389/fmars.2022.907192.

Silvano, R. A. M., & Valbo-Jørgensen, J. (2008). Beyond fishermen's tales: Contributions of fishers' local ecological knowledge to fish ecology and fisheries management. *Environment, Development and Sustainability*, 10(5), 657. https://doi.org/10.1007/s10668-008-9149-0

Stange, K. (2017). Knowledge production at boundaries: an inquiry into collaborations to make management plans for European fisheries (PhD dissertation). *Wageningen University and Research*. https://doi.org/10.18174/402072

Steins, N. A., Kraan, M., van der Reijden, K. J., Quirijns, F. J., van Broekhoven, W., & Poos, J. J. (2020). Integrating collaborative research in marine science: Recommendations from an evaluation of evolving science-industry partnerships in Dutch demersal fisheries. *Fish and Fisheries*, 21(1), 146–161. https://doi.org/10.1111/faf.12423

Steins, N. A., Mackinson, S., Stephenson, R. L., Mangi, S. C., Pastoors, M. A., Baker, M., Ballesteros, M., Brooks, K., Calderwood, J., McIsaac, J., Neis, B., Ogier, E., & Reid, D. G. (n.d.). A will-o'-the wisp? On the utility of voluntary contributions of data and knowledge from the fishing industry to marine science (under review). *Frontiers in Marine Science.*

Steins, N. A., Mattens, A. L., & Kraan, M. (2022). Being able is not necessarily being willing: Governance implications of social, policy, and science-related factors influencing uptake of selective gear. *ICES Journal of Marine Science.* https://doi.org/10.1093/icesjms/fsac016

Stephenson, R. L., Paul, S., Pastoors, M. A., Kraan, M., Holm, P., Wiber, M., Mackinson, S., Dankel, D. J., Brooks, K., & Benson, A. (2016). Integrating fishers' knowledge research in science and management. *ICES Journal of Marine Science*. https://doi.org/10.1093/icesjms/fsw025

Thompson, S. A., Stephenson, R. L., Rose, G. A., & Paul, S. D. (2019). Collaborative fisheries research: The Canadian Fisheries Research Network experience. *Canadian Journal of Fisheries and Aquatic Sciences*, 76(5), 671–681. https://doi.org/10.1139/cjfas-2018-0450

Turnhout, E., Metze, T., Wyborn, C., Klenk, N., & Louder, E. (2020). The politics of co-production: Participation, power, and transformation. *Current Opinion in Environmental Sustainability*, 42(2018), 15–21. https://doi.org/10.1016/j.cosust.2019.11.009

van Stralen, M. R., & Troost, K. (2021). Ervaringskaart relatieve stabiliteit van sublitorale mosselbanken in de Waddenzee. Bureau MarinX rapport 2021.198.

Verweij, M. C., van Densen, W. L. T., & Mol, A. J. P. (2010). The Tower of Babel: Different perceptions and controversies on change and status of North Sea fish stocks in multi-stakeholder settings. *Marine Policy*, 34(3), 522–533. https://doi.org/10.1016/j.marpol.2009.10.008

Visseren-Hamakers, I. J., Razzaque, J., McElwee, P., Turnhout, E., Kelemen, E., Rusch, G. M., Fernández-Llamazares, Á., Chan, I., Lim, M., Islar, M., Gautam, A. P., Williams, M., Mungatana, E., Karim, M. S., Muradian, R., Gerber, L. R., Lui, G., Liu, J., Spangenberg, J. H., & Zaleski, D. (2021). Transformative governance of biodiversity: Insights for sustainable development. *Current Opinion in Environmental Sustainability*, 53, 20–28. https://doi.org/10.1016/j.cosust.2021.06.002

PART V
Closing remarks

The conclusion in Chapter 14 offers a renewed conception of transdisciplinarity in marine science in the light of contemporary challenges that require a new culture of reflective and diverse marine transdisciplinarity. The concluding remarks highlight key aspects to be considered in future research agendas to develop new transdisciplinary approaches, as outlined from the beginning of this book, in the co-production of culturally sensitive knowledge, as well as in the assessment of socio-ecological interactions that need to be incorporated into management and governance systems in order to make an effective transition towards sustainable oceans and seas.

By synthetizing the main aspects of the contributions of this book, the conclusions aim to move one step closer to achieving sustainability. By providing critical reflections based on the development of transdisciplinary research experiences that shed light on the underlying problems for their development, the book concludes with recommendations for overcoming these difficulties. The book shows that it is not only a question of integrating the different knowledge systems to guarantee a democratic process in the just and equitable transition towards sustainability (social, environmental, and economic) but also to include the diverse moral and ethical angles involved in the inherent complexity of present-day socio-ecological problems. This implies an epistemological transformation, in the theoretical and methodological approach, that can only be achieved through cooperation between disciplines and by building bridges between science and society. The conclusions of this book point to this breakthrough.

DOI: 10.4324/9781003311171-18

14

TOWARDS A NEW CULTURE OF REFLEXIVE AND DIVERSE MARINE TRANSDISCIPLINARITY

Sílvia Gómez and Vera Köpsel

Contemporary transdisciplinary challenges

> Transdisciplinarity in marine science is a research paradigm that aims at merging different ways of knowing the sea to produce a knowledge pool grounded in science, empirical assessment, and social and cultural experiences which should inform actionable political decision-making.

Since the "participatory turn" in the 1990s, the engagement of non-scientists has been a key element of transdisciplinary research both in terms of participation in the scientific process itself and in research governance (Jasanoff, 2003, Wynne, 2007). From the 2000s onwards, new approaches to transdisciplinarity signal a shift from a scientific-theoretical orientation towards problem-oriented research (Hirsch-Hadorn et al., 2008; Mobjork et al., 2010) highlighting the necessary pragmatic, practical, and "real-world" engagement of research on complex environmental problems (Jahn et al., 2012). Hence, transdisciplinarity pivots at the science-society-policy interface, undermining the contextually rooted problems towards transformative political change. Reflexivity addressing research practice is an important step in this consecution (Jahn et al., 2012). The practice of reflexivity ensures the testing of knowledge in the real world by constructing a space for "joint problem solving" with the participation of all societal agents. Knowledge emerges from iteration and reflection for the creation of a common pool of knowledge, and as a way of assessing the quality of knowledge (Gibbons et al., 1994) – ensuring a democratic process.

Investigating real-world problems to master socio-ecological complexity requires us to think of them as a whole, taking into account their culturally embedded context dependency (Scholz et al., 2006). Therefore, case studies facilitate an empirical approach from the viewpoint of local practice and everyday experiences with a view to producing a large-scale disruptive process towards socio-ecological transformation.

DOI: 10.4324/9781003311171-19

Despite the long history of the terminology of transdisciplinarity, a common and standardised definition of transdisciplinarity is still lacking (Jahn et al., 2012; cf. Chapter 1). At the same time, as argued by Brum Bulanti et al. in Chapter 2, marine management policies that are still based on traditional approaches show clear signs of failure. This double shortcoming reflects weaknesses in the evolution of the transdisciplinary approach itself, at the same time as environmental problems persist or worsen, making an urgent orientation towards problem-solving research necessary.

In the 1970s, Rosenfeld (1992), focusing on the research process, pointed out that "transdisciplinary refers to a process in which representatives of different disciplines work jointly for extended periods to produce novel and shared conceptual frameworks that have the promise of generating transcendent theoretical avenues" (1992: 1351). In the early 2000s, transdisciplinarity subsumed interdisciplinary collaboration by blurring disciplinary boundaries, and was seen as a process of exchange over, as Godemann (2008) calls it, a "common knowledge base" that bridges science and society by involving all social actors participating in the production of problem-oriented knowledge:

> Transdisciplinarity is a critical and self-reflexive research approach that relates societal with scientific problems; it produces new knowledge by integrating different scientific and extra-scientific insights; its aim is to contribute to both societal and scientific progress; integration is the cognitive operation of establishing a novel, hitherto non-existent connection between the distinct epistemic, social-organizational, and communicative entities that make up the given problem context.
>
> *(Jahn et al., 2012: 9)*

The contemporary challenge is to ensure democracy, equity, and justice in addressing the claims of all social groups, taking into account cultural diversity that may integrate other principles of thought that do not always coincide with those of Western society. As Petriello and his co-authors underline in Chapter 7 of this book, governance systems, technological solutions to climate change, and environmental protection measures intersect with gender, ethnicity, class, and indigenous claims. Responses to complex socio-ecological problems often do not take into account whether they match or counteract the beliefs and values of affected human populations or the multifaceted connections between them and their environment. The question of the appropriate cooperation of science, social actors, business, and public administration in this new scenario opens up new challenges in the constantly evolving transdisciplinary approach. Although the need for transdisciplinary studies is well-recognised in theory, a constant review of transdisciplinary approaches is therefore needed to facilitate their implementation in the most appropriate way.

The chapters in this book have identified the weaknesses and strengths of transdisciplinarity, as well as pointed out issues to be addressed in future research agendas in order to take a step further towards achieving a transition towards social, economic, and ecological sustainability of the seas and oceans (see Figure 14.1). The chapters are embedded in current debates on transdisciplinarity to shed light on new directions to help achieve this goal.

The main conclusion of this book is that the key transdisciplinary principle is to recognise the value of the "common good", to think in plural, trust, and acknowledge the knowledge of the other by using inclusive methodologies that subscribe to dialogical and relational ontologies and epistemes. As highlighted throughout this book, transdisciplinarity is a research paradigm that aims at the appropriate merging of different ways of knowing to produce a knowledge pool grounded in science, empirical assessment, social and cultural experiences, and practices that should inform actionable political decision-making.

The social and the ecological are melted into a unique purpose, that of preserving and conserving our oceans and seas to keep both them and us alive. The conceptual, methodological and practical considerations outlined in the different chapters of this book (see Table 14.1) clearly highlight the necessity of integrating diverse types of knowledge, diverse practices of living off and with the ocean, and diverse ways of managing its resources in order to create conditions that respect the needs of human and non-human communities interwoven with the marine. In order to achieve this goal, we think that the following cornerstones should be integrated in future approaches to transdisciplinarity in the marine domain.

Who takes part in transdisciplinary approaches and how?

Among the most common questions that arise from the contributions in this book is: Who comes into play in transdisciplinary processes? What role do different actors take on in the co-production of new knowledge? What is the value of their participation, the establishment of relationships between the participating parties, and the translation of research results into actions for the socio-ecological transformative turn? The multifaceted image of the term "transdisciplinarity" explored in Chapter 1 illustrates how the uses of the concept can vary depending on the disciplines involved, the research topic addressed, the research objectives, the complex socio-ecological problems addressed, the types of knowledge involved, as well as its understanding of society and its involvement in transforming the environment. All these determinants can be placed at the interface of a cross-scale linkage (from the most local to the global) involving multiple actors (see Figure 14.1). The contributions in this book address key points such as who asks the research question(s), what questions are in focus, and for what purpose we shape different approaches to transdisciplinarity. It is thus not about setting a standard or agreed-on definition, but about understanding the patterns of transdisciplinarity and the general meaning as a perspective of inquiry in the pursuit of cooperation between knowledges. Janico argues that "transdisciplinary vision offers hope for humanity with its advocacy of an open-minded rationality that embraces natural science, social science, and humanities" (2012: 221). In addition, we claim that society taking part at different levels of marine reality forms a key component of transdisciplinarity practice, and

TABLE 14.1 Book chapters and key insights

Chapter	*Type*	*Focus topic*	*Key insight(s)*
1 Grünhagen et al.	C	Understandings of transdisciplinarity in the marine sciences	Definitions of transdisciplinarity are multifaceted across all marine fields, scientists and stake-holders. This can promote collaboration and knowledge exchange.
2 Brum Bulanti et al.	C	Societal constructions of the ocean now and in the past	"Humanizing" conceptualisations of the ocean fosters discussions about power relations, cultural discourses and representations of the marine.
3 K. Collins	C	How theories of space can help contextualise ocean knowledges	Societal views of the sea and shaped by underlying beliefs, values, cultural backgrounds, and ethics. Deconstructing them is key to problem-solving.
4 P. Buchan	C	Democratisation of the human-ocean relation via marine citizens	Marine citizens can diversely enrich the peer community of ocean research, but institutional and personal barriers to knowledge sharing exist.
5 Carrick, Fitzsimmon and Gray	M	Critical reflection of a stakeholder process that used Bayesian Belief Networks (BBN)	BBNs are a useful tool for improving stakeholder participation by visually representing qualitative data, but their complexity can be a pitfall.
6 McAteer and Flannery	M	The impacts of professionalising marine citizen science	Professionalisation reduces the capacity of citizen science projects to empower local knowledge and to transform marine management processes.
7 Petriello et al.	M	Knowledge co-production with Intuit communities in Canada	Knowledge co-production benefits from iterative self-reflection and the deconstructions of meanings, purpose, processes, and outcomes.
8 Schaber et al.	M	Assessing stakeholders' views on sustainability via a quantitative survey	Sustainability is defined differently across stake-holders and groups; finding and using common ground can improve fisheries management.
9 J. Sá Cuoto	E	Ethnography can help consider fishers' know-ledge in sustainable policies in Portugal	Co-production of transdisciplinary ocean knowledge needs valuable local, embodied, and empirical knowledge from communities, and so does sustainable ocean management.

10 Ramos et al.	E	Collecting local ecological knowledge from fishers in Uruguay	Incorporating local knowledge into research provides initiatives for effective management and to promote dialogue between stakeholders.
11 Machado et al.	E	Using a participatory methodology to under-stand impacts of an oil spill on fishers in Brazil	Collaborative research principles constitute an important tool in the study of environmental disasters, which are complex and transdisciplinary.
12 Florido-del-Corral and Abbot-Jiménez	E	Analysing a political stakeholder process around a marine reserve in Conil, Spain	If plural stakeholder knowledges are not effectively integrated into a political process, participants lose trust. Transdisciplinarity can help to address socio-ecological challenges.
13 Steins et al.	C	Reflecting long-standing col-laboration with fishers, ways forward for marine transdisci-plinarity are discussed	Not more scientific knowledge will solve marine management problems, but transdisciplinary approaches to both science and management that transform our marine governance systems.

C = conceptual; M = methodological; E = empirical/case study.

Source: own compilation based on this book

so do the institutions involved. Beyond a theoretical, epistemological, and methodological approach, transdisciplinarity is a research attitude that is internalised by researchers, as "when transdisciplinarity is considered as a way of being, it is inseparable from personal life and extends far beyond the professional activities of a researcher" (Rigolot, 2020: 1).

Transdisciplinarity is not a stage of research, a process, but a research perspective to address and understand complex issues, a new research paradigm. Understanding transdisciplinarity in this way means producing new concepts and opening up new approaches that bring together all disciplines in the definition of the subject of study where all prisms of the socio-ecological front are included. To practice transdisciplinarity, we need sufficiently broad concepts that imply the involvement of different actors and their knowledge committed to the challenge of solving complex socio-ecological problems. Depending on the conceptualisation of the real-world problem to be addressed, the social aspect will be incorporated in the research, underpinning one or another understanding of transdisciplinarity. Modulable in its content in relation to the actors referred to in the socio-ecological relationships, the disciplines evoked to address the research questions, and the understanding of the reality described, transdisciplinarity must be embedded in new concepts. New concepts imply targeting new research problems and objectives in which different disciplines are integrated and which go beyond their own internal disciplinary logic.

Therefore, relying on new ontologies and epistemes of knowledge or, as Wheaton et al. (2021) highlighted, "conflicts of knowledge", and "contextuality" considering a co-learning process at the local level is a constant negotiation. A negotiation is here viewed as a mutual learning process and appreciation of the difference, and it is not contradictory to producing a common ground of knowledge when "unbridgeable differences such as incommensurability of certain knowledges or knowledge communities where integration by consensus is not a viable or preferred option" (Pohl, 2021: 23). On the contrary, as underlined in Chapter 7, indigenous knowledge negotiation can become assimilation that reproduces colonial practices, while indigenous voices are silenced and lost in the confusion of scientific jargon and specific technical language divorced from society. The practices of knowledge co-production described by Petriello et al. (Chapter 7) brought to the table the precariousness and weaknesses of collaborative work when conducting research among indigenous peoples. The chapter rightly appeals to humanise collaborative projects. In addition, J. Sá Cuoto highlights in Chapter 9 the need to preserve fisherfolk's knowledge as key knowledge for facing challenges such as climate change.

Collaboration with the participation of the affected social agents becomes paramount to producing sound problem-solving actions, especially in the face of environmental disasters such as the oil spill described by Machado et al. in Chapter 11. As concluded by the chapter authors, "conceptual theoretical models that emerge from participatory processes, can contribute to a better understanding of the research results and favour greater integration of the knowledge acquired with

FIGURE 14.1 Infographic – Strengths, challenges, practices, and new directions of Marine Transdisciplinarity.

Source: Based on all chapters in this book. © Vera Köpsel

intervention actions, generating more co-participatory and co-designed solutions" (p. 221 of this book).

Although timidly introduced, wide-ranging conceptions have begun to proliferate, encompassing social and environmental purposes and describing current issues to be addressed at the science-society-policy interface. They present themselves as a promising alternative for a transdisciplinary perspective following a holistic approach. The notion of ocean health, for example, places the need to understand how health and marine environments and people are intertwined at the centre of the problematisation of the marine socio-ecological crisis. The concept encapsulates concerns about sustainability, human well-being, and ocean health, which are inextricably linked. It captures new ideas about conservation in which the relationship between humans and the marine environment is the driver of the central research questions in which society is viewed as part of the problem but also part of the solution. As highlighted by Brum Bulanti et al. in Chapter 2, considering this kind of conceptualisation of reality implies new ways of producing theory, methods, and practices from different worldviews and forms of knowledge, even overcoming limits and obstacles of the disciplinary frameworks themselves, building what Franke et al. (2020) call meta-disciplines such as "ocean and human health" (Franke et al., 2020). This is an emerging line in a lot of universities and research centres pioneering in transdisciplinary approaches (Lloret et al., 2020).

How do we relate to others?

As formulated in Chapter 2 of this book, the way we relate to each other becomes the critical focus for understanding contemporary socio-environmental problems. This implies being critical of the dominant knowledge production paradigms which are predominantly based on the ontology and epistemology of Western societies, underpinned by the nature-culture dualism that undermines binary boundaries. This worldview, however, fosters classifications and categorisations that attribute roles which polarise social and gender relations, reproducing asymmetries and inequalities of power. A renewed epistemological, ontological, and paradigmatic approach that takes a more dialogical, integrative perspective of diverse and multi-scalar knowledge can ensure that all dimensions at stake are considered. The contemporary socio-ecological crisis is the consequence of the disregard of some of the cogs in the socio-ecological system. The failure of management policies in fisheries, for instance, reflects this unbalance (Gómez and Maynou, 2021) that occurs when transdisciplinarity is dominated by few as summarised in by Steins et al. in Chapter 13 and observed in its practice by McAteer and Flannery in Chapter 6. Knowledge research in the field of fisheries is already well-researched, as is reflected by the contributions in this book (see Chapters 9, 10, 12, and 13). Based on longstanding collaboration with fishers, Chapter 13 illustrates very well this experience of knowledge co-production with fishers and the fishing industry.

Knowledge production and values reflect our relational position towards the environment, whether hierarchical or equitable, reciprocal or defined according to an order of precedence that guides the importance of factors in this relationship (which can be situated between the most anthropocentric to the most ecocentric). As described by Collins in Chapter 3, the reasons to support conservation may vary, but it is the position of social actors in all processes of knowledge creation in a socio-ecological relationship that sheds light on the ethical and moral orientation of values and beliefs. As highlighted in Chapter 3, "understanding the foundations of these beliefs is crucial for understanding how decisions are made"

(p. 54). Concretising the abstract theoretical assumptions that science has about real-world problems helps to unmask the power relations and inequalities that can prevent alleged nature-based solutions from producing strong dissent because of the social consequences these solutions might produce (see Chapter 5). As stated in Chapter 3, acknowledging the uncertainty of scientific knowledge post-normal science assumes that by incorporating local knowledges, traditional scientific knowledge is supplemented with ethical, moral, and cultural values central for ensuring democracy.

Knowing the knowledge systems

As highlighted by Steins et al. in Chapter 13, knowledge pool production means relying on each other's expertise, disciplines, stakeholders, industry, and administration. Blurring disciplinary boundaries to find spaces of interaction and exchange does not mean abandoning one's disciplinary identity but, on the contrary, reinforcing it and getting to know it better, as suggested by Florido-del-Corral and Abbot-Jiménez in Chapter 12. It is necessary for each of us to hold an internal knowledge of our own expertise, but also of others in order to know our role in relation to them and to suitably trespass disciplinary boundaries. We can get to this point via the comprehension of and cooperation with other professionals to avoid adulteration of concepts, perspectives, methodologies, and theories originally produced in the frame of other disciplines than our own. Borrowing perspectives from other disciplines that are inappropriately reinterpreted can produce a significant distortion by trivialising power relationships and reproducing, rather than changing, the inequalities they are intended to redress. Considering social aspects in environmental assessments does mean not only to invite stakeholders to participate in our research, but to recognise their knowledge in the results (see Chapter 6). Recognising the potentials of each disciplinary expertise whilst cooperating with professionals from different backgrounds instead of trying to produce transdisciplinary assessments from single dominant disciplinary perspectives may help to bridge different ontologies. Chapter 12 shows us clearly how misunderstandings and incomprehension in knowledge transfer and risk sharing can lead to failure in transdisciplinary projects. Instead, researchers should deliberately deal with disciplinary differences to fuel feedback processes since outputs and inputs are co-dependent (Blythe et al., 2017).

At the forefront of misunderstandings remains the recognition of the value of different kinds of knowledge. By the standards of Western science, objectivity is what attributes value and legitimacy to knowledge. Experiential knowledge contains judgement based on perceptions, feelings, beliefs, and, ultimately, variable subjectivities according to interests that traditionally place it as supplementary to the production of "real", objective scientific knowledge. In line with Wheaton et al. (2021), the chapters in this book show us that knowledge production is – and always has been – politically and historically at the interface of the field of political economy. Moreover, the subjective dimension of the objective valorisation of concerns and problems can reveal mismatches between abstract knowledge produced according to large-scale global political and economic values and local-scale ecological, socio-cultural, and economic specificities. This can fuel claims for rights, dissent, and conflict (see Chapter 3). The disqualification of subjective knowledge can unfold an escalation of misunderstandings between the different social actors involved in the production of knowledge for policymaking.

The key to transdisciplinarity, as Semenyuk (2019) points out, is not so much the integration of knowledge per se, but "the integration of more or less distant objective-subjective and methodological branches of knowledge" (ibid.: 4). Between "soft" (social scientific) and "hard" (natural scientific) disciplines, an integration implies approaching divergent epistemologies. Here, methodological design can crucially influence the degree of permeability between the social and natural sciences. Greater or lower permeability of knowledge may tip the balance towards one or the other knowledge system by reproducing existing hierarchies, for example when experiential knowledge is used as complementary to validate traditional Western scientific knowledge. Notwithstanding, also the contrary effect may be produced when knowledges are integrated in a balanced way by making other ways of knowing known and respected (Noël-Knapp et al., 2019).

Participatory processes are underpinned as those methodological inquiry systems best suited to merge qualitative (subjective) and quantitative approaches. As shown by Carrick, Fitzsimmons and Gray in Chapter 5, the combination of participatory models and flexible methods such as PAR allows subjectivity to penetrate the rigidity of quantitative methodologies by integrating "soft" science (narratives, emotions, perceptions, beliefs, values) into "hard" science, thus helping to give social and cultural meaning to quantitative assessments. Nonetheless, as concluded in Chapter 5, flexible methodologies do not equal malleable or manipulable approaches that co-opt dissenting voices during participatory and reflexive processes by trying to situate supposedly socially embedded solutions on their ground through the mere process of consultation or communication with stakeholders. Despite this, in some cases, consultation or the use of local knowledge to complement and shed light on scientific assertions may help to introduce data that otherwise would remain unattained. In Chapter 10, for instance, Ramos et al. highlight the value of using local knowledge to identify factors that may influence the vulnerability of fishing communities.

Shaping the politics of participation

McAteer and Flannery describe in Chapter 6 how "politics of participation", as it is called by Kok et al. (2021), drive the dynamics of transdisciplinary processes. The power to decide who may participate in producing knowledge obliges to define the role of not only scientists and science in relation to society but also actors involved in relation to science and to who they are. Depending on this definition the direction of society's engagement in research will vary. Designing and implementing well-moderated participatory processes is the way of bridging science and society with the political will to produce a transformative turn at the science-society-policy interface.

When it comes to determining who may participate in a research process, Kok et al. (2021) point out the necessity to politicise the inclusion of actors as it is a conscient political act. If we do not consciously define who are our stakeholders, we risk (re)producing inequalities and reproducing interests. Moreover, it is necessary to combine pragmatic choices weighting representation with deliberation (Kok et al., 2021). In Chapter 8, Schaber et al. state aptly that

> it is very difficult to gauge whether a stakeholder speaks for him- or herself, or on behalf of the whole group. In fact, they may relate to several groups, and groups may be diverse themselves. Since the inclusion of everyone is usually not possible, some kind of bias is very likely to occur.
>
> *(p. 168)*

Making ourselves aware of this potential bias and working around it as well as possible is an important responsibility that we, as the scientists who engage, carry in the participatory processes we initiate.

Despite these social limits that revert to the position of facilitators to build bridges (Gómez and Maynou, 2021; Kok et al., 2021), funding agencies often restrict the requirements for engagement processes, the identification of participants, and the definition of transdisciplinarity. The institutionalisation of citizen science discussed in Chapter 6 shows how this disempowers local knowledge and reinforces top-down governance by neutralising bottom-up procedures. This bears the danger of reducing the role of society to data collection, education, and awareness-raising, with citizen science understood as acquiring responsibility rather than acting as a catalyst for citizens' rights and duties (see Chapter 4). This reveals the hierarchy of knowledge, disqualification, and mistrust of non-academic knowledge, while social aspects are considered and integrated for the sole purpose of "ticking the participation box". This logic follows a process of knowledge transfer or communication. Defining a common understanding of who the societal actors involved are and what roles they occupy is a frequent problem in transdisciplinary research. The same is true for the term "participatory" or "consultative", as highlighted by Grünhagen et al. in Chapter 1. At the boundary between "inclusion" and "transformation", stakeholder participation can be defined as part of "consultative transdisciplinarity" or "participatory transdisciplinarity" research (Mobjork, 2010).

Stakeholder engagement is often viewed from the participants' passive position. The roles of stakeholders are reduced to receiving information from science, or sharing their perspectives which are ultimately masked under administrative decisions or attenuated by research funding agencies. In the practice of co-management models, Gómez et al. (2021: 8) note that

> the lack of agility hampered by the multiple bodies between down and top often results in concealing power relations, which neutralises the conflict as a way of claiming rights. The agreement can be reduced to a socialisation process that achieves social capital, mutual knowledge, and trust, neutralising the effect of fighting for equity and rights.

Significant in defining the role of the actors involved is the way in which they are referred to as "stakeholders", "social actors", "practitioners" or "citizens" when producing "citizen science". Beyond the names and roles assigned to actors is their potential as activists to bridge science and action, with attention to the power and access to decision-making. It is not enough to build trust and understanding among participants; it requires the commitment of science and society, but also of public administration and business to take transformative action. This also implies an integration of the value of the common good.

The potential of ethnography in transdisciplinary approaches

Throughout the chapters in this book, the potential of ethnography to build bridges between ontologies, cultural systems, and the realms of science and society is repeatedly underlined. In Chapter 9, J. Sá Cuoto highlights the potential of ethnography to capture the knowledge of fisherfolk that is at risk of disappearing under neoliberal models in fisheries. Ethnography enables the recovery of social and cultural skills to re-establish successful socio-ecological relationships in the face of the uncertainties of climate change. Knowledge is lost through the process of alienation of fishers in relation to fisheries production. In Chapter 11, Machado et al. show how the application of anthropological concepts and ethnographic methods to collect epidemiological knowledge in the context of culturally embedded everyday life allows for a "dialogue of knowledge" (p. 209).

The ability of social anthropology to capture the ethical components encapsulated in the subjectivities (aspirations, frustrations, feelings, emotions) entailed in the ontologies of social groups makes it a promising discipline to consider in future transdisciplinary projects. As already pointed out by Noël-Knapp et al. (2019: 8), "documentation of different ways of knowing are common in cultural ethnographies arising in anthropology". The exercise of transdisciplinarity emerges from social anthropology as it attempts to "transcend the narrow scope of disciplinary worldviews" since presenting itself as a "general science of humans" (Pohl et al., 2021). When joining forces in transdisciplinary processes, the non-natural sciences of anthropology, ethnology, social sciences, and humanities carry great potential for better understanding the human activities and societal dynamics at play in managing our marine spaces and, at the same time, for developing the applied solutions we need for bridging science and society in times of an ocean in crisis.

References

Blythe, J. et al. (2017) Feedbacks as a Bridging Concept for Advancing Transdisciplinary Sustainability Research. *Current Opinion in Environmental Sustainability*, 26–27: 114–119. http://dx.doi.org/10.1016/j.cosust.2017.05.004

Franke et al. (2020) *Operationalizing Ocean Health: Toward Integrated Research on Ocean Health and Recovery to Achieve Ocean Sustainability*. One Earth 2.

Gibbons, M., Limoges, C., Nowotny, H. et al. (1994) *The New Production of Knowledge*. Sage, London

Godemann, J. (2008) Knowledge Integration: A Key Challenge for Transdisciplinary Cooperation. *Environmental Education Research*, 14(6): 625–641, https://doi.org/10.1080/13504620802469188

Gómez, S., Carreño, A., Lloret, J. (2021) Cultural Heritage and Environmental Ethical Values in Governance Models: Conflicts Between Recreational Fisheries and Other Maritime Activities in Mediterranean Marine Protected Areas. *Marine Policy*, 129: 104529. https://doi.org/10.1016/j.marpol.2021.104529

Gómez, S., Maynou, F. (2021) Balancing Ecology, Economy and Culture in Fisheries Policy: Participatory Research in the Western Mediterranean Demersal Fisheries Management Plan. *Journal of Environmental Management*, 291: 112728. https://doi.org/10.1016/j.jenvman.2021.112728

Hirsch Hadorn, G., Hoffmann-Riem, H., Biber-Klemm, S., et al. (Eds.) (2008) *Handbook of Transdisciplinary Research*. Springer Science

Jahn, T., Bergmann, M., Keil, F. (2012) Transdisciplinarity: Between Mainstreaming and Marginalization. *Ecological Economics*, 79: 1–10. https://doi.org/10.1016/j.ecolecon.2012.04.017

Jasanoff, S. (2003): Technologies of Humility: Citizen Participation in Governing Science. *Minerva*, 41: 223–244

Juanico, Melton B. (2012) The Long Road Towards Transdisciplinarity. In Virginia A. Miralao, Joanne B. Agbisit (Eds.) *Philippine Social Sciences: Capacities, Directions, and Challenges*. Philippine Social Science Report 2012

Kok, K.P.W., Gjefsen, M.D., Regeers, B.J., Broerse, E.W.J. (2021) Unraveling the Politics of 'Doing Inclusion' in Transdisciplinarity for Sustainable Transformation. *Sustainability Science*, 16: 1811–1826. https://doi.org/10.1007/s11625-021-01033-7

Lloret et al. (2020) The Roses Ocean and Human Health Chair: A New Way to Engage the Public in Oceans and Human Health Challenges. *International Journal of Environmental Research and Public Health*, 17: 5078. https://doi.org/10.3390/ijerph17145078 www.mdpi

Mobjörk, M. (2010) Consulting Versus Participatory Transdisciplinarity: A Refined Classification of Transdisciplinary Research. *Futures* 42: 866–873. https://doi.org/10.1016/j.futures.2010.03.003

Noël-Knapp, C., Reid, R.S., Fernández-Giménez, M.E., Klein, J.A., Galvin, K.A. (2019) Placing Transdisciplinarity in Context: A Review of Approaches to Connect Scholars, Society and Action. *Sustainability*, 11: 4899. https://doi.org/10.3390/su11184899

Pohl, C. et al. (2021) Conceptualising Transdisciplinary Integration as a Multidimensional Interactive Process. *Environmental Science and Policy*, 118: 18–26

Rigolot, C. (2020) Transdisciplinarity as a Discipline and a Way of Being: Complementarities and Creative Tensions. *Humanities & Social Sciences Communications*, 7: 100. https://doi.org/10.1057/s41599-020-00598-5

Rosenfeld, P.L. (1992). The Potential of Transdisciplinary Research for Sustaining and Extending Linkages Between the Health and Social Sciences. *Social Science and Medicine*, 35(11): 1343–57.

Scholz, R.W. et al. (2006) Transdisciplinary Case Studies as a Means of Sustainability Learning. Historical Framework and Theory. *International Journal of Sustainability in Higher Education*, 7(3): 226–251. https://doi.org/10.1108/14676370610677829

Semenyuk, E.P. (2019). The Information Effect of Transdisciplinarity in the Concept of Sustainable Development. *Scientific and Technical Information Processing*, 46(1): 1–13. https://doi.org/10.3103/S0147688219010027

Wheaton, B., Te Aramoana Waiti, J., Olive, R., Kearns, R. (2021) Coastal Communities, Leisure and Wellbeing: Advancing a Trans-Disciplinary Agenda for Understanding Ocean-Human Relationships in Aotearoa New Zealand. *International Journal of Environmental Research and Public Health*, 18: 450. https://doi.org/10.3390/ijerph18020450.

Wynne, B. (2007). Public Participation in Science and Technology: Performing and Obscuring a Political-Conceptual Category Mistake. *East Asian Science, Technology and Society: an International Journal*, 1: 99–110.

INDEX

Note: Page numbers in *italics* indicate a figure and page numbers in **bold** indicate a table on the corresponding page.